Energy, Environment and Economic Transformation in China

I0124912

China has achieved rapid economic growth since the market-oriented reform in 1978 and became the second largest economy in the world in 2010. However, the growth model in China is still extensive in nature and may be characterized by high energy consumption and heavy environmental pollution, among other factors. In fact, China has become number one in the world for carbon emission, since 2007, and energy consumption, since 2009. This book endeavours to analyse whether such energy driven and environmentally restricted economic growth can be sustainable in China in the long run.

The book describes the basic facts of energy consumption and environmental pollution in China from the perspectives of industries, regions and energy-types. It also outlines the evolution of energy and environmental policies implemented in China. In particular, the author makes use of the environmental activity analysis model to assess the sustainable transformation of economic model in Chinese industries and regions. This model captures the negative externalities of pollutants and estimates accurately the environmental total factor productivity. The possibilities of win-win development and double dividend are also forecast.

This book proposes new methods to measure the environmental total factor productivity, evaluate the process of low carbon transformation, quantify the structural bonus, estimate the abating cost and forecast how to achieve win-win development. Researchers may find these methodologies useful for measuring other pollutants and for analysis in other countries.

Shiyi Chen is Professor of Economics at Fudan University and a visiting scholar at Humbolt University in Berlin, Germany. He is the Director of Leading Group of Ecology, Environment, Humanities and Social Sciences Research at Fudan University, Co-Director of Shanghai-Hong Kong Development Institute (CUHK-Fudan) and Director of Fudan Policy Lab for Sustainable Development. He holds a PhD in Econometrics at Kyungpook National Univerity, Republic of Korea. His research interests are in energy, environment and sustainable development, economy and finance in contemporary China, applied econometrics and so on. He has published in English journals such as *Quantitative Finance*, *Journal of Forecasting*, *China Economic Review*, *The World Economy*, *Energy Policy* and top Chinese journals.

Energy, Environment and Economic Transformation in China

Shiyi Chen

Routledge
Taylor & Francis Group

LONDON AND NEW YORK

First published 2013
by Routledge
2 Park Square, Milton Park, Abingdon, Oxfordshire OX14 4RN

Simultaneously published in the USA and Canada
by Routledge
711 Third Avenue, New York, NY 10017

First issued in paperback 2014

Routledge is an imprint of the Taylor and Francis Group, an informa business

British Library Cataloguing in Publication Data
A catalogue record for this book is available from the British Library

Library of Congress Cataloging in Publication Data
Chen, Shiyi.
  Energy, environment and economic transformation in China / by
Shiyi Chen.
    p. cm.
  Includes bibliographical references and index.
  1. Air—Pollution—Economic aspects—China.   2. Carbon dioxide
mitigation—China.   3. Industries—Environmental aspects—China.
  4. Environmental policy—China.   5. Sustainable development—China.
  I. Title.
  HC430.A4C44 2013
  363.738'70951—dc23   2012035800

ISBN 978-0-415-50953-4 (hbk)
ISBN 978-1-138-91035-5 (pbk)
ISBN 978-0-203-37462-7 (ebk)

Typeset in Times New Roman
by RefineCatch Limited, Bungay, Suffolk

To my wife, Mei Chen, and our son, Xiaoyu Chen

Contents

Figures

Tables

Acknowledgements

I would like to take this opportunity to acknowledge everyone who has made a contribution to my study. First and foremost, I would like to thank Professor Jun Zhang, Kiho Jeong and Wolfgang Härdle, who once were my advisors and still encourage me to research ambitiously now. I owe a great debt to President Yuliang Yang, Vice President Xiaoman Chen and Shangli Lin, and Professor Cheng Wang, Trevor Davies and Boqiang Lin, from whom I have gained enormous encouragement and support. Thanks also go to Yongling Lam, Heather Cushing and Olwyn Hocking, my commissioning editor, manager and producer at Routledge.

This monograph contains several revised versions of previously published papers. The permission to reproduce materials published in the following papers was granted by the *Economic Research Journal*: Chen, S. (2010) Green Industrial Revolution in China – A Perspective from the Change of Environmental Total Factor Productivity (1980–2008), Issue 11, 21–34; Chen, S. (2010) Energy-Save and Emission-Abate with its Impact on Win-Win Development in Chinese Industry: 2009–2049, Issue 3, 129–143; Chen, S. (2009) Energy Consumption, CO_2 Emission and Sustainable Development in Chinese Industry, Issue 4, 41–55; by *Social Science in China*: Chen, S. (2011) Marginal Abatement Cost and Environmental Tax Reform in China, *Social Sciences in China*, Issue 3, 85–100; by the *China Economic Quarterly*: Chen, S. (2011) Estimates of Sub-industrial Statistical Data in China (1980–2008), 10(3), 735–776; by *The Journal of World Economy*: Chen, S. (2011) Reduction Pattern of Carbon Intensity and Its Explanation in China, Issue 4, 124–143; Chen, S. (2010) Shadow Price of Industrial Carbon Dioxide: Parametric and Nonparametric Approach, Issue 8, 93–111. I appreciate the support of these journals.

Funding has been received from the National Natural Science Foundation of China (project 71173048), the National Social Science Foundation of China (key project 12AZD047), the Ministry of Education (New Century Excellent Talent Plan and project 11JJD790007), Shanghai (Ling-Jun Talent Plan and leading academic disciplines project of B101) and Fudan University (Zhuo-Shi Talent Plan).I express my sincere thanks for their support. Part of work in this book has been accepted by the 2012 International Symposium on Econometric Theory and Applications (SETA, 2012) and the 2010 Econometric Society World Congress (ESWC, 2010) and I presented these myself at both conferences. I am much obliged for the invitations from the organizers.

I am very grateful to my grandfather, Yin Chen, for his enlightenment and support. I miss my grandmother, Chengxun Ding, forever. My gratitude also goes out to my parents. Finally, I owe by far the greatest debt to my wife, Mei Chen, and our son, Xiaoyu Chen, for their inspiration, encouragement and support.

1   Introduction

1.1 Energy, environment and economic growth in China

Since its economic reform from central planning to markets in 1978, China's economic performance has been remarkable, and industry is the biggest growth-driven sector when compared with the agricultural and tertiary sectors. It is well known that the Chinese government opted for a heavy-industry-oriented development strategy to catch up with the developed world for a long time before the reform. This strategy, based on China's comparative disadvantages, has resulted in the persistence of a dual economy. The higher than average rate of growth of industry depends on massive inputs of production resources, by using 'price scissors' to transfer resources from agriculture to industry. Though the industrial development strategy was adjusted during the reform period to emphasize both light and heavy industry evenly, the extensive growth model, with the characteristics of high growth, high investment, high energy use and heavy environmental pollution emission, has not changed greatly.

The depletion of fossil fuels and consequent increase in environmental pollution indicates, in particular, that this extensive model of economic development is not sustainable, and a transformation towards a low-carbon model is inevitable. Especially in recent decades, this transformation has been far from simple in China, because its rapid economic growth has depended heavily on high energy consumption and heavy waste emission. As depicted in Figure 1.1, industrial energy consumption and carbon dioxide (CO_2) emission have seen continuous increase since the reform. In an attempt to speed up economic transformation, the central government has introduced a number of energy-saving and emission-abating policies and regulations. In particular, during the ownership rights reform period from 1996 to 2002, China employed the policy of furlough (*xiagang*) and aimed to 'grasp the large and let go of the small' (*zhuada fangxiao*) and shut down about 84,000 small energy and emission-intensive enterprises, causing the first stagnancy or decline of energy use and CO_2 emission, as shown in the Figure. However, the phenomenon of heavy industrialization reappeared after 2002. Figure 1.1 shows that the consumption of primary energy and carbon emission had such an unprecedented increase after 2002 that it almost reversed the process of economic transformation already happening in China. Now, China

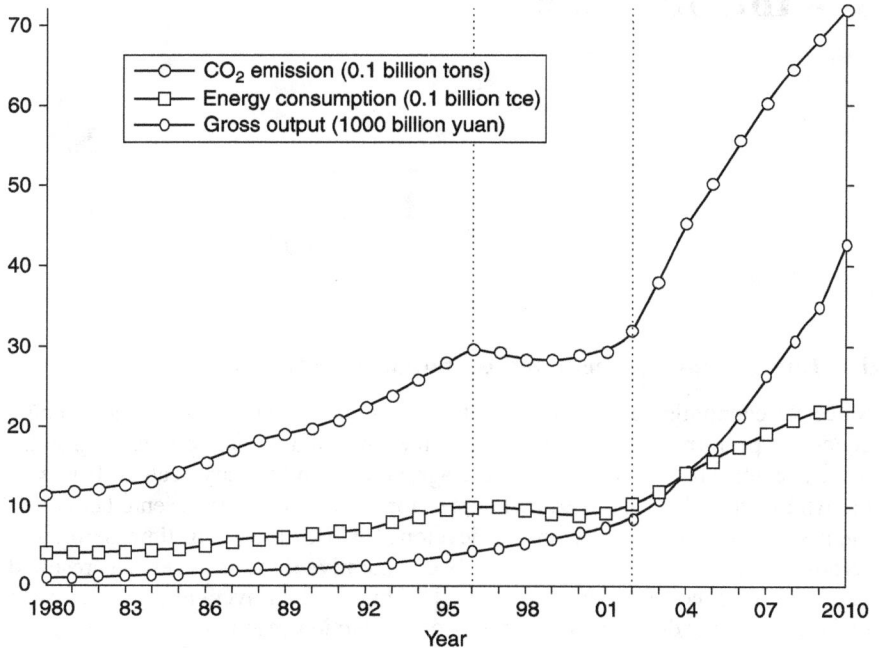

Figure 1.1 Industrial gross output, energy consumption and CO_2 emission in China (1980–2010).

became the world's largest consumer of energy and emitter of CO_2 in absolute terms, followed in both by the US, but only the second largest GDP producer, behind the US. In 2010, industrial energy consumption and CO_2 emission achieved the peak of 2.3 billion tons of coal equivalent (tce) and 7.2 billion tons, respectively, while industrial gross output attained the absolute value of 42.6 trillion of RMB. In fact, during the period of the eleventh Five-Year Plan and the beginning of the twelfth Five-Year Plan, the central government also proposed official quantitative abating targets for energy and carbon intensity, respectively; but the results have yet to be seen, as its effects need further observation.

1.2 Total factor productivity, economic transformation and what do we learn from descriptive statistics of industrial and regional input and output panel data?

Since the pioneering work of Solow, rising total factor productivity (TFP) has been considered as a key engine of economic growth, along with quantitative increases in factors of production. Within the analytical framework of neoclassical growth accounting, a rising share of TFP in total output growth signals the transformation towards a development model based on quality rather than quantity (Solow, 1957; Kim and Lau, 1994; Krugman, 1994; Young, 1994). In particular, when the contribution of TFP growth to output growth exceeds that of the quantitative inputs of all factors, it can be said that the development model has been transformed from being extensive and unsustainable to being intensive and sustainable. This will be the basic economic theoretical framework used in this book to assess whether the traditional economic growth model has been transformed to become sustainable or not. Of course, in the literature, there is still some debate over the contribution of TFP growth to China's output growth during the period of economic reforms since 1978, including whether and when its development model has already made this transformation. Chen, Jefferson and Zhang (2011) provide a relevant survey. In existing literature, however, measures of TFP have been usually calculated as the Solow residual in a growth accounting framework that only includes the traditional inputs of capital and labour, neglecting both the energy inputs required to sustain economic growth and their environmental impacts.

As described above, the estimation of the production function and then TFP requires the data on input as well as output quantities. To comprehensively analyse economic transformation, it is necessary for us to take both the energy inputs and the environmental factors into account, along with the traditional factors. Thus, the construction of new input and output databases that include the energy and environmental variables is extremely important for the analytical objectives of this book. As Jorgenson and Stiroh (1999, 2000) denoted, growth differs widely among industrial sectors or different regions. For the economy as a whole, negative growth in one sector or region can offset positive growth in another; therefore it is essential to disaggregate estimates of economic growth to the sub-industrial or provincial level to find the true pattern behind the aggregation. Following this approach, this book avoids the limitations of an aggregation analysis by breaking down Chinese economic analysis into industrial sectors and provincial levels, which should provide us with a new understanding of the forces driving the transformation of China's economy. In particular, we constructed two types of database: two-digit sub-industrial and provincial panel data over the reform period is the basic research sample employed in the following chapters. In the Appendix, we introduce the principles behind the construction of the panel databases and report part of the input and output data. Here, we just briefly introduce the two types of database and, based on their descriptive statistics, assess whether we can see something useful in relation to productivity growth and economic transformation from both industrial and regional viewpoints.

This research concentrates on the study of Chinese industry, rather than the whole economy, because the output, energy consumption and carbon emissions of Chinese industry account for most of the national total, and the analysis of it is bound to give us an overall picture of the aggregate economy. The two-digit sub-industries are classified according to the new version of the National Standard of Industrial Classification (GB/T4754), revised in 2002 in China. This information is constructed into the panel data for 38 sub-industries between 1980 and 2010, which will be used as the sample in most of the chapters, except Chapters 2 and 8, and it is expected to enhance the information available when analysing the microeconomic performance for each sector. The codes and names for the 38 sub-industrial sectors are in Table A.1 in the Appendix; they belong to three bigger categories: mining, manufacturing, utilities (electric power, gas and water production and supply). Table 1.1 reports the descriptive statistics for the main input and output variables used in this book.[1] For example, the output variable is gross industrial output value (GIOV) with the unit of 100 million RMB at 1990 prices, rather than value-added, due to the possible inclusion of intermediate inputs, such as energy, into the analysis. The capital stock and CO_2 emissions cannot be obtained directly and need to be estimated. The capital stock is estimated by using a perpetual inventory approach, depreciated at constant 1990 prices of investment in fixed assets, and CO_2 emissions are estimated from three types of primary energy use, according to their different carbon emission coefficients. The labour input is the annual average number of employed workers (unit: 10 thousand workers) and energy input is total energy consumption with a unit of 10 thousand tce. To investigate the impact of energy on growth further, we divided all sectors into light and heavy industry according to the ranking of the absolute quantities of energy consumption in 2004.[2] That is, the light industry corresponds to the top half of sectors with a lower energy consumption, and the heavy industry to the lower half of sectors with a larger value of energy use.

Table 1.1 brings into focus the contrast between light and heavy industry. The principal feature we obtain from this table is that the mean levels of capital stock and energy consumption in heavy industry are much higher than in light industry, respectively, with much smaller differences in the means of GIOV and labour. The averaged CO_2 emissions in heavy industry are 28.1 times those in light industry. Obviously, high investment and energy consumption do not lead to

Table 1.1 Descriptive statistics of main sub-industrial variables used (1980–2010)

Variables	Light Industry				Heavy Industry			
	mean	*s.d.*	*min*	*max*	*mean*	*s.d.*	*min*	*max*
GIOV (100 million RMB)	1106	2653	17	34,329	3526	8239	44	99,694
CO_2 Emission (10 thousand tons)	552	597	16	3106	15,503	35,239	158	305,967
Capital Stock (100 millionRMB)	339	358	13	3308	1698	2396	43	24,592
Labour (ten thousand workers)	131	109	10	691	354	258	18	1279
Energy Consumption (10 thousand tce)	433	349	37	1918	4784	7149	113	60,946

equally higher output growth and employment, but they emit far higher levels of CO_2. Also, there is a higher degree of variability within the heavy sub-industries for output and inputs, especially for energy and emissions, compared with the light sub-industries. Based on such statistical information, heavy industry in China holds the obvious characteristics of higher investment, higher energy consumption, higher carbon emission and higher variance in performance levels, but less output, which implies that there may be lower TFP and then more extensive growth in heavy industry than in light industry.

In Chapter 8, we compile data from a panel of 31 provinces to assess the low-carbon economic transformation from a regional perspective. The names of input and output variables, including energy consumption, electricity consumption and multiple environmental pollution emissions (such as CO_2, COD, SO_2, waste water and waste gas) and their sample period are outlined in Table 1.2.[3] Unlike the traditional classification of eastern, middle and western regions, we divided all the provinces into low and high energy and emission regions, according to the classifying index built by ourselves[4], to check the effect of different energy and emission levels on regional economic transformation. According to the classification, the low energy and emission region in this chapter includes 16 provinces, in which Beijing, Guangdong and Shanghai are ranked the top three, the value of their index being less or close to 0.5; the high energy and emission region includes 15 provinces, in which Guizhou and Ningxia are ranked the last two, the value of their index being greater than 2. The descriptive statistics of regional variables reported in Table 1.2 reveal similar information to the industrial Table 1.1. The regional value-added in the low energy and emission region is 2.1 times that of the high energy and emission region, but the inputs of capital, labour, energy and electricity in the former region are only 1.7, 1.4, 1.2 and 1.5 times the size of those in latter region. The environmental emissions (such as CO_2, waste water, waste gas, COD) in the low energy and emission region are also only 1.2, 1.8, 1.1 and 1.3 times those of the high energy and emission region; the SO_2 emission in the former region is even less than that in the latter region. The investment in the treatment of industrial and environmental pollutions in the low energy and emission group is higher than that in the high energy and emission group, though the area of afforestation in the former region is less than that in the latter. Such statistical information also implies that the low energy and emission region may have the higher TFP and therefore a more intensive growth model relative to the high energy and emission region. There also exists an abnormal phenomenon in Table 1.2. For instance, the standard deviation of all the variables except waste gas and afforestation in the low energy and emission region is higher than that in the high region. When looking at the maximum statistics, although Shandong is in the low energy and emission region, it has the highest CO_2 emission and energy consumption in 2009 and the highest SO_2 emission in 2005. Similarly, Guangdong is in low energy and emission region but also has the largest waste water emission and electricity consumption in 2009. In Chapter 7, we will investigate the impact of abnormal energy and emission distribution on regional low-carbon economic transformation.

Table 1.2 Descriptive statistics of main provincial variables used

Variables and Sample Period	Low Energy and Emission Region				High Energy and Emission Region			
	mean	s.d.	min	max	mean	s.d.	min	max
Regional GDP (100 million RMB, 1980–2009)	4188	5527	25	36,035	2012	2390	63	15,511
CO_2 Emission (10 thousand tons, 1980–2009)	12,157	11,821	63	88,713	9973	10,156	550	67,624
Waste Water Emission (10 thousand tons, 1985–2009)	171,780	123,500	4	694,010	94,125	68,050	9181	299,020
Waste Gas Emission (100 million cu.m, 1985–2009)	5500	5413	1	35,089	5257	6267	248	54,598
COD Emission (10 thousand tons, 2000–2009)	48	30	0.8	111	39	26	3.2	112
SO_2 Emission (10 thousand tons, 2000–2009)	64	49	0.1	200	80	44	3.2	156
Capital Stock (100 million RMB, 1980–2009)	7361	10,474	24	68,612	4243	5391	168	36,152
Labour Forces (10 thousand workers, 1980–2009)	2264	1582	100	6081	1640	1066	147	4945
Energy Consumption (10 thousand tce, 1980–2009)	5124	4923	40	33,855	4289	3947	269	24,962
Electricity Consumption (100 million kwh, 1990–2009)	643	669	5	3610	436	361	42	2344
Investment in the Treatment of Industrial Pollution (100 million RMB, 1998–2009)	12.2	12.4	0.0007	65.6	8.1	7.7	0.1	44.4
Investment in the Treatment of Environmental Pollution (100 million RMB, 2004–2009)	110	96	0.2	449	62	46	5	223
Area of Afforestation (hectare, 2004–2009)	75,602	85,971	1117	416,130	219,530	174,530	5385	861,930

1.3 What do the statistical indicators tell us about the economic transformation?

Drawing on the constructed sub-industrial panel data, we further calculate several economic indicators of the 38 sub-industries. To see their varying trends and differences among the sub-industries easily, Figure 1.2 depicts their weighted averaging values for the industry as a whole, and for light and heavy industry, in which the weights are each sub-industrial gross output value and the light and heavy industry are classified in terms of the same criteria as those in Table 1.1.

In Figure 1.2, sub-figure (a) describes the industrial value-added per capita, that is, labour productivity, which is used internationally to assess the industrialization level and measure the economic benefits and production efficiency during the course of industrialization. Obviously, since the mid-1990s, industrial labour productivity has increased rapidly, leading to a continuously rising industrialization level; corresponding with the reappearance of heavy industrialization after 2002, the output per capita in heavy industry even exceeded that in light industry, and the momentum of heavy industrialization still exists currently in China.

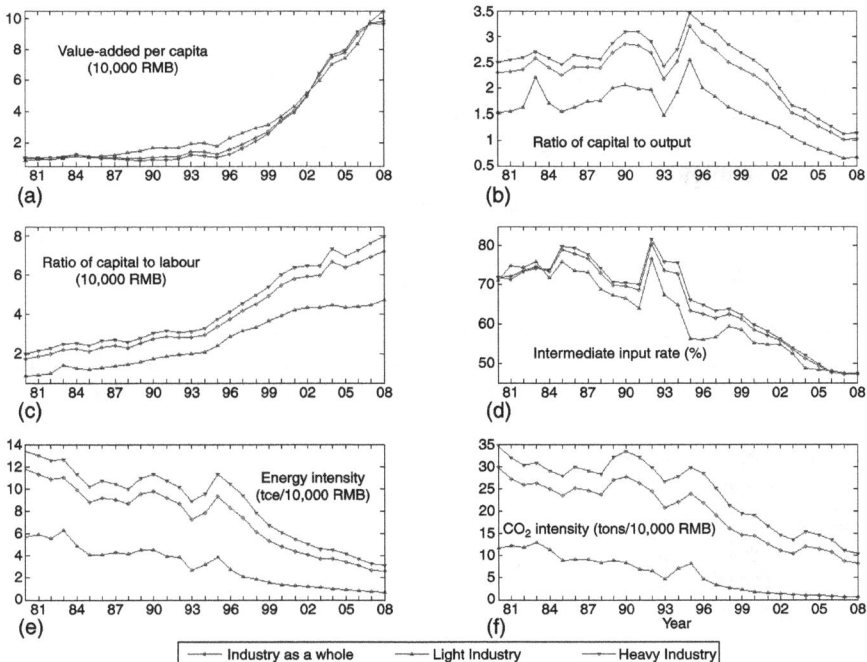

Figure 1.2 Trend of main economic indicators for light industry, heavy industry and aggregated industry.

Zhang (2002) utilizes the ratio of capital to output to describe the capital deepening phenomenon in the process of industrialization in China. He finds that the ratio of capital to output reversed the long-run decreasing trend and began to increase in 1994. He argues that this phenomenon is because the speed of capital formation will be finally dominated by the law of diminishing marginal returns and the input factors driven economic growth model will be unsustainable in the long term. Thus, the rising ratio of capital to output could be regarded as the important characteristic of extensive economic growth. Young (1994) also denotes that the excessive capital deepening that pushes economic growth in eastern Asia is not sustainable. However, sub-figure (b) of Figure 1.2 shows a different trend of the ratio of capital to output. That is, the ratio varies very little before 1993 but experiences a decreasing trend after 1995. Li and Zhu (2005) find the same phenomenon and think that the better indicator for estimating the capital deepening is not the ratio of capital to output but that of capital to labour, because the capital deepening implies that more capital and less labour are utilized during the multiple-input-combination production process. Like labour productivity, the ratio of capital to output is in fact the reciprocal of capital productivity. Sub-figure (b) reveals that Chinese industrial capital productivity increased from 1995, and more rapidly in light industry than in heavy industry, which tells us that light industry had higher productivity growth and was experiencing more rapid economic transformation, consistent with the information revealed by Table 1.1. Fisher-Vanden and Jefferson (2008) also find that industrial capital productivity began to rise in China from the late 1990s, following long-term decrease. But they denote that this cannot cure the investment hunger, and this excessive investment – still existing now in China – further worsens allocative efficiency.

If the ratio of capital to labour is used to measure the capital deepening, as shown in sub-figure (c), Chinese industry experienced a continuous capital deepening process, and the ratio in heavy industry was much higher than that in light industry; the difference between them became larger from the beginning of this century. Consistent with Chen *et al.* (2011), the standard perception of industrialization is a general shift in relative importance from light to heavy industry. Light industry is of great importance normally at the early stage of industrialization and labour-intensive in nature, with a relatively low ratio of capital to labour; while heavy industry is at the middle or late stage and capital-intensive, with a relatively high ratio of capital to labour. Zheng *et al.* (2009) also find that, relative to advanced economies, the ratio of capital to labour is still low in China. Therefore, China is still in the middle of industrialization, and the capital-, energy- and pollution-intensive industrial sectors (such as chemical products, cement, iron and steel) will still play a fundamental role in the economy, both at the present and in the foreseeable future. As the organic composition of capital, the ratio of capital to labour reflects the factor endowment of each industrial sector, and its increasing trend also reveals the actual industrialization process from a labour-intensive to capital-intensive industrial structure. Solow (1957) once outlined that the growth of output per capita consist of two parts: the growth of capital to labour ratio and the growth of TFP, known as the Solow residual. Comparing sub-figure (a) with (c), the higher growth rate of value-added per capita than capital to labour ratio implies that the TFP improves for industry as a whole in China; and the phenomenon that the growth rate of output per capita is similar between light and heavy industry, and the growth rate of capital to labour ratio in light industry is lower than that in heavy industry, indicates that light industry should experience a higher TFP growth and more effective economic transformation than heavy industry, which is also consistent with Table 1.1.

Sub-figure (d) depicts the share of industrial intermediate input to industrial gross output, called the industrial intermediate input rate. The decreasing trend of the industrial intermediate input rate since the beginning of the 1990s implies that the industrial value-added rate is increasing and, correspondingly, the industrial production efficiency is improving. The higher intermediated input rate in heavy industry than that in light industry indicates that light industry has a higher industrial value-added rate, more efficient production and a higher class development model than heavy industry. The industrial intermediate input rate is used to measure the level of resources consumption and materials utilization and is also the sustainable development indicator, together with both indicators of energy intensity and carbon dioxide emission intensity in sub-figure (e) and (f), respectively, to assess whether the economic development model is transformed or not.

As shown in Figure 1.1, Chinese industrial energy consumption and CO_2 emissions have risen sharply since the start of this century. However, both sub-figure (e) and (f) show that industrial energy intensity and CO_2 intensity have actually fallen, especially after 1995, indicating that Chinese efficiency of energy utilization and emission reduction is in fact rising, and, of course, that the energy-using and emission-abating efficiency in light industry is far higher than that in heavy industry.[5] Although efficiency is increasing, we cannot be too optimistic, because the value of energy and emission intensity is still too large in Chinese industry as compared with other countries. For example, the energy intensity in 2004 was 0.99 for China, 0.23 for the US, 0.19 for Germany, 0.11 for Japan and only 0.65 for India (unit: kg oil equivalent/1 dollar of GDP). The CO_2 emission intensity (tons CO_2/1 million dollars of GDP) in 2004 was 2755 for China and only 549 for the US.[6] This shows that the task of saving energy, reducing emissions and transforming the economic development model is still large and challenging in China.

1.4 The main framework

This book will study from different viewpoints the economic transformation in China and its links to high energy consumption and heavy environmental pollution emissions. There are 11 chapters, and we now briefly introduce the research content for each chapter.

In the Introduction, the facts of energy, environment and economic growth are outlined to illustrate the importance of this research. Specifically, the subindustrial and provincial input and output panel databases are constructed to further describe the traditional growth characteristics in China and the necessity of transforming the economic development model within the constraints of energy and environment. Based on three-dimensions (i.e. industries, provinces and energy types) decomposition for CO_2 emission in China between 1995 and 2007, Chapter 2 finds that, to achieve CO_2 abatement, it is necessary to transform the growth model from being capital driven, to promote energy and capital productivity, adjust energy and industrial structures, and so on. In the case of Shanghai, it is particularly necessary to develop nine newly defined high-tech industries, focus on CO_2 abatement in the transportation industry, change the growth model driven by government investment, and substantially reform SOEs in Shanghai, among other things.

Based on the decomposition of carbon intensity, Chapter 3 finds that the decline of energy intensity, or the improvement of energy productivity, is the most important factor to drive the abatement of carbon intensity in Chinese industry. The adjustment of the energy structure and industrial structure also plays a role during the process of carbon intensity reduction. The former factor is called the direct effect but the latter is an indirect one. By using parametric and nonparametric directional distance functions, Chapter 4 estimates the shadow price of industrial carbon dioxide for 38 sectors between 1980 and 2008. The measurements show that the shadow price for light industries is higher than that for heavy industry, and the estimated shadow price increases over time. The estimates of shadow price make it possible to determine the carbon tax rate, the pricing of pollution permit trade and calculate the traditional productivity index. Chapter 5 reviews the evolution of energy and environmental policies in China and then carries out green growth accounting, based on the translog production function. In the production process, both energy consumption and CO_2 emission are treated as input factors. The results tell us that China's industry seems to have transformed its growth model, with productivity as the main driving force in many sectors. In addition to productivity growth, energy and capital have also been driving industrial growth during the sample period, while labour and emissions made a lower or even negative contribution to it. Some heavy industries are still characterized by extensive growth and must improve their energy-save and emission-abate technology to support the sustainable development of overall Chinese industry. In Chapter 6, we quantify the structural change effect and find that structural reform has contributed to TFP and output growth substantially but has experienced a decreasing effect over time. Empirical analysis reveals that the reforms in factor markets, energy structure, ownership structure, size structure and foreign funding structure contributed substantially to the overall trend and the sectoral heterogeneity of structural change effect during the industrial transformation process.

Chapter 7 employs the directional distance function (DDF) that includes carbon dioxide emission as undesirable output to estimate the environment restricted total factor productivity and its decomposition in Chinese industry between 1980 and 2008. The results reveal that the actual productivity growth, allowing for the negative externality of emission, is much lower than traditional measures, indicating that the overall transformation of the industrial development model, promoted by the increasing shares of productivity on output growth, is still at its early stage. However, from the mid-1990s to the beginning of this century, the measure of actual productivity, technical change and efficiency in heavy industry significantly exceeded that in light industry due to effective energy-saving and emission-abating regulations implemented by central government. Though hindered by the energy and emission restriction, China is experiencing the effective but gradual low carbon transformation of a development model in industry. Based on the slacks-based measure DDF theoretical mechanism, Chapter 8 builds the dynamic indicator to evaluate the low carbon economic transformation of Chinese provinces since the reform. The measurement indicates that Chinese low-carbon transformation underwent very poor development during 1986–1990 and the beginning of this century, then performed very well in the 1990s and recent years. The evaluation indicator of low-carbon transformation produced from this economic model takes both the growth quality and speed into account and can be employed as the more appropriate alternative to a conventional GDP criterion. The heterogeneity and inconsistency of the evaluated indicator for each province reveals that Chinese low-carbon transformation is still within the unstable early stage, and specific energy and environmental policies should be implemented to support the long-run process of great transformation in China.

The Porter hypothesis states that environmental governance may lead to win-win opportunities, that is, improve productivity and reduce undesirable output simultaneously. Chapter 9 proposes a dynamic activity analysis model to forecast the possibilities of win-win development in Chinese industry between 2009 and 2049. The evidence reveals that energy-saving and emission-abating activity will result in both the improvement in the net growth of potential output and the steadily increasing growth of total factor productivity. This favours the Porter hypothesis. Environmental taxation reform is one of the most efficient means to help China realize the required reduction of carbon intensity. But there exists the heated controversy over its economic impact and carbon tax pricing. Chapter 10 measures the marginal abatement cost (MAC) of carbon dioxide and uses it as the pricing base of carbon tax, and then to forecast its economic and ecological impact in China in the following ten years. The main results include: (1) the measured MAC of CO_2 increases over time and differs across industrial sectors, and so the carbon tax rate forecasted from this should follow the same pattern; (2) to fully reflect the institution value of environmental taxation, the tax rate should be high enough to influence the emitters' behaviour and stimulate them subconsciously towards environmental protection; (3) carbon tax has a negative impact on output but the influence is extremely small. Carbon tax plays an obvious role in abating carbon intensity within most sectors, but in some heavy sectors it is not enough, and other environmental regulations are necessary.

Chapter 11 summarizes the studies and the corresponding conclusions in this book. The critical policies deduced from the studies are also summarized. Such policies to push the economic transformation forward include the reform of the energy price system, the reform of environmental taxation, the reform of economic structure, and the application of a new evaluation indicator of economic transformation instead of only using traditional GDP. In the Appendix, the basic principles to construct the sub-industrial panel data are introduced. That is, the sub-industries must be reclassified and recombined to match one another, the scope of all the industrial variables must be adjusted to the same statistical content, and some missing data should be added rationally. The three dimensional database for six industries from 30 provinces in China between 1995 and 2007 and the provincial panel data for 31 provinces over the reform period are also discussed briefly. Some of the variables are included in the Appendix, at the end of this book.

2 Industrial and regional composition of energy-induced CO₂ emissions

2.1 Introduction

Since the economic reform, China has achieved rapid economic growth, but this has also caused serious resources waste and heavy environmental pollution. The CO_2 emission, the main component of air pollution, is picked out as an example here. Because CO_2 emission mainly results from the combustion of fossil fuels, as outlined in Chapter 1, the data on CO_2 emission cannot be obtained directly and needs to be estimated from three types of primary energy (i.e. coal, oil and gas) according to their respective CO_2 emission coefficients. Thus, precisely speaking, the CO_2 emission information used in this book should be called the energy-induced CO_2 emission.

The CO_2 emission of China was relatively low in the 1960s and 1970s, similar to that of Germany and Japan, then increased quickly after the reform, finally exceeding that of the US in 2008 to become the largest CO_2 emitter. The financial crisis brought the rare opportunity for each country to transform its extensive economic growth model, and energy-saving and emission-abating regulation is becoming an important method to push economic transformation forward. Even in the US, to achieve economic recovery and lead the new round of economic growth, the government has invested considerably in new energy and low-carbon techniques that have been chosen as the new growth point in US for the future. This chapter first describes CO_2 emission in China as a whole, its industrial and regional distribution, and then analyses its important influential factors. This analysis will be compared with one for the city of Shanghai, to reveal the emitting heterogeneity of CO_2 in China.

2.2 The industrial and regional composition of CO₂ emission in China

In this section, we first describe the CO_2 emission of China as a whole. Due to the availability of the data, the CO_2 emission used in this chapter is estimated from the end-use of energy consumption.[1] Figure 2.1 shows the change of industrial composition of end-use energy consumption in China. The industries discussed in this chapter cover six sectors: agriculture, industry, construction, transport,

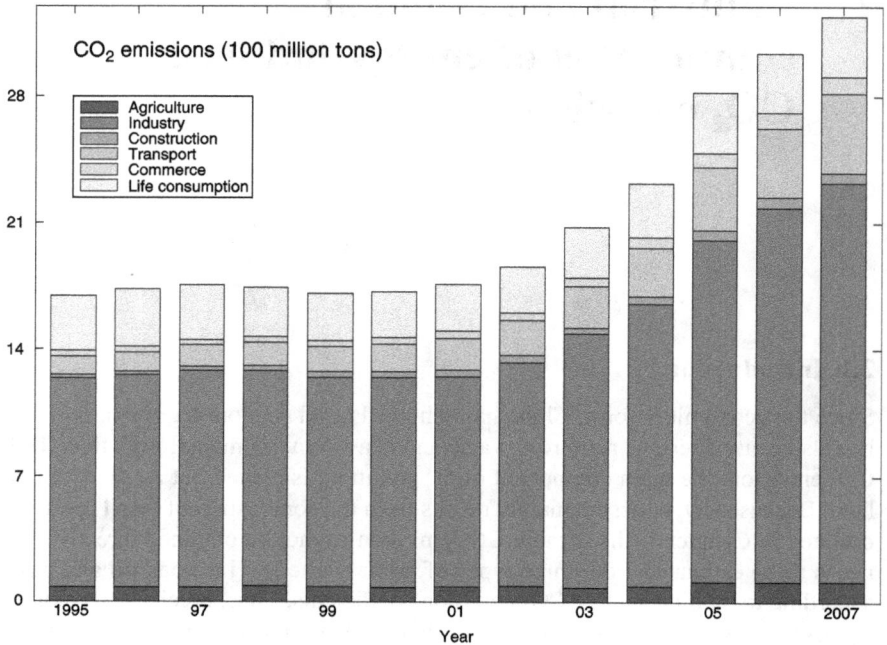

Figure 2.1 Industrial composition of CO$_2$ emission in China (1995–2007).

commerce and the household sector; the first five belong to the production sector.[2]

Obviously, the production sector (including agriculture, industry, construction, transport and commerce) emits most of the CO$_2$, and its share of total CO$_2$ emission increased from 82 per cent in 1995 to a very high 90 per cent in 2007. Accordingly, the emission share of life activities consumption in the household sector decreased from 18 per cent to 10 per cent. Within the production sector, the emission share of agricultural, construction and commercial industries did not vary much and remained at about 4.4 per cent, 1.7 per cent and 2.3 per cent, respectively; industry was the top emitter, with a share which maintained a very high and steady value of 68 per cent; the transport industry was the only one where the emission share increases continuously, from 6 per cent in 1995 to 14 per cent in 2007, which reflects the rapid development of transport infrastructure and the car industry during this period from another angle.

Figure 2.2 exhibits the change of CO$_2$ emission intensity of different industries in China. Such industries as agriculture, construction and commerce had lower CO$_2$ intensity, being about 0.5 tons CO$_2$ per 10,000 RMB of GDP; the household sector had a relatively high CO$_2$ intensity of 0.8 tons per 10,000 RMB of GDP; however, the production sector, and its components of industry and transport, had much higher CO$_2$ intensity. In particular, the continuous decrease of CO$_2$ intensity

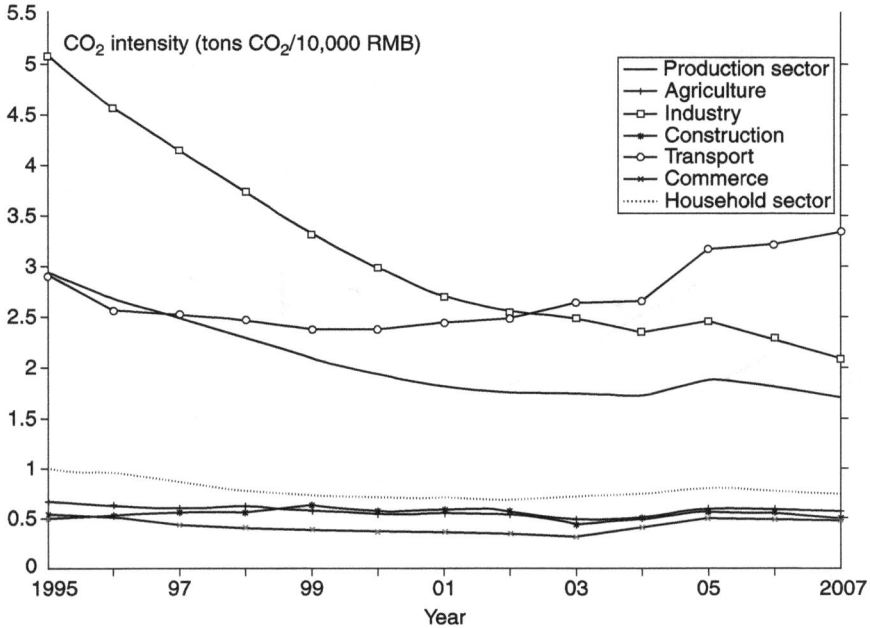

Figure 2.2 Change of CO$_2$ intensity for different industries in China (1995–2007).

in the production sector was due to a larger reduction of CO$_2$ intensity in industry, which was greater than the increasing trend of CO$_2$ intensity in the transport industry. As the only industry where emission intensity has risen continuously and more rapidly since 2003, the transport industry should be especially emphasized in energy-saving and emission-abating processes in the future.

From regional viewpoints,[3] the eastern region had more rapid economic growth and more CO$_2$ emission, followed by the middle region; the western region, though containing most provinces, had the lowest CO$_2$ emissions. As for their respective emission intensity, as shown in Figure 2.3, the CO$_2$ emission intensity for China as a whole and the three regions all experienced an overall deceasing trend, which is only level or somewhat rising during the period of the tenth Five-Year Plan (2001–2005), corresponding with the reappearance of heavy industrialization at that stage. Of course, though emitting the highest levels of CO$_2$, the eastern region had the lowest emission intensity and its carbon productivity was higher than that of the middle and western regions.

2.3 The industrial composition of CO$_2$ emission in Shanghai

In this section, China's largest and most advanced city – Shanghai – will be studied to investigate its features of end-use energy-induced CO$_2$ emission and compare

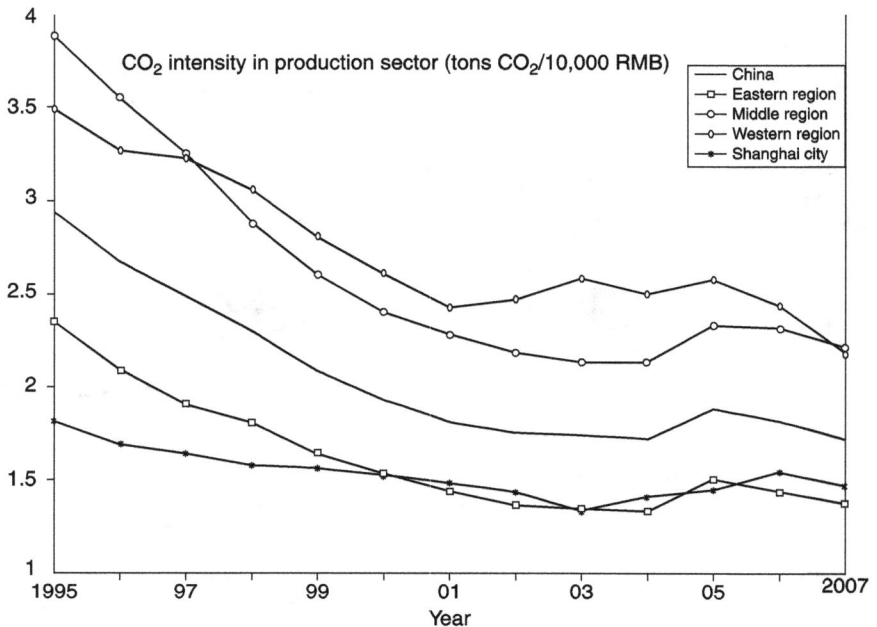

Figure 2.3 Change of CO_2 intensity in the production sector for different regions (1995–2007).

these with other regions and the averaged levels for China. Figure 2.4 addresses the change in the industrial composition of CO_2 emission in Shanghai and Figure 2.5 describes the emission intensity of different industries based there.

As shown in Figure 2.4, the CO_2 emission in Shanghai rose from 39.87 million tons in 1995 to 106.63 million in 2007, and its share of total emission in China increased also, from 2.4 per cent to 3.3 per cent. Compared with Figure 2.1, the national CO_2 emission turned out to be standstill and even fell during the period of the ninth and tenth Five-Year Plan, while the CO_2 emission of Shanghai increased quickly. The same trend for both Shanghai and China was the sharp rise of CO_2 emission after 2003. There exists a big difference in the industrial composition of carbon dioxide emission for Shanghai and China. Industry is the largest emitter for both of them. In 1995, the proportion of industrial CO_2 emission to total emissions in Shanghai was the same as that in China, being about 69 per cent; since then, it remained almost constant in China and began to continuously decrease in Shanghai, to less than 50 per cent by 2007, due to the shrinking in the speed of Shanghai's industrial development at that stage. The share of CO_2 emission from the transport industry in Shanghai's tertiary industry was higher than that of China (26 per cent and 10 per cent, respectively), and grew more rapidly (from 15 per cent in 1995 to 36 per cent in 2007), which partly reflected the development

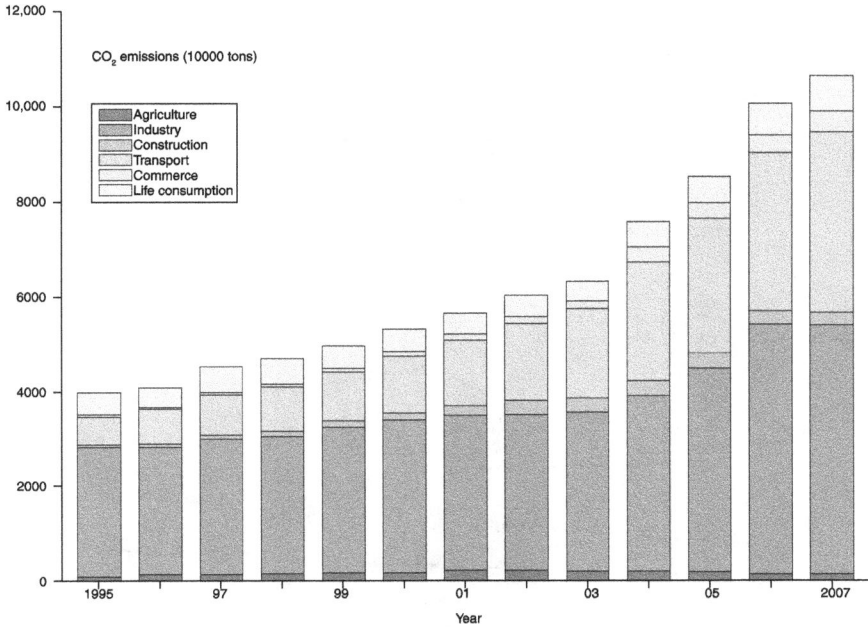

Figure 2.4 Industrial composition of CO_2 emission in Shanghai (1995–2007).

direction of tertiary industry driven by the construction of an international ship-ping and trade centre in Shanghai. The proportion of CO_2 emission in the house-hold sector experienced a decreasing trend in both Shanghai and China, but the size of this proportion was smaller in Shanghai than in China. Specifically, the end-use energy-induced CO_2 emission of the household sector in China accounted for 14 per cent of total emission on average, which ranks it the second among six industries studied in this chapter and just behind industry, as shown in Figure 2.1. But the emission share of the household sector in Shanghai was only 9 per cent on average, behind industry and the transport sector (Figure 2.4).

When comparing Figures 2.2 and 2.5, Shanghai had a lower carbon dioxide intensity in both production and household sectors than China, indicating higher carbon productivity. In both Shanghai and China, the carbon intensity in the production sector fell throughout, only rising a little around 2005 and 2006, and then continued to fall; while in the household sector, intensity fell first and then reduced extremely slightly, the absolute value of which is far smaller than that in the production sector. Within the production sector, only the transport industry had an obvious growth of carbon intensity; other industries vary slightly, at most, increasing a little. The bigger differences between Shanghai and China are addressed in section 2.4. The CO_2 intensity of the transport industry in Shanghai

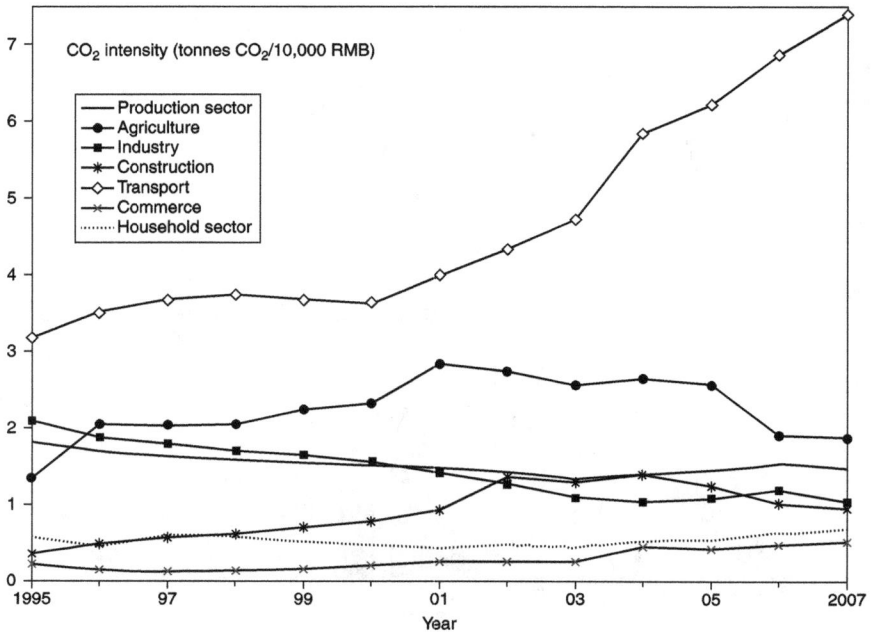

Figure 2.5 Change of CO_2 intensity for different industries in Shanghai (1995–2007).

was much higher than that in China and increased sharply from 2001 (from 3.2 tons per 10,000 RMB of GDP in 1995 to 7.4 in 2007), indicating that the transport industry merits attention in carbon reduction measures in the future. The industrial CO_2 intensity in China decreased considerably (from 5.1 tons per 10,000 RMB of GDP in 1995 to 2.1 in 2007); while that in Shanghai was much lower than that in the national industry in absolute value (on average 1.5 and 3.1 tons/10,000 RMB, respectively), but its growth was lower than national growth. In addition, the carbon intensity of the agricultural and construction industry in Shanghai was higher than the average in China, too. Seen from Figure 2.3, the CO_2 intensity of the production sector in Shanghai was similar to the level in the eastern region, lower than both the average level for China and also that of the middle and western regions. They experienced similar trends, falling first, then bouncing, and finally falling again. Similar to what is shown in Figure 2.2, the CO_2 intensity of Shanghai decreased the least during the period of the ninth and tenth Five-Year Plan.

2.4 The factors driving the different emission patterns between Shanghai and China

As described above, China and Shanghai have exhibited different CO_2 emission patterns. In this section, we try to identify the different influential factors from the perspectives of industries, provinces (or regions) and energy types.[4] To this end, we collect four dimensional data from the official Yearbook for three types of primary energy (coal, oil and gas) used by six industries (agriculture, industry, construction, transport, commerce, household sector) in 30 provinces between 1995 and 2007. This is only used in this chapter, and is totally different from the data used in other parts of this book. The variables include end-use energy consumption, estimated carbon dioxide and capital stock, gross output value, total residents, resident income per capita, and so on. All value-type variables are depreciated at 1995 price level.

To link the analysis in this chapter to the overall theme of economic transformation in this book, we specify a newly influential factor of capital scale to capture the extensive characteristics of Chinese economic growth, that is, capital deepening. In sum, the influential factors we discuss in this section consist of capital scale, energy intensity, industrial structure, energy structure and household sector, and so on. The decomposition results show that, in both China and Shanghai, the most important explanatory variables are capital deepening and energy intensity, followed by industrial structure and energy structure. The household sector has the weakest explanatory power and should not be the focus of future carbon reduction; this is different from developed countries that emphasize the emission abatement in areas related to household life. In the following, we will introduce the important factors driving the different CO_2 emission patterns in both China and Shanghai.

2.4.1 Capital deepening

The capital deepening factor, represented by stock scale, had the largest positive effect on CO_2 emissions in both China and Shanghai, indicating that the rapid increase of CO_2 emission is still closely related to the extensive growth model driven by capital deepening in China. Zhang (2002) regards the capital deepening phenomenon represented by capital scale as the important feature of the traditional growth model in China. Zhang (2003) denotes that, relative to the labour force, fixed capital investment grows too fast in China. Qin and Song (2009) show that excessive capital investment or investment hunger, accompanied by the command plan, still exists in China now. If capital deepening is measured by the ratio of capital to labour, as shown in Figure 2.6, three regions in China all experienced a continuous capital deepening process. The capital per capita in the eastern region rose from 23,000 RMB in 1995 to 95,000 RMB in 2007, much higher than the similar ratio in the middle and western regions. But the ratio of capital to labour in Shanghai was far higher than the average level for the eastern region, implying that Shanghai had a different investment model from the other regions.

In fact, unlike the Wenzhou Mode and South Jiangsu Mode, there exists a particular investment mode in Shanghai, the so-called Shanghai Mode, that is dominated by the Shanghai government as it allocates resources and develops the economy. For example, Shanghai invested a total of 2.6 trillion RMB during the period of 15 years between 1992 and 2007, in which it invested 0.39 trillion, 0.96 trillion and 1.3 trillion RMB, respectively, in the periods of the eighth, ninth and

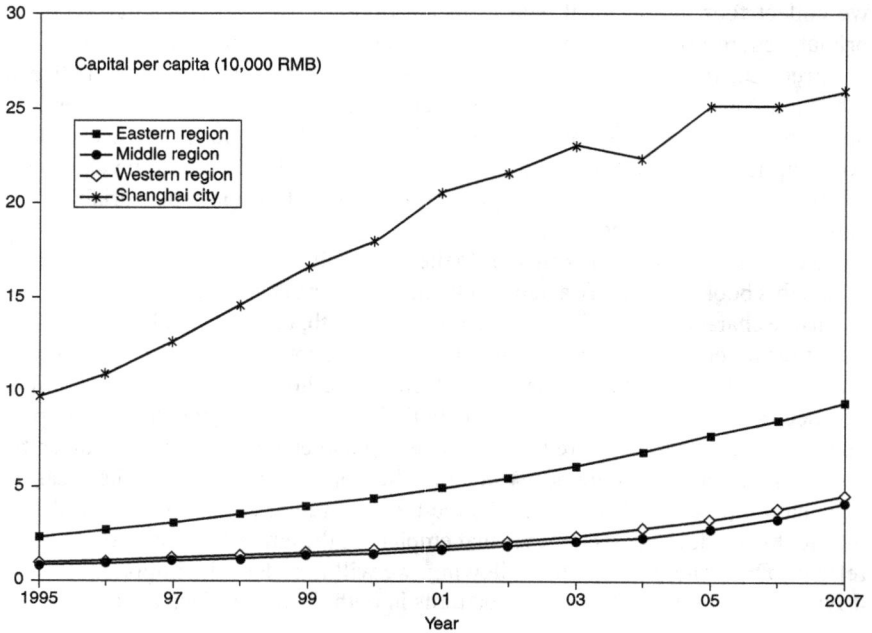

Figure 2.6 Ratios of capital to labour of three regions in China (1995–2007).

tenth Five-Year Plan; this played the critical role in developing the Pu-Dong area and transforming the city of Shanghai from an industrial and commercial city to an economic centre in China. But the drawbacks of government driven investment mode are obvious, too. In this case, the main market players were normally state-owned enterprises (SOEs), ownership reform was behind the average level in China, and investment efficiency was not high; these greatly hindered the development of market power. The report 'The transformation of economic growth model in Shanghai', proposed by the Shanghai Academy of Social Science in 2005, showed that, since the 1990s, the marginal effectiveness of economic investment driven by local government was in fact decreasing, and that investment efficiency has deteriorated continuously. The big influence of the capital scale factor described in this sub-section implies that transforming the capital-driven extensive growth model to become sustainable is the only way for China and Shanghai to save energy and reduce environmental emission substantially in the future.

2.4.2 Energy intensity

In both China and Shanghai, the decrease of energy intensity led to the largest emission reducing effect, which then effectively outweighed the rapid increase of CO_2 emission resulting from capital deepening. The difference between Shanghai and China is that, during 2001–2007, the energy intensity in Shanghai ceased falling and even began to rise a little; that has resulted in an increase in CO_2 emission.

Specifically, during the period of the ninth Five-Year Plan (1996–2000), energy intensity played the largest role in abating CO_2 emission at the national level; that corresponds with the substantial SOEs reform, the shutdown of many energy- and emission-intensive small firms and effective implementation of environmental regulations. Though the SOEs reform and environmental regulation were behind the national level, the decrease of energy intensity in Shanghai (from 0.76 tce per 10,000 RMB GDP in 1995 to 0.65 in 2000) still caused a relatively large CO_2 emission reduction. During the period of the tenth Five-Year Plan (2001–2005), the reappearance of heavy industrialization and rapid urbanization caused the lowest decrease of energy intensity and then reduction of CO_2 emission at the national level. Accordingly, 2003–2005 were the only three years in which Chinese energy consumption elasticity exceeded the value of 1. From 2001, the factor of energy intensity started to push the CO_2 emission in Shanghai, totally different from the case of China. This is maybe because the six pillar industries established in this period were all big energy consumers.[5] At this stage, many big SOEs such as the Shanghai Automotive Industry Corporation, Baosteel, the Jinshan Petrochemical Group, the Shanghai Electric Group and many shipbuilding corporations developed very rapidly; that drove heavier industrialization in Shanghai than in China, leading to an increase of energy intensity in Shanghai from the lowest 0.58 tce/10,000 RMB to 0.69 in 2006.

During the period of the eleventh Five-Year Plan, central government officially proposed binding targets to save energy intensity by the end of 2010 by 20 per cent relative to 2005, and that increased the impact of energy intensity on the reduction of CO_2 emission at the national level. But, in Shanghai, the enterprises in the six pillar industries accounted for more than 60 per cent of the total industrial gross output and it is difficult to close all of them. Therefore, in 2006 and 2007, the energy intensity in Shanghai still played a positive role in pushing CO_2 emission higher. The Shanghai government realized the negative impact of the six pillar industries on sustainable development and changed the new manufacturing strategy to develop nine advanced technique industries such as those developing new energy techniques, civil aviation manufacture, advanced manufacturing equipment, bio-pharmaceutics, electronic information, new energy vehicles, ocean engineering equipment, innovative uses of materials, software and information services, to replace the traditional six industries. But the results need further observation. Though the energy intensity in China and Shanghai has decreased overall, its absolute value is still large as compared with other economies and has more room to fall. Because Shanghai does not produce primary energy and imports almost all the energy that the economic growth needs, its degree of dependence on energy imports is more than 90 per cent. Thus, it is very critical for Shanghai to decrease its energy intensity and then to increase energy utilization efficiency during the process of economic transformation.

2.4.3 Industrial structure

The industrial structure has had more impact on CO$_2$ emission reduction than the energy structure, which is discussed in the following sub-section. The change of energy structure drives the reduction of carbon dioxide emission in both China and Shanghai. However, the evolution of industrial structure has had a negative impact on carbon emission reduction in China, but has had a positive impact on emission reduction in Shanghai.

Figures 2.7 and 2.8 describe the evolution of three industries for China as a whole and for Shanghai. In China, the GDP share of primary industry decreased continuously, broadly consistent with the general characteristics of the structural transformation process documented in the literature of transitioning economies. The GDP share of tertiary industry rose first and then started to fall from the beginning of this century; this is different from the standard structural evolution pattern of tertiary industry where output share should keep increasing. This may be due to the continuous rise of GDP share of the second industry (from 55 per cent in 1995 to 69 per cent in 2007), which is different from the experience of industrialized economies which tend to have a hump-shaped pattern. It is not difficult to understand the positive impact of industrial structure on carbon emission in China, because the secondary industry contains the largest carbon emitters. This indicates that China's industrial structural adjustment is still in the early phase and needs updating further. It is not the case in Shanghai. The GDP share of primary

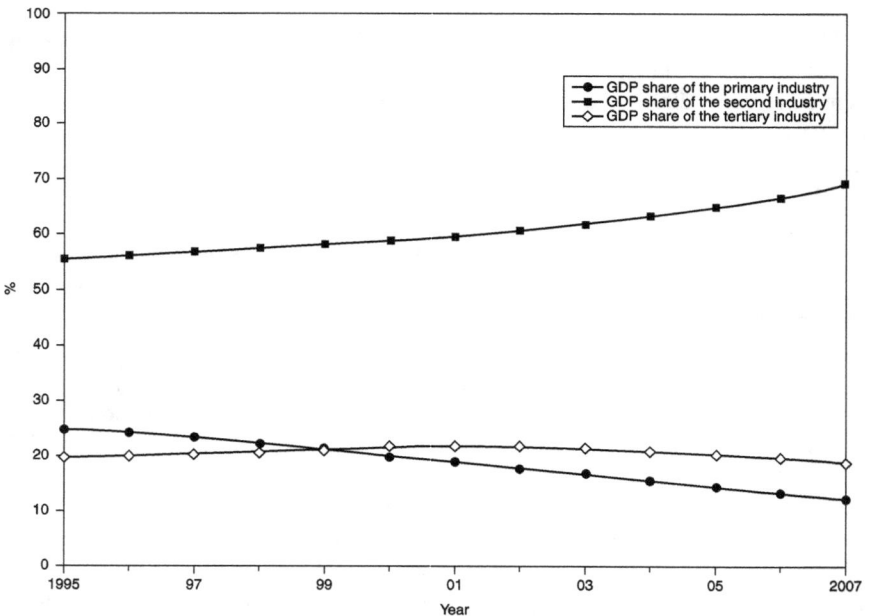

Figure 2.7 Structural change of three industries in China (1995–2007).

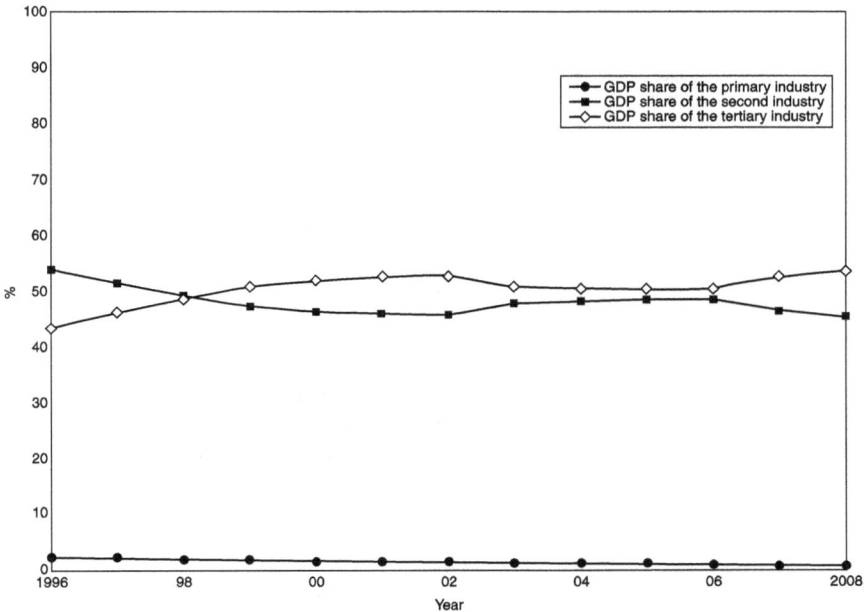

Figure 2.8 Structural change of three industries in Shanghai (1996–2008).

industry in Shanghai is only 1–2 per cent, far less than 19 per cent of national level; thus it does not greatly influence emission abatement. But tertiary industry in Shanghai developed very quickly; its output share has exceeded secondary industry since 1999 and accounted for more than 50 per cent of total GDP in Shanghai. The majority of primary energy consumption in tertiary industry in Shanghai, represented by the transport industry is cleaner oil rather coal, totally different from the situation across China; this structural evolution is beneficial to CO_2 emission abatement. Of course, the development of tertiary industry in Shanghai is not enough; its GDP share is still similar to secondary industry in recent years (about 50 per cent in 2003–2006, and 54 per cent in 2008). Therefore, to achieve the goal of substantial emission reduction, it is necessary to radically develop the tertiary industry in Shanghai by building international financial, shipping, trade and economic centres before the deadline of 2020 set by central government.

2.4.4 Energy structure

The change of energy structure in Shanghai has had the same positive influence as in China on the abatement of CO_2 emission, but to a greater degree. Figures 2.9 and 2.10 exhibit the composition by type of primary energy consumption in China and Shanghai. For China as a whole, coal is the main consumed energy, accounting

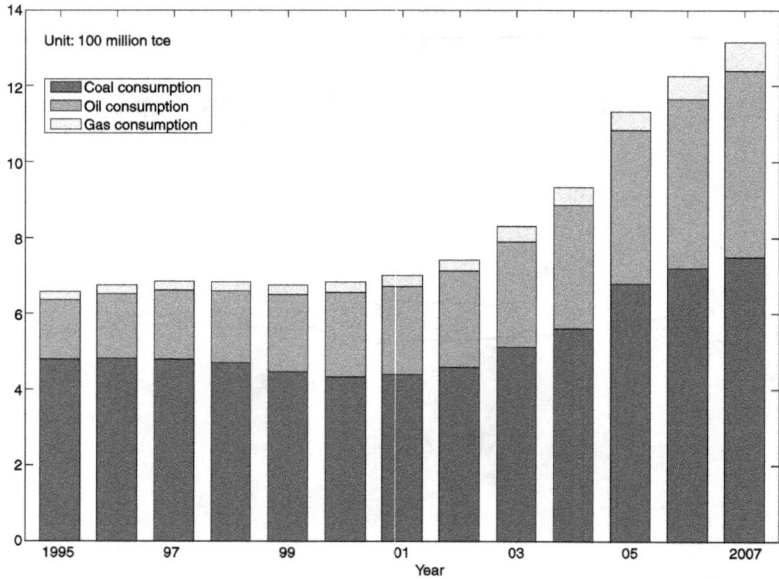

Figure 2.9 Three primary energy consumptions and their composition in China (1995–2007).

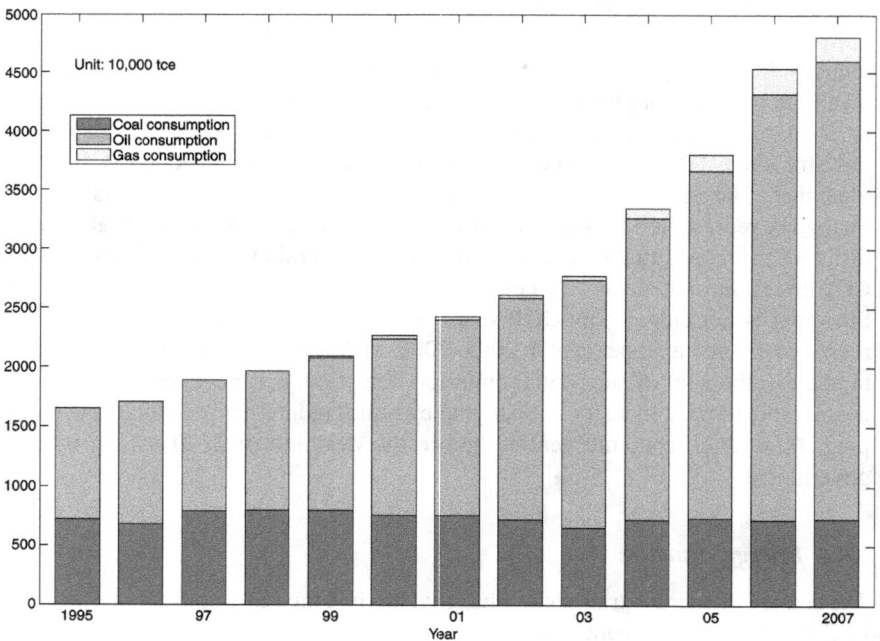

Figure 2.10 Three primary energy consumptions and their composition in Shanghai (1995–2007).

for about 64 per cent of total energy consumption on average, but its share decreased from 73 per cent in 1995 to 57 per cent in 2007. The consumption shares of oil and gas all rise. Because the carbon emission coefficient of coal is higher than that of oil and gas, the evolution of the energy structure in China is helpful for the reduction of CO_2 emission. The energy structure in Shanghai is totally different from that in China. The average share of oil is as high as 68 per cent and that of coal is only 30 per cent. And the consumption share of oil rose from 57 per cent in 1995 to 81 per cent in 2007, but the share of coal decreased accordingly, from 43 per cent to 15 per cent. The change of energy structure in Shanghai is bound to abate emission more than that in China.

Figure 2.11 depicts the type composition of energy consumption for six industries in China. Overall, among these, agricultural, construction and commercial industries consume far less energy; the main energy consumers are industry, transport and households. Though the absolute increase is large, the share of coal in industry, transport and household consumption is decreasing. For example, coal share in the largest consumer of industry fell from 77 per cent in 1995 to about 72 per cent in 2001 and 2007. The coal share in the household sector decreased more, from 90 per cent in 1995 to 76 per cent and 57 per cent in 2001 and 2007, respectively. The newly developing industry of transport has a larger proportion of oil consumption, rather coal; and its oil consumption share increased

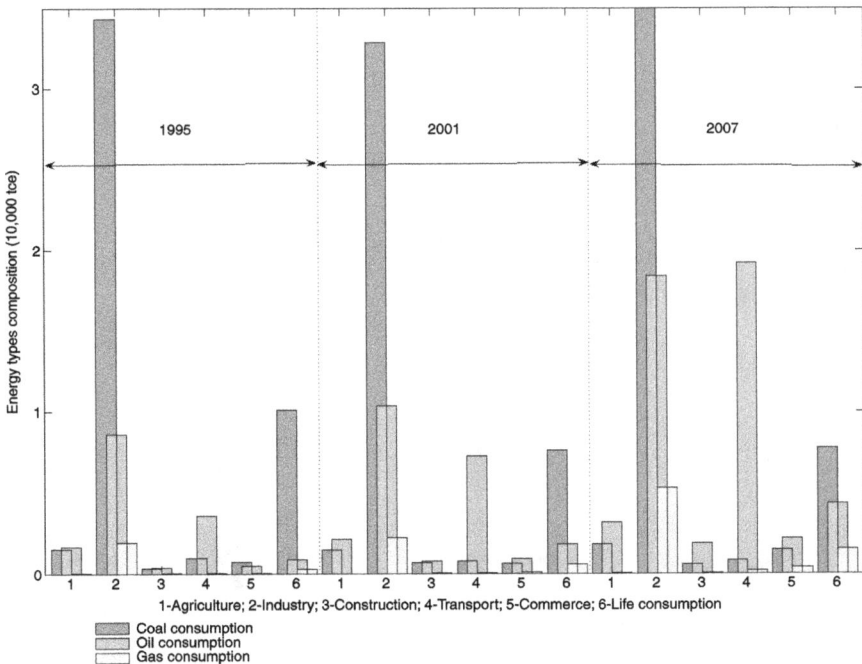

Figure 2.11 Types composition of energy consumption for six industries in China.

fastest (from 79 per cent in 1995 to 90 per cent and 95 per cent in 2001 and 2007) while its coal share decreased (from 21 per cent to 10 per cent and 4 per cent, respectively). The change of energy structure, where coal share falls and oil share rises, is helpful for the reduction of carbon emission. Because the adjustment of the energy structure is restricted by the endowment of energy resources, its effect on CO_2 emission is relatively small. For instance, China and the US are the top two coal consumers in the world, and the quantity of electricity generation based on burning coal makes them the top two greenhouse gas emitters, too. Due to the relatively rigid energy structure, with high coal consumption, it is difficult to reduce the emissions and their intensity by adjusting the energy structure in the short run

Figure 2.12 displays the type composition of energy consumption for six industries in Shanghai. Similar to China, the agricultural, construction and commercial industries use less energy and the main energy consumers are industry, transport and household life. The share of coal in the industry and household sectors decreased, too. For example, the coal share in the largest consumer, industry, decreased from 49 per cent in 1995 to 47 per cent in 2001 and 27 per cent in 2007. The coal share of life activities consumption fell more, from 82 per cent in 1995 to 47 per cent and 19 per cent in 2001 and 2007, respectively. Corresponding

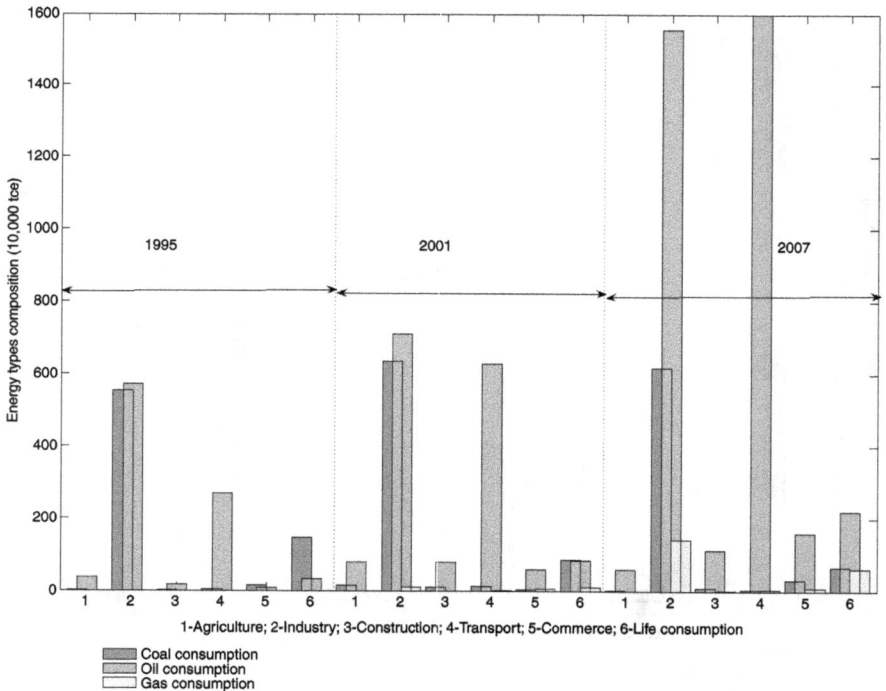

Figure 2.12 Types composition of energy consumption for six industries in Shanghai.

with this, the oil consumption share in the industry and household sectors has increased to 67 per cent and 63 per cent in 2007 (the values in China are 21 per cent and 32 per cent). Unlike China, the oil consumption share in the newly developing transport industry in Shanghai is extremely high, keeping constant at about 99 per cent; on the contrary, the coal and gas share in the transport industry is very low and hardly varies, which could tell us why the oil consumption proportion in Shanghai is so high. Overall, the change of energy structure is good for emission reduction. With the increasing share of clean energy and renewable energy, the adjustment of the energy structure will play a larger role in abating CO$_2$ emission.

2.5 Summary and comments

During the period of the ninth Five-Year Plan, carbon dioxide emission in China experienced a standstill and even decreased, but that in Shanghai was different from other provinces and grew by about 5.7 per cent; during the period of the tenth Five-Year Plan, both China and Shanghai increased CO$_2$ emission by 10 per cent; between 2006 and 2007, the growth rate of CO$_2$ emission in China reduced to 7 per cent but that in Shanghai rose further to 12 per cent. The decomposition results can be summarized as follows.

In both China and Shanghai, the production sector influenced the CO$_2$ emission much more than the household sector; this is different from the situation in developed countries in which the important sector to abate emission is the household one. According to the newly specified capital scale factor, capital deepening is mainly responsible for the rise of CO$_2$ emission in China and especially Shanghai. Thus, it will be very difficult to substantially save energy and reduce carbon emission without the transformation of the economic model, particularly changing factors such as being capital driven to becoming productivity driven. Energy intensity is the important factor to reduce CO$_2$ emission, by overcoming the rise of emission due to capital deepening. The reciprocal of energy intensity is economic productivity. Because almost all the energy consumption in Shanghai comes from other provinces in China or is imported from other countries, improving energy utilization efficiency or energy productivity becomes particularly important for Shanghai in the future.

The change of industrial structure has a greater effect on CO$_2$ emission change than energy structure. The industry and transport industry should become the main industries to abate emission. In Shanghai, the transport sector should be emphasized more because its energy intensity is far higher than that in industry. The strategy to develop tertiary industry is helpful for the reduction of CO$_2$ emission in both China and Shanghai. In Shanghai, tertiary industry needs developing further by building international financial, shipping, trade and economic centres. The important way to do so in Shanghai is to transform the government driven investment mode and focus on the substantial development of market forces. As for the energy structure, the coal and oil share in Shanghai is 30 per cent and 68 per cent on average, better than that in China (the corresponding values being

64 per cent and 32 per cent). Compared with developed countries, however, the gas consumption share in Shanghai is very low, at only 2 per cent. In Shanghai, the plan in the near future is to increase the gas share to 10 per cent by such measures as increasing gas resources, developing the gas market and generating electricity by using gas rather the traditional coal.

3 How can industrial CO_2 emission intensity be reduced?

3.1 Introduction

To control greenhouse gas emission and transform the traditional economic growth model, China decided to reduce the carbon dioxide (CO_2) emission per GDP (i.e. CO_2 intensity) by 40–45 per cent between 2005 and 2020. This was the first time that the Chinese government proposed quantitative abatement target of CO_2 emission. Though it is only a relative target to reduce the emission per GDP rather the absolute emission level, some researchers still argue that it is too high to be achieved. This chapter aims to discover the factors driving the decline of energy-induced CO_2 intensity in China between 1980 and 2008 using the decomposition technique. It will also derive corresponding policy suggestions from the decomposition that will be necessary to achieve the required reduction of carbon intensity by 2020. In accordance with this research objective and the applicability and availability of the database, for the remainder of this book (with the exception of Chapter 8), we will investigate total energy consumption and its induced CO_2 emission, rather than the end-use which was examined in Chapter 2, and focus on the energy consumption and CO_2 emission in industry.

3.1.1 Absolute CO_2 emission in China

It is well known that the economic growth in China has been achieved through a high level of investment, energy consumption and waste emission, which are still extensive in nature. As shown in Figure 3.1, national CO_2 emission in China increased from 0.15 billion tons in 1953 to a peak of 3.5 billion tons in 1996. During this time, emission growth increased steadily, the only exception being the Great Leap Forward (1958–1961). After 1996, CO_2 emission started declining and, on some occasions, came to a standstill, over five to six years. It then increased sharply to 6.7 billion tons in 2008. In comparison with other countries, as shown in Figure 3.2, CO_2 emission in China was low in the 1960s and 1970s, similar to that in Germany and Japan, but then became higher and higher, exceeding Germany, Japan and India and approaching the same level as the US. China has became the largest emitter of CO_2 in absolute terms in 2008 (CEACER, 2009).

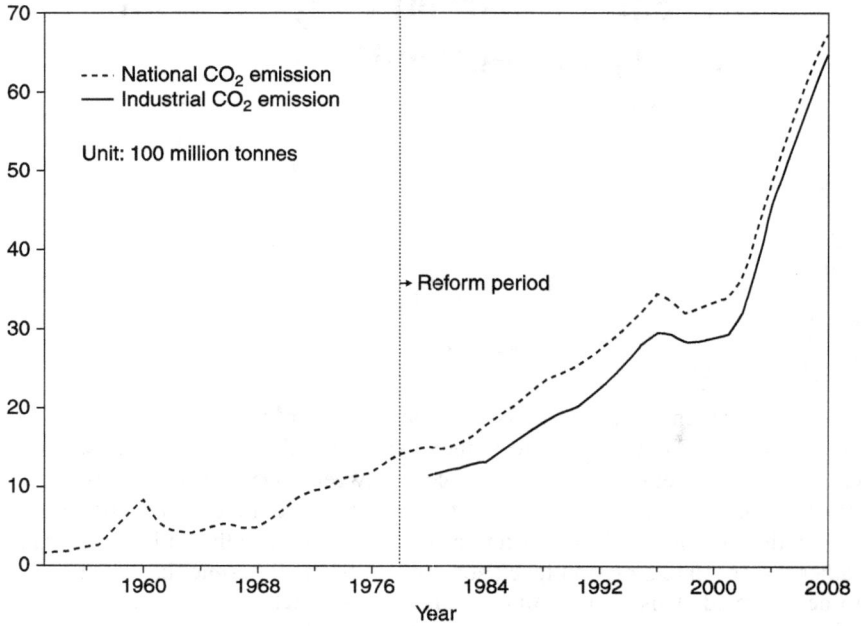

Figure 3.1 CO$_2$ emissions in China (1953–2008).

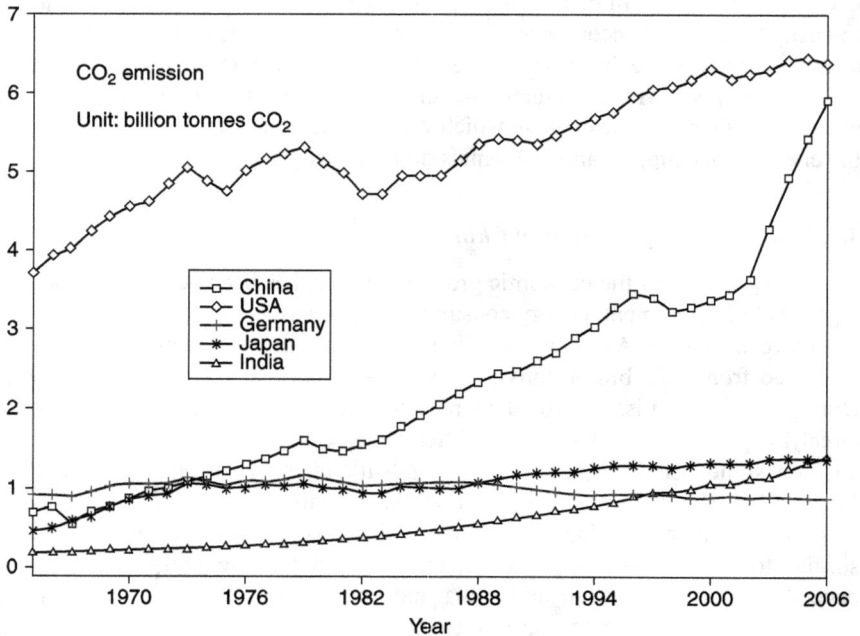

Figure 3.2 CO$_2$ emissions compared with four nations (1965–2006).

As denoted above, industrial CO_2 emission will be the main subject of this chapter because industry is the main emitter; main factors include the burning of fossil fuels and the manufacture of cement, lime, iron and steel. As illustrated in Figure 3.1, during the reform period, industrial output accounted for only 40 per cent of national output but industrial CO_2 emission constituted the majority of the total national CO_2 emission – the share being 84 per cent on average and more than 90 per cent since the beginning of this century. It is clear that the industriali-zation and urbanization of China will carry on. As a result, energy- and emission-intensive industrial sectors will continue to play a fundamental role in the economy. Therefore, a detailed analysis of carbon emission in industry is crucial for our understanding of emission abatement in China as a whole.

3.1.2 CO₂ intensity in China

Although industrial CO_2 emission is growing in China (except for the period between 1997 and 2001), Figure 3.3 reveals that industrial CO_2 intensity basically followed a decreasing trend during the whole reform period[1]. Industrial CO_2 intensity is defined as CO_2 emission per unit of industrial value-added, the reciprocal of which is carbon productivity. Thus, the decline of CO_2 intensity is equivalent to the improvement of carbon productivity, which implies that Chinese industrial CO_2 abatement is substantial and efficient during the reform period and

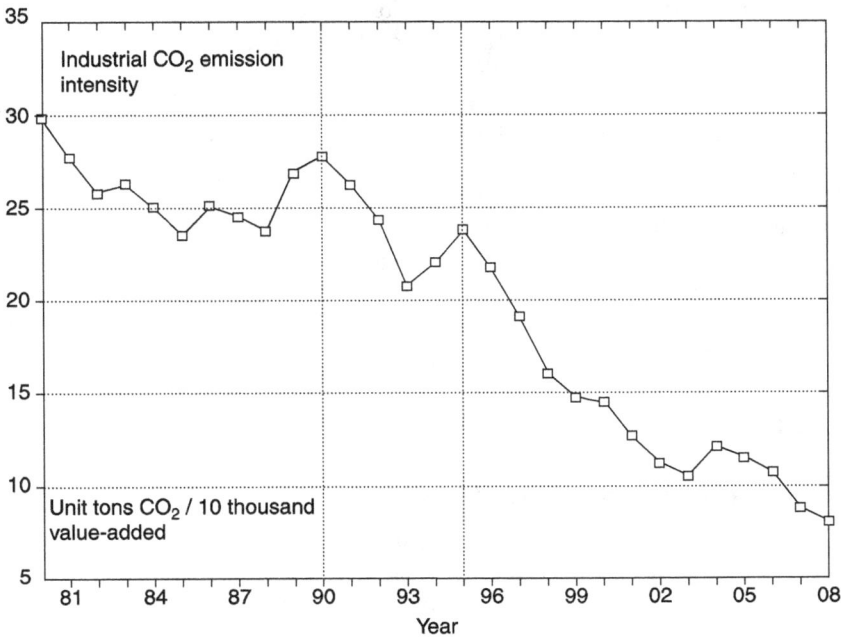

Figure 3.3 Industrial CO_2 intensity (1980–2008).

the industrialization process in China is also the low carbonization process simultaneously.

As mentioned above, because CO_2 emission mainly results from the combustion of fossil fuels, the abatement of CO_2 is therefore closely related to the consumption of fossil energy. In response to shortages of energy, the Chinese government formulated an energy-saving policy in 1980; this, was coupled with the fast development of light industry represented by town and village enterprises (TVEs), and an early decline in CO_2 intensity can be seen in Figure 3.3. The early reform of the energy industry was implemented by encouraging the production of energy. The resulting rapid development of energy- and emission-intensive small enterprises (such as the coal mines) relieved the tight energy supply caused by the long-term command economy but caused serious coal resources waste and environmental pollution. Thus, CO_2 intensity began to rise in the late 1980s and the government had to clean up the coal market in 1989. It was not until the concept of sustainable development became accepted worldwide in the early 1990s that the consciousness of environmental protection took root in China. The government once again emphasized the energy policy aim of 'saving and developing energy simultaneously' and began to restrict the development of energy enterprises instead of the encouragement given in the 1980s. This lead to the decline of CO_2 intensity again in the 1990s, as shown in Figure 3.3, especially evident in the largest reduction rate of CO_2 intensity (56 per cent) from 1995 to 2003, which basically corresponded to the reform of State-Owned Enterprises (SOEs) by 'grasping the large and letting go of the small' (*zhuada fangxiao*). The phenomenon of heavy industrialization re-emerged after the turn of the century. CO_2 intensity rose to 12 tons per 10 thousand value-added in 2004, but continued to decrease after 2004. This may have been due to the implementation of new energy-saving and emission-abating policies, such as the 'Outlines of China Medium and Long Term Energy Saving Plan 2004–2020' enacted in 2004, the quantitative goal to save energy intensity by 20 per cent in the eleventh national Five-Year Plan proposed in 2006, and the 'China National Plan for Coping with Climate Change' released in 2007. What was driving the decline of energy-induced CO_2 intensity during the reform period? And what policy implications could be derived to achieve future abatement as scheduled? These two questions will be answered through the analysis in the following parts of this chapter.

3.2 Literature review

3.2.1 Survey of decomposition techniques

The variables to be decomposed in the energy and environmental area are normally energy consumption, energy intensity, energy elasticity, and carbon emission or carbon intensity in additive/multiplicative decomposing forms. The decomposition techniques normally include index decomposition, input and output structural decomposition, and the nonparametric distance function based decomposition.

Ang and Zhang (2000) survey 124 papers published before 1999, 109 of which use index decomposition. Ang *et al.* (2003) contains a complementary survey of decomposition literature after 1999.

Index decomposition is represented by Laspeyres index and Paasche index decomposition and was extensively used in the 1970s and 1980s; see Doblin (1988), Park (1992) and Ang (1993), among others. The Laspeyres index approach is also extended to decompose the structural effect from labour productivity, namely the shift-share approach (Timmer and Szirmai, 2000). Boyd *et al.* (1987) propose another arithmetic mean Divisia index (AMDI) decomposition approach, and use it to analyse the industrial energy consumption in the US. Liu *et al.* (1992) further put forward the adaptive weighting Divisia index (AWDI) decomposition approach, which became popular in the 1990s; see Greening *et al.* (1998), Fisher-Vanden *et al.* (2004, 2006), Liu (2006) and Fan *et al.* (2007b).

Before 1995, two imperfections of decomposition methods remained to be resolved; they were the existence of decomposition residual (the largest residual term was for Laspeyres index decomposition and the lower residual for AWDI decomposition) and the calculation difficulty due to zero value. Then, Sun (1998) proposed a modified Laspeyres index decomposition approach to completely decompose the residual according to the principle of 'jointly created and equally distributed'. This approach was adopted to decompose the energy consumption in China by Zhang (2003) and Steenhof (2006), among others. Ang and Choi (1997) and Ang *et al.* (1998), respectively, put forward the modified Divisia decomposition approach in additive and multiplicative forms, namely the Logarithmic Mean Divisia Index (LMDI) decomposition method, and considerably improved two of the imperfections. As the Fisher index required, the ideal index should satisfy three tests (i.e. time-reversal, circular, and factor reversal tests); and only the modified Laspeyres decomposition method and LMDI method are able to pass these tests. That is because the modified Laspeyres method is normally used in additive decomposition to undertake incremental decomposition, while the LMDI approach could take both the additive and multiplicative decomposition forms, and so the latter is used more extensively in application (Ang, 2004). Many papers choose the LMDI approach to decompose the target variables (Wang *et al.*, 2005a; Liu *et al.*, 2007). Other complete decomposition methods such as mean rate-of-change index (MRCI) by Chung and Rhee (2001) and the Shapley method by Albrecht *et al.* (2002) are still new and hardly used in the literature.

Input and output structural decomposition makes use of comparative static technique to decompose the structural effect from energy consumption or carbon dioxide emission, which could be regarded as the detailed version of Laspeyres decomposition method (see surveys by Rose and Casler, 1996). Constrained by the availability of I-O tables, this approach can only undertake a periodwise analysis. Many studies use the structural decomposition methods; see, for example, Chang *et al.* (2008), Guan *et al.* (2008), Kahrl and Roland-Holst (2009), Weber (2009), Wood (2009) and Zhang (2010). The third decomposing method in terms of nonparametric output distance function can decompose energy productivity into technical efficiency and technical progress; see Wang (2007).

3.2.2 Survey of decomposition of CO$_2$ emission or intensity in China

The main literature on factors decomposition of Chinese CO$_2$ emission is reviewed below. Wang *et al.* (2005a) decomposes Chinese CO$_2$ emission in 1957–2000 and finds that the most important factor in reducing CO$_2$ emission is the reduction of energy intensity. The secondary factors are energy structure, investment in renewable energy and economic growth. Fan *et al.* (2007b) analyse the influential factors of Chinese carbon intensity in 1980–2003 through decomposition and find that the decline of carbon intensity results mainly from the decline of energy intensity and then the change of energy structure. Chang *et al.* (2008) decompose the factors influencing Taiwan CO$_2$ emission in 1989–2004, and find the important factors are change of energy intensity and energy structure, export level and domestic final demand, among others. Zhang (2010) concludes that the structural variables in the supply side, measured by sectoral value-added share, are important influential factors by using structural decomposition. In 1992–2002, the rapid growth of manufacturing industry boosts carbon emission, while in 2002–2005, the share decline of carbon extensive sectors leads to the decrease in CO$_2$ emission.

3.3 LMDI decomposition technique

This chapter chooses the multiplicative Logarithmic Mean Divisia Index (LMDI) method to decompose the overall industrial CO$_2$ intensity in China, symbolized by *CI*, from two dimensions of 38 industrial sectors and 3 types of primary energy (coal, oil and gas), represented by the subscripts *i* and *j*, respectively (that is, $i = 1,2, \ldots, 38; j = 1,2,3$). The variables *Y* and *C* are defined to represent the overall industrial value-added and industrial CO$_2$ emission. The symbols of C_{ij}, E_{ij}, EC_{ij} and ES_{ij} represent CO$_2$ emission, energy consumption, CO$_2$ emission coefficient and consumption structure of energy types for *i*th industrial sector and *j*th energy type. The four variables, E_i, Y_i, EI_i and S_i, are the energy consumption, industrial value-added, energy intensity and industrial structure (industrial value-added share) of the *i*th sector. The overall industrial CO$_2$ intensity can be expressed equivalently as:

$$CI = \frac{C}{Y} = \frac{\sum_{i=1}^{38}\sum_{j=1}^{3} C_{ij}}{Y} = \sum\sum_{ij} \frac{C_{ij}}{E_{ij}} \cdot \frac{E_{ij}}{E_i} \cdot \frac{E_i}{Y_i} \cdot \frac{Y_i}{Y}$$

$$= \sum\sum_{ij} EC_{ij} \cdot ES_{ij} \cdot EI_i \cdot S_i \tag{3.1}$$

Define the following symmetrical logarithmic weighting function:

$$L(a,b) = \begin{cases} \dfrac{a-b}{\ln a - \ln b} & a \neq b \\ a & a = b \end{cases} \tag{3.2}$$

Therefore the chain index of overall industrial CO$_2$ intensity can be decomposed into the following four influential terms by using LMDI multiplicative approach:

$$RCI = CI^t/CI^{t-1} = RCI_{ec} \cdot RCI_{es} \cdot RCI_s \cdot RCI_{ei}$$

$$= \exp\left[\sum\sum_{ij} \frac{L(CI_{ij}^t, CI_{ij}^{t-1})}{L(CI^t, CI^{t-1})} \ln\left(\frac{ES_{ij}^t}{ES_{ij}^{t-1}}\right)\right]$$

$$\cdot \exp\left[\sum\sum_{ij} \frac{L(CI_{ij}^t, CI_{ij}^{t-1})}{L(CI^t, CI^{t-1})} \ln\left(\frac{S_i^t}{S_i^{t-1}}\right)\right]$$

$$\cdot \exp\left[\sum\sum_{ij} \frac{L(CI_{ij}^t, CI_{ij}^{t-1})}{L(CI^t, CI^{t-1})} \ln\left(\frac{EI_i^t}{EI_i^{t-1}}\right)\right] \tag{3.3}$$

where t and $t-1$ are adjacent time points. *RCI* represents the overall development index of CO$_2$ intensity. RCI_{ec}, RCI_{es}, RCI_s and RCI_{ei} are decomposed four chain factor development index; that is, the carbon emission coefficient index, energy structure index, industrial structure index and energy intensity index, respectively. Because the carbon emission coefficient of three types of primary energy is assumed to be fixed when calculating CO$_2$ emission, the RCI_{ec} term on the right side of equation (3.3) reduces to 1 in fact, and the number of final decomposed terms is only three.

The LMDI approach leads to perfect decomposition without the produce of residual term, as proved by the following algebraic calculation and manipulation:

$$RCI/(RCI_{ec} \cdot RCI_{es} \cdot RCI_s \cdot RCI_{ei})$$

$$= \frac{CI^t}{CI^{t-1}} /\exp\left[\sum\sum_{ij} \frac{L(CI_{ij}^t, CI_{ij}^{t-1})}{L(CI^t, CI^{t-1})} \ln\left(\frac{EC_{ij}^t ES_{ij}^t S_i^t EI_i^t}{EC_{ij}^{t-1} ES_{ij}^{t-1} S_i^{t-1} EI_i^{t-1}}\right)\right]$$

$$= \frac{CI^t}{CI^{t-1}} /\exp\left[\sum\sum_{ij} \frac{L(CI_{ij}^t, CI_{ij}^{t-1})}{L(CI^t, CI^{t-1})} \ln\left(\frac{CI_{ij}^t}{CI_{ij}^{t-1}}\right)\right]$$

$$= \frac{CI^t}{CI^{t-1}} /\exp\left[\frac{\sum\sum_{ij}(CI_{ij}^t - CI_{ij}^{t-1})}{L(CI^t, CI^{t-1})}\right]$$

$$= \frac{CI^t}{CI^{t-1}} /\exp\left[\frac{CI^t - CI^{t-1}}{L(CI^t, CI^{t-1})}\right] = \frac{CI^t}{CI^{t-1}} /\exp\left[\ln\frac{CI^t}{CI^{t-1}}\right] = 1 \tag{3.4}$$

The choice of decomposition techniques depends on the features of the database available. The panel data for 38 industrial sectors between 1980 and 2008, employed in this book, make the year to year decomposition possible by using the index decomposition instead of input and output structural decomposition only suitable for periodwise analysis. The LMDI decomposition method in multiplicative form, as introduced previously in this section, is finally chosen to construct the chain development index, preferably analysed in this chapter rather than any versions of Laspeyres index decomposition that could only undertake the incremental decomposition in additive form. As opposed to all the literature reviewed in Section 2.2, the main contribution of this chapter is that it looks at the largest number of sectors

(38 sectors) as one dimension to decompose CO_2 intensity in the longest sample period (29 years), making it possible to investigate the varying pattern of target variables over the whole reform period within the sector of Chinese industry. Unlike other studies, this chapter focuses on explaining the varying patterns of intensity and structural factors according to historical experience and economic policies, rather than providing just description of various decomposed factors.

3.4 Factors decomposition of industrial CO_2 intensity in China

Figure 3.4 depicts the trend of chain development index of CO_2 intensity and its influential factors by decomposition in Chinese industry (averaged over all sectors). Obviously, the index of CO_2 intensity fluctuated very much but exhibited negative growth in many years (that is, it exhibits positive growth for only seven years among almost 30 years, and since 1995 was only positive in 2004). Figure 3.4 also reveals that the majority of the varying pattern of the CO_2 intensity index could be explained by the index of energy intensity because there is much overlapping of the two index lines to reflect their similar volatility. The indices of energy structure and industrial structure have a weaker correlation with, and explanation of, the variation of carbon intensity index than that of energy intensity. Seen from the year to year decomposition, and its average over four subperiods and the whole period reported in Table 3.1, the index values of CO_2

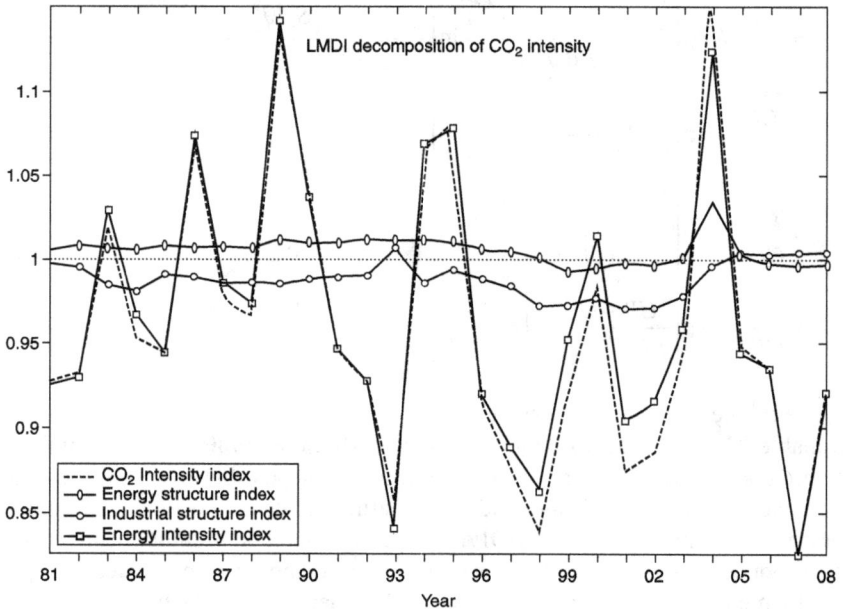

Figure 3.4 Industrial CO_2 intensity index and its influential factors through decomposition (1981–2008).

Table 3.1 Industrial CO₂ Intensity Index and its LMDI Decomposition in China (1980–2008)

Periods	CO₂ Intensity Index	Factor Indexes through LMDI Decomposition			Periods	CO₂ Intensity Index	Factor Indexes through LMDI Decomposition		
		Energy Structure	Industrial Structure	Energy Intensity			Energy Structure	Industrial Structure	Energy Intensity
1981/1980	0.9270	1.0052	0.9966	0.9254	1998/1997	0.8388	0.99999	0.9719	0.8631
1982/1981	0.9314	1.0079	0.9946	0.9292	1999/1998	0.9186	0.9925	0.9722	0.9520
1983/1982	1.0187	1.0060	0.9843	1.0288	2000/1999	0.9829	0.9939	0.9763	1.0130
1984/1983	0.9518	1.0051	0.9804	0.9659	2001/2000	0.8741	0.9976	0.9700	0.9033
1985/1984	0.9414	1.0076	0.9900	0.9437	2002/2001	0.8850	0.9963	0.9706	0.9153
1986/1985	1.0663	1.0061	0.9887	1.0719	2003/2002	0.9368	1.0003	0.9778	0.9578
1987/1986	0.9764	1.0070	0.9843	0.9851	2004/2003	1.1549	1.0332	0.9958	1.1226
1988/1987	0.9656	1.0061	0.9859	0.9735	2005/2004	0.9471	1.0016	1.0027	0.9431
1989/1988	1.1352	1.0108	0.9848	1.1404	2006/2005	0.9340	0.9966	1.0021	0.9352
1990/1989	1.0333	1.0094	0.9881	1.0361	2007/2006	0.8238	0.9957	1.0032	0.8247
1991/1990	0.9433	1.0088	0.9887	0.9457	2008/2007	0.9202	0.9962	1.0035	0.9205
1992/1991	0.9286	1.0115	0.9897	0.9276	1980–1990	0.9883	1.0069	0.9877	0.9938
1993/1992	0.8545	1.0104	1.0062	0.8405	1990–1995	0.9615	1.0101	0.9915	0.9600
1994/1993	1.0624	1.0104	0.9852	1.0673	1995–2004	0.9206	0.99995	0.9781	0.9412
1995/1994	1.0794	1.0096	0.9933	1.0764	2004–2008	0.9502	1.0045	1.0015	0.9445
1996/1995	0.9132	1.0051	0.9878	0.9198	1980–2008	0.9545	1.0048	0.9877	0.9618
1997/1996	0.8775	1.0044	0.9834	0.8884					

intensity and energy intensity are much closer, but those of CO$_2$ intensity and the energy/industrial structure are far from each other[2] During the whole reform period, the development index of industrial CO$_2$ intensity reduced by 4.55 per cent on average, within which the energy intensity index reduced by 3.82 per cent, the industrial structure index reduced by 1.23 per cent but the energy structure index rose by 0.48 per cent. The detailed explanation of three influential factors on the variation of CO$_2$ intensity will be addressed next.

3.4.1 Energy structure effect

As illustrated by Figure 3.5, coal constitutes the main primary energy consumption in China, followed by oil and gas. In primary energy consumption in Chinese industry, coal accounts for 74.2 per cent on average, making China one of the few countries to rely on coal. That explains why the CO$_2$ intensity is so high in China when compared with other countries. Evidently the CO$_2$ emission coefficient of coal is higher than that of oil and gas. It follows that the variation of energy structure has an impact on the carbon intensity.

As shown in Figure 3.5, the share of coal has gradually risen from 68 per cent at the early reform period to a peak of 78.4 per cent in 1995; this corresponded to the energy policy of encouraging energy production implemented in 1980s to solve the shortage caused by the long-run planned economy. Similar to the industrial reform to increase operating autonomy in urban areas, energy intensive industry (such as coal mining, petrochemical and electric power enterprises) has undertaken the reform of contract responsibility system, one after the other. Among them, the coal industry is the one that has experienced the most thorough reforms, and where competition is fiercest because the State decentralized authority to local governments, which then relaxed the entry barriers for coal

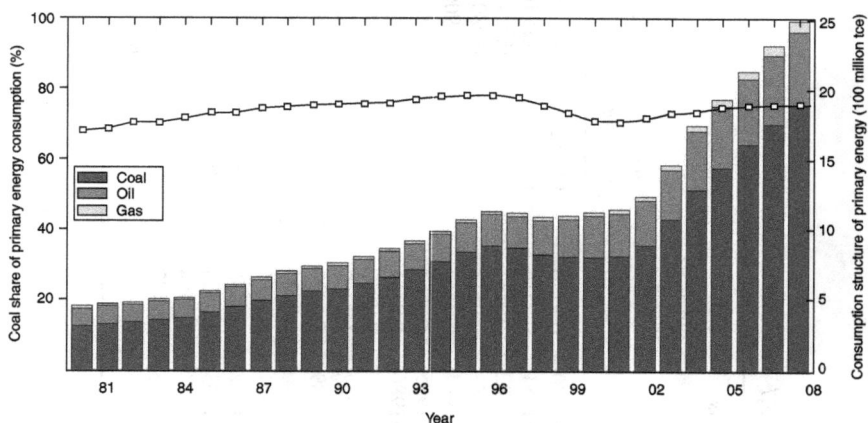

Figure 3.5 Structure of primary energy consumption in Chinese industry (1980–2008).

enterprises and encouraged townships, communes and brigades (*shedui*), individuals and foreign investors to mine coal. As a result, there was a surge in small coal mining all over the country for a time. In 1994, the dual-track price of coal was first phased out of the state plan and then fully liberalized, as opposed to the situation for product oil and electric power. It was then that the coal market was fully developed.

However the production of coal eventually ran out of control. The destructive exploitation and the lack of environmental protection measures caused serious damage and waste of coal resources and environmental pollution. As shown in the sub-periods of Table 3.1, from 1980 to 1995, energy structure was the only factor to drive CO_2 intensity in reverse, the latter reducing by 1.17 per cent–3.85 per cent but the former rising by 0.69 per cent–1.01 per cent. From 1995 to this century, because the Chinese government closed down many small energy- and emission-intensive enterprises,[3] and started to restrict the development of energy enterprises instead of the encouragement provided in the 1980s, the share of coal consumption decreased for the first time during the whole period – with the lowest being 70.89 per cent in 2000. Thus, this was the only phase (1995–2004) when the energy structure had a positive effect on the reduction of CO_2 intensity and the indices of the three factors decrease simultaneously, though the value of the reduction rate of the energy structure index was very small, as shown also in Table 3.1. As heavy industrialization appeared again in this century, coal share started to rise, attaining the highest level (75.9 per cent) in 2008 (see Figure 3.5). Accordingly, the energy structure index increased by 0.45 per cent and that led to the final positive growth effect of the energy structure during the whole period and hindered the CO_2 intensity from reducing.

Of course, the overall influence of the energy structure on CO_2 intensity is comparatively small because the adjustment of the energy structure is restricted by the energy endowment in each country. For instance, China and US are the top two coal consumers in the world, and the burning of coal makes the two countries the top two CO_2 emitters as well. Therefore, because of an energy structure where coal reservation and consumption are the majority, it is difficult to abate CO_2 emission and its intensity by adjusting the energy structure in the short run.

3.4.2 Industrial structure effect

As shown in Figure 3.4 and Table 3.1, compared with the positive impact of the energy structure on CO_2 intensity in only one sub-period, the effect of industrial structure on the reduction of CO_2 intensity is overall positive, with the only exception at the last sub-period of heavy industrialization. The industrial structure is the main component of the economic structure. Adjustment in the industrial structural indicates that the production factors, such as capital, labour and energy, are re-flowed and reallocated among industrial sectors with different techniques, efficiency and profit. This leads to the changing of output share in different sectors. According to neoclassical growth theory, structural adjustment is an important source of sustainable growth and a radical way to transform the development model. Timmer and Szirmai (2000) refer to the positive effect of structural adjustment on economic growth as the structural bonus hypothesis.

In this chapter, industrial structural adjustment is characterized as the flow of production factors between light and heavy industries. Following Zhang *et al.* (2009), all sectors are divided into low and high energy groups according to the ranking of energy consumption in 2004, because heavy industry is normally assumed to use more energy and vice versa. We also classify all the sectors into low and high emission groups in term of the ranking of CO_2 emission in 2004. That is, the low emission group corresponds to the top half of sectors with the lower value, and the high emission group to the last half of sectors with the larger value of CO_2 emission Figure 3.6 describes the varying trend of value-added share for low and high energy and emission groups, respectively.

Since the advent of economic reform, China has revised its economic strategy from focusing on heavy industry in order to catch up with the developed world and have a more balanced approach to developing both light and heavy industries. This has led to the rapid growth of light industry in the following two decades, including TVEs, private enterprises and foreign funded enterprises. Thus, the value-added share of the low energy and emission group increased continuously in the 1980s and 1990s and that of the high energy and emission group decreased symmetrically, as illustrated in Figure 3.6. The same negative growth of both the

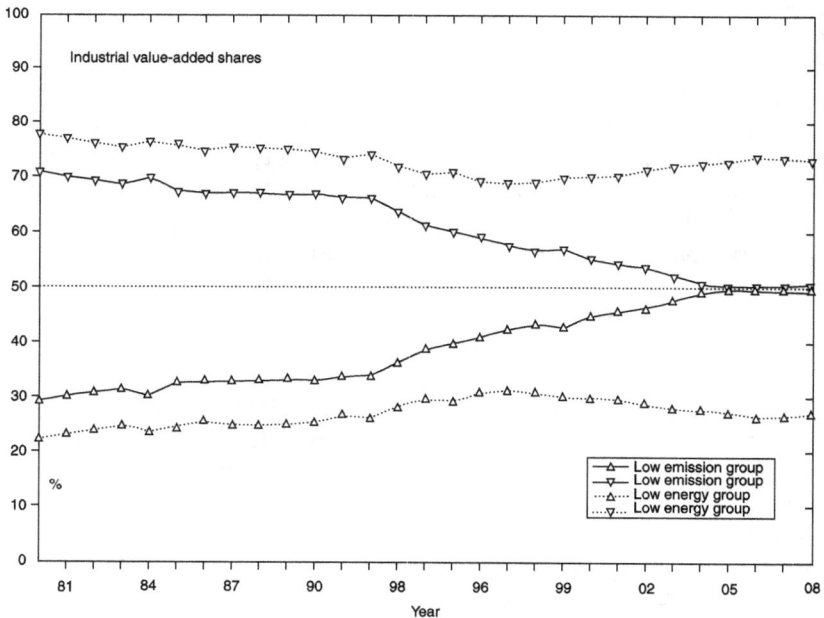

Figure 3.6 Change of output structure between low and high energy and emission groups (1980–2008).

industrial structure index and the CO_2 intensity index, revealed by Figure 3.4 and Table 3.1, implies that factors reallocation to more efficient sectors drives the reduction of carbon intensity and that a structural bonus exists.

Most importantly, since the SOEs' ownership reform encouraged 'grasping the large and letting go of the small' (particularly closing tens of thousands of small-scale emission intensive enterprises such as coal mines and power generation, as stated in the previous sub-section) and the furlough policy (*xiagang*) was implemented in the mid-1990s, the value-added share of the heavy emission group fell faster, imposing the biggest positive effect on the reduction of carbon intensity, though it is not obviously reflected in accordance with the low and high energy groups. In fact, in this phase, the development index of the industrial structure reduced by 2.19 per cent. Heavy industrialization reappeared after 2003. Figure 3.6 shows that the value-added share in the high emission group stopped the downward trend for almost 20 years and stayed level to that of the low emission group. The output share in the high energy group is shown to somewhat rise and go above 70 per cent. As shown in Figure 3.4 and Table 3.1, the index of the industrial structure reveals a positive growth by 0.15 per cent for the first time and its effect on the reduction of CO_2 intensity starts to be negative.

In sum, the effect of the energy structure and the industrial structure on CO_2 intensity reduction is not big but has a reverse influence – the former being negative and the latter, positive. Thus, one approach to new industrialization is to adjust the industrial structure, in which the important factors are continuously reallocated from energy and emission intensive groups to light industry with advanced technology, low energy consumption and CO_2 emission.

3.4.3 Energy intensity effect

Similar to the findings in most literature on decomposing CO_2 emission or its intensity, this chapter also finds that the decline of energy intensity is the main force to drive the reduction of CO_2 intensity, as the values of both the energy and the CO_2 intensity index are very close (see Table 3.1), and both index lines overlap in most periods (a significant gap is only visible in the mid and late 1990s as a result of a dramatic change in energy structure and industrial structure; see Figure 3.4). Since energy consumption causes CO_2 emission, and the abatement of CO_2 intensity radically depends on the decline of energy intensity or the promotion of energy efficiency (defined as 'the reciprocal of energy intensity'), there is an intimate relationship between emission abatement and energy saved.

In 1953, after three years of land reform, the national energy consumption was only 0.05 billion tons of coal equivalent (tce). In 1978, during the early reform period, it was 0.57 billion tce. In 2008, it had risen to 2.85 billion tce. Figure 3.7 depicts the trend of industrial energy consumption and its intensity, in sub-figure (a), and the sub-industrial contribution to industrial energy intensity reduction between 1980 and 2008 in sub-figure (b). Here, the weights to compute the aggregated industrial energy intensity and sub-industrial contribution are all gross industrial output value. Industrial energy consumption is 0.45 and 2.5 billion tce

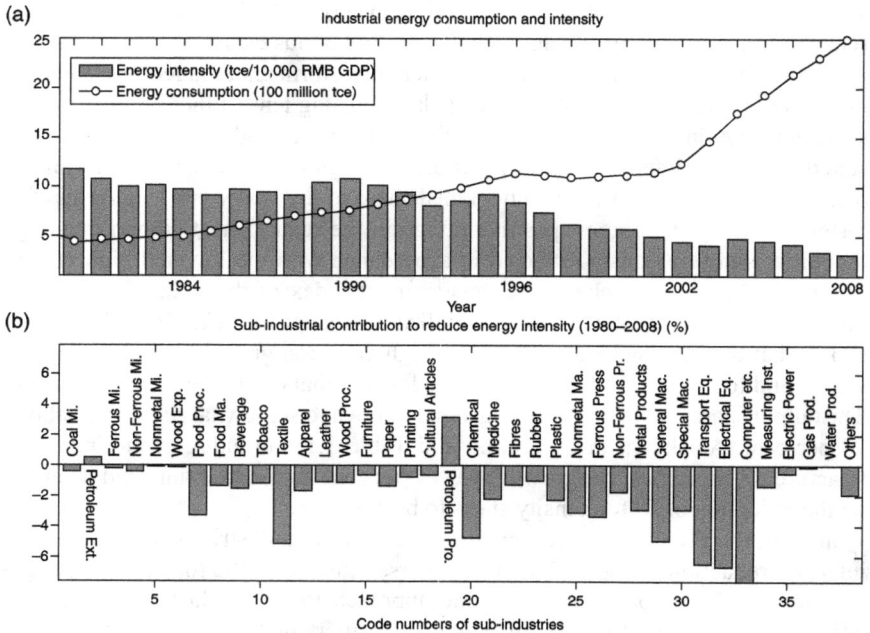

Figure 3.7 Industrial energy consumption, energy intensity and its sub-industrial contribution (1980–2008).

in 1980 and 2008, respectively, accounting for 74 per cent and 87 per cent of national consumption. This implies a sharp increase in the demand for energy in Chinese industry. Accordingly, the high shares of 76.1 per cent and 96.3 per cent of industrial CO_2 emission to national emission in 1980 and 2008, revealed in Figure 3.1, do not seem strange. The higher coal share in primary energy consumption and higher CO_2 emission coefficient contribute mostly to this fact.

Figure 3.3 and Table 3.1 have divided the whole reform period into four sub-periods according to the varying pattern of industrial CO_2 intensity. The change of energy consumption and its intensity in sub-figure (a) of Figure 3.7 displays a similar pattern. Before 1995, energy consumption rose steadily and energy intensity fell by 2.28 per cent on average, with fluctuation. As stated in the previous two sub-sections, during the radical ownership reform period from 1996 to the advent of this century, industrial energy consumption reversed the previous increasing trend and was at a standstill, and industrial energy intensity showed the biggest decrease of the whole period (from 9.14 tce per 10 thousand value-added in 1995 to 4.15 in 2003, with a reduced rate of 7.68 per cent on average), indicating a significant improvement of energy efficiency corresponding to the dramatic adjustment of energy and industrial structure. The re-emergence of heavy industrialization after 2003 is also shown in the variation of energy consumption and intensity. It could be attributable to the fanatical expansion of housing and car industries, the rapid urbanization, accelerated exports of energy and emission intensive products after the access into the World Trade Organization (WTO), the continuous and massive infrastructure investment, and the new entry of private capital into heavy industries due to the low price of natural resources. As described in sub-figure (a), industrial energy consumption rose at an unprecedent rate, and industrial energy intensity temporarily rose, too, during the long-term decreasing process. Due to this, the eleventh national Five-Year Plan in 2006 put forward the quantitative target to reduce energy consumption per GDP by 20 per cent between 2006 and 2010.

Sub-figure (b) in Figure 3.7 depicts the respective contribution of all 38 sub-industries to the energy intensity reduction of China's aggregated industry in 2008 relative to the year of 1980. To produce 10,000 RMB of industrial value-added, China's industry consumed 11.7 tce in 1980 and only 3.2 tce in 2008, the accumulated reduction rate being about 73 percentages. Specifically, manufacture of communication equipment, computers and other electronic equipment fell by 9 per cent and contributed the most to the total reduction. Such sectors as the manufacture of medicines, metal products, general purpose machinery, special purpose machinery, transport equipment, electrical machinery and equipment contributed between 2 and 6 per cent, individually, and caused the accumulated reduction of industrial energy intensity by 26 per cent. More than half of the energy intensity reduction was achieved by advanced technical industrial sectors, indicating that developing advanced technical industries benefits not only the information technique revolution but also the green industrial revolution for the future in China. Light industries are also the main contributors of industrial energy intensity reduction. For instance, the manufacture of textiles contributed 5 per cent of the energy intensity reduction. The sectors such as the processing of food from agricultural products, and the manufacture of foods, beverages, tobacco, textile wearing apparel, footwear and caps, leather, fur, feather and related products also contributed between 1 and 3 per cent, individually. Of course, traditional heavy industries also contributed much to the energy intensity reduction of aggregated industry in China. For example, the manufacture of raw chemical materials and chemical products reduced the energy intensity by almost 5 per cent. Other sectors such as the manufacture of plastics, the manufacture of non-metallic mineral products, smelting and pressing of ferrous metals, smelting and pressing of non-ferrous metals, the manufacture of chemical fibres, and the manufacture of paper and paper products also contributed between 1 and 3 per cent of the energy intensity reduction. In sum, 17 sectors in the heavy industries reduced the industrial energy intensity by an accumulative 21 percentages. Among 38 sectors, only two sectors played a negative role in reducing the energy intensity. They are the extraction of petroleum and natural gas, and the processing of petroleum, coking and processing of nuclear fuel, which increased the aggregated industrial energy intensity by 0.5 per cent and 3.1 per cent, respectively, and should merit more attention to increase their energy productivity.

In other words, though fluctuating, industrial energy intensity experienced an overall decreasing trend, indicating the continuous improvement of utilization efficiency of energy. Many studies find such factors historically driving the promotion of energy efficiency as the technology advances, including the adoption of new energy technologies and energy-saving and emission-abating technologies, R&D expenditure, reform of energy prices, ownership reform, and an industrial structural adjustment from the high to low emission group (Garbaccio et al., 1999b; Fisher-Vanden et al., 2004, 2006; Mukherjee and Zhang, 2007; Fisher-Vanden and Jefferson, 2008). Zhang et al. (2008) also note that the nature of the low carbon economic development model is to promote energy efficiency, and the core is technology innovation.

As shown in Figure 3.8, energy intensity in China reduces the quickest when compared with other economies, only rising a little after 2002. However, the absolute value of energy intensity in China is still too large and has much room to fall. For instance, in 2004, the Chinese energy intensity was 99,000 tons of oil equivalent per US$10,000 of GDP (price level of the year 2000), much higher than that of the US, Germany, Japan and Hong Kong (respectively, 23,000, 19,000, 11,000 and 14,000) and even higher than the value of 65,000 in India, a developing country. Figure 3.9 shows the energy consumption among different economies, and tells us that China is facing more and more pressure to reduce energy use. After reform, energy consumption in China increased rapidly and was almost close to that of the US. In 2006, energy consumption in China reached 1696.8 million tons oil equivalent (toe), only lower than the 2326.4 of the US, and far

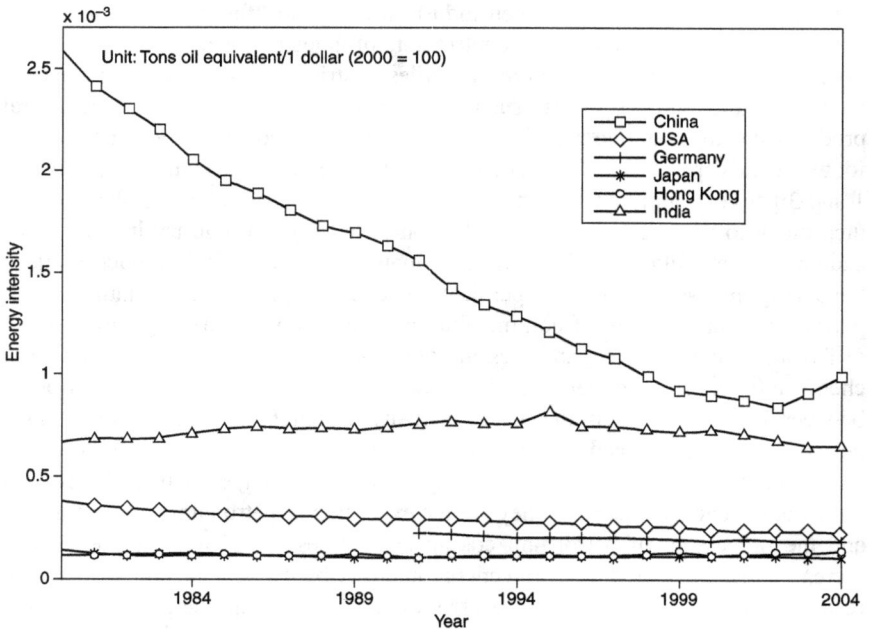

Figure 3.8 Energy intensity among different economies (1980–2004).

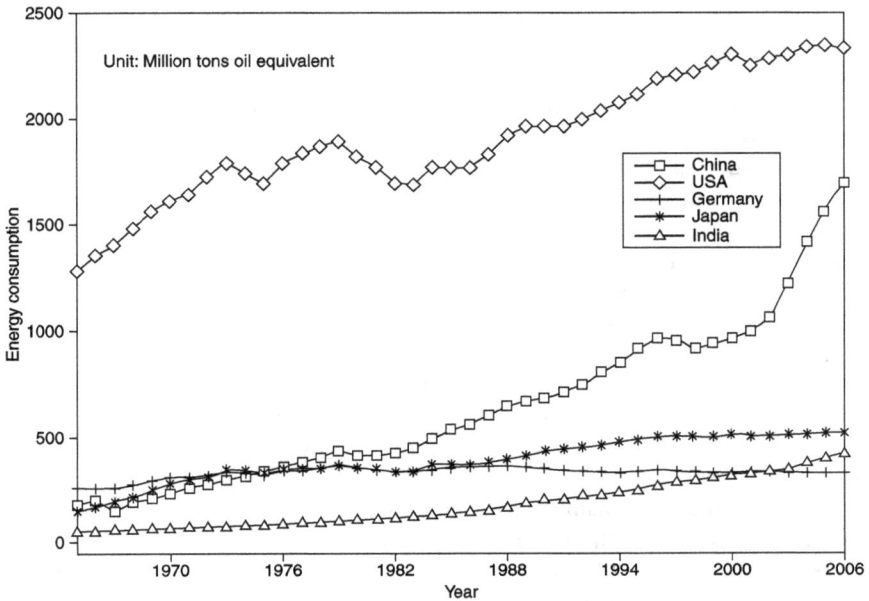

Figure 3.9 Energy consumption among different economies (1965–2006).

higher than that in Japan, India and Germany (520.3, 423.2 and 328.5, respectively). From the combination efficiency of energy processing, conversion, storage, transport and end use, different energy types and structure will produce different combination efficiency. Though the current energy utilization efficiency in China has increased by almost 10 per cent relative to 1980, according to Wu and Cheng (2006), it is too low as compared with other countries. For example, in 2004, the energy utilization efficiency in China was 36.48 per cent, much lower than 49.9 per cent, 51.43 per cent and 65.45 per cent in the US, Japan and France, indicating that China also has much room to increased its energy utilization efficiency.

3.5 Summary and comments

Calculating further using the data behind Figure 3.3, the CO_2 intensity of Chinese industry in 2008 was 8.1 tons CO_2 per 10,000 industrial value-added, while the values in 1980, 1995 and 2004 were 29.9, 23.9 and 12.1, respectively. That means it has abated by 72.9 per cent over the past 28 years, 66 per cent over the past 13 years and 33 per cent over the past four years. Based on such historical abating speed in industry, it is highly possible for China to realize the abating target of national CO_2 intensity proposed in 2009, that is, 40–45 per cent between 2005 and 2020. As shown in Figure 3.7, industrial energy intensity, the driving force behind

CO_2 intensity found in this chapter, is 4.5 tce in 2005 and decreases by 29 per cent to 3.2 tce in 2008, the shrinking magnitude of which also exceeds the required 20 per cent save of national energy intensity from 2006 to 2010 proposed in the eleventh Five-Year Plan. Such intensity indexes reveal that the industrial sector has saved energy and abated emission better than the whole state over recent years, though it does consume energy and emit CO_2 the most at the absolute level. Of course, such an optimistic conclusion reached above may be due to the calculation of industrial CO_2 and energy intensity by using the weighting average of nearly 40 industrial sectors, which is possibly different from the information behind the aggregation industry data directly provided by the official yearbooks.

As CEACER (2009) demonstrates, between 1980 and 2000, China quadrupled its economy but only doubled its energy consumption; that is, the elasticity coefficients of energy consumption was about 0.5, contributing mostly to the reduction of CO_2 intensity. However, during the period of the tenth Five-Year Plan (2001–2005), the re-emergence of heavy industrialization resulted in the elasticity coefficients of energy consumption being over 1 successively from 2003 to 2005, forcing the central government to put forward a quantitative energy-save target for the first time, namely, to reduce energy intensity by 20 per cent during the period of the eleventh Five-Year Plan (2006–2010). If China can achieve the 20 per cent reduction of energy intensity during each of the Five-Year Plans before 2020, it can continue to quadruple the economy in 2020 while only doubling energy consumption, as it did before the twenty-first century, leading to the successful achievement of the abating target of CO_2 intensity by 2020.

In terms of the factors decomposition of CO_2 intensity for 38 industrial sectors between 1980 and 2008, some conclusions and corresponding policy implications can be obtained.

The most important driving force for the reduction of Chinese CO_2 intensity is energy intensity. According to the decomposing results, to achieve the goal of successful abatement of CO_2 intensity, it is necessary to substantially reduce energy intensity or improve energy efficiency from now on. Though energy efficiency in China keeps improving, it is still much lower than that in advanced countries and has much room for further improvement.

The following policies can contribute to the improvement of energy efficiency: levy an environmental tax soon (such as a 'carbon tax') and form the market mechanism of energy pricing to reflect the scarcity of resources and cost of environmental governance; continuously increase R&D expenditure through multi-channels on energy-saving technology and decarbonizing technology; establish technology markets and innovation loci devolved from state-dominated systems to firms, research institutes and universities; and employ advanced technology to renovate traditional heavy industry and rapidly develop light industry, promoting new industrialization by drawing on the green and information technological revolution.

Because the heavy industries do not contribute great deal to the industrial energy intensity reduction – and several heavy sectors have even increased energy intensity – it is very important to improve the energy productivity of heavy industries. Of course, advanced technical industry and light industry play a critical role in reducing industrial energy intensity and this should be further emphasized in the future.

Energy structure and industrial structure are also factors to influence CO_2 intensity abatement in the future. Although the room for reducing carbon intensity by adjusting energy structure is not great, due to the restrictions of resources endowment, in the long run, there is the possibility of changing the current energy structure dominated by coal to a cleaner energy structure by developing clean energy (hydro, nuclear, wind and solar), new energy and renewable energy. Structural adjustment represented by industrial structure is always the radical approach to transform the development model. China should develop the light and advanced industrial sectors faster than the energy-and-emission intensive sectors to promote the abatement of carbon dioxide intensity in the future.

4   Measurement of CO$_2$ shadow price

4.1 Introduction

Since the energy crises, environmental disasters and growing awareness of global climate warming in the 1970s, people begin to realize the importance of energy and environment in sustainable development. Linked with this, researchers also began to introduce the variables of energy and environment into economic development studies. As shown in the literature, it's easier to handle the energy variable in the analysis of economic growth. The researchers normally regard energy consumption as one kind of the intermediate inputs, such as the famous KLEM model proposed by Jorgenson *et al.* (1987). However, it is very complicated to work with the environmental variables because there exists no market pricing of environmental pollution and it is difficult to estimate the its cost in the production process. Accordingly, the environmental factors are often neglected, intentionally or unconsciously, by the researchers in the field of economic growth analysis for a long time, and the pricing of environmental pollution has become one of the puzzles in environmental economics. To overcome the scarcity of environmental price information, the shadow price of environmental pollution started to be measured occasionally in the literature. Now more and more people have come to realize the importance of the shadow price of environmental pollution in environmental regulation and green growth accounting. For example, the shadow price of environmental pollution can be used as the reference pricing of an environmental tax rate and pollution emission trade. It could also be utilized to build the environmental demand function and undertake green national income accounting. It makes it possible, too, for us to calculate the traditional productivity indices that depend on price information. Overall, the shadow price has become an important concept in ecological and environmental economics nowadays. In this chapter, we will attempt to measure the shadow price of carbon dioxide among different industrial sectors and discuss some applications of the estimated shadow price.

4.2   Methods to measure the shadow price in the literature

There are two kinds of methods to measure the shadow price of environmental pollution in the literature. The first one is the parametric method. For instance,

Aigner and Chu (1968) and Schmidt (1976) estimated the production function based shadow price early on. Pollak *et al.* (1984) and Gollop and Roberts (1985) estimated the shadow price by using the cost function. Pittman (1981, 1983) measured the shadow price originally by using the distance function proposed by Shephard (1970). Since then, more and more researchers have estimated the shadow price by employing the distance function. In the case of one desirable output and another undesirable output with negative externalities, the theoretical formula to calculate the relative shadow price of two outputs could be derived in terms of the dual relationship between the distance function and cost function. According to the less desirable output, the relative shadow price can be explained as the opportunity cost to reduce one unit of undesirable output; this is also called the marginal abating cost of undesirable output. The value of the relative shadow price equals the negative slope of the corresponding point on the production technique frontier. As opposed to the cost function, the distance function has another advantage in that it does not need the price information of inputs and outputs.

According to the different distance functions, the parametric methods to estimate the shadow price could be further divided into the following three types: the shadow price estimation based on the Shephard input distance function (Hailu and Veeman, 2000; Lee, 2005); that based on the Shephard output distance function (Färe *et al.*, 1993; Coggins and Swinton, 1996; Vardanyan and Noh, 2006; Rezek *et al.*, 2007);[1] and that based on the directional distance function. Within the Shephard-type distance function, the desirable and undesirable outputs are not differentiated and treated as if they increase in the same proportion. The directional distance function could identify the negative externalities of environmental pollution that are different from the desirable output and takes the increase of desirable output and the decrease of undesirable output into account simultaneously. Thus, the directional distance function is more appropriate for handling the variables of environmental pollution and is used to calculate the shadow price by more and more researchers, including Chung (1996), Murty and Kumar (2002), Aiken and Pasurka Jr (2003), Färe *et al.* (2005), Vardanyan and Noh (2006), Murty *et al.* (2007) and Cuesta *et al.* (2009).

The second approach is the data envelopment analysis (DEA) based nonparametric method. The directional distance function can also be used under this nonparametric framework to calculate the shadow price of environmental pollution; see Boyd *et al.* (1996), Boyd *et al.* (2002), Lee *et al.* (2002), Bellenger and Herlihy (2009), Kaneko *et al.* (2010) and Tu (2009). Färe and Grosskopf (1998) discuss both the parametric and nonparametric methods to measure shadow price. Overall, in measuring the shadow price of environmental pollution, the parametric method appears earlier and is used more, due to its greater convenience to compute the shadow price by differentiating the parametric function. The nonparametric method appeared later, was first used to estimate the distance function, and has been used to calculate the shadow price in recent years. This chapter will use both the parametric and nonparametric methods to estimate the directional distance function, based on each, to compute the carbon dioxide shadow price of 38 industrial sectors in China over the reform period.

4.3 Methodology used in this chapter

In this section, we will introduce both the parametric and nonparametric methods to estimate the directional output distance function and calculate the shadow price.

Firstly, let i = 1,2,. . .,38, representing 38 industrial sectors in this study; t = 1,2,. . .,29, corresponding to each year during the sample period (1980–2008); j,j'=1,2,3,4, successfully representing four input vectors $\mathbf{x}$ such as capital stock, labour, energy consumption and intermediate inputs be used in this chapter. Desirable and undesirable outputs are industrial gross output and CO_2 emissions, symbolized by $\mathbf{y}$ and $\mathbf{b}$, respectively.[2]

4.3.1 Parametric measurement of shadow price

Chung (1996) firstly specifies the parametric directional output distance function by using the translog function. Due to no requirement to assume the strong disposability of output variable, this specification is very appropriate for us to introduce the undesirable output of carbon dioxide reasonably. Following Chung (1996) and Färe *et al.* (2005), the parametric translog directional output distance function to estimate shadow price in this chapter is specified as follows:

$$
\ln[1+\vec{D}_o(\mathbf{x}^i,\mathbf{y}^i,\mathbf{b}^i;1,-1)] = \alpha_0 + \sum_{j=1}^{4}\beta_j\,\ln\mathbf{x}_j^i + \frac{1}{2}\sum_{j=1}^{4}\sum_{j'=1}^{4}\beta_{jj'}\,\ln\mathbf{x}_j^i\,\ln\mathbf{x}_{j'}^i
$$

$$
+\gamma_1\ln\mathbf{y}^i+\frac{1}{2}\gamma_{11}\ln^2\mathbf{y}^i+\eta_1\ln\mathbf{b}^i+\frac{1}{2}\eta_{11}\ln^2\mathbf{b}^i+\varphi\ln\mathbf{y}^i\ln\mathbf{b}^i
$$

$$
+\sum_{j=1}^{4}\delta_j\,\ln\mathbf{x}_j^i\,\ln\mathbf{y}^i+\sum_{j=1}^{4}\phi_j\,\ln\mathbf{x}_j^i\,\ln\mathbf{b}^i \qquad (4.1)
$$

To measure the shadow price, it is necessary to estimate all the 28 coefficients in expression (4.1). Using $\ln[1+\vec{D}_o(\cdot)]$ rather than $\ln[\vec{D}_o(\cdot)]$ to ensure the domain of the logarithm to be positive because the value of $\vec{D}_o(\cdot)$ equals 0 when the unit is on the technique frontier. Chung (1996) argues that the coefficients could be estimated by minimizing the deviation between all the units and the production frontiers. The optimization problem could be stated as follows:

$$
\min \sum_{i=1}^{38}\ln[1+\vec{D}_o(\mathbf{x}^i,\mathbf{y}^i,\mathbf{b}^i;1,-1)] \qquad (4.2)
$$

s.t.

$$
\ln[1+\vec{D}_o(\mathbf{x}^i,\mathbf{y}^i,\mathbf{b}^i;1,-1)]\geq 0 \qquad i=1,\ldots,38 \qquad (i)
$$

$$
\frac{\partial\ln[1+\vec{D}_o(\mathbf{x}^i,\mathbf{y}^i,\mathbf{b}^i;1,-1)]}{\partial\ln\mathbf{y}^i}\leq 0 \qquad i=1,\ldots,38 \qquad (ii)
$$

$$\frac{\partial \ln[1 + \vec{D}_o(\mathbf{x}^i, \mathbf{y}^i, \mathbf{b}^i; 1, -1)]}{\partial \ln \mathbf{b}^i} \geq 0 \qquad i = 1, \ldots, 38 \qquad \text{(iii)}$$

$$\gamma_1 - \eta_1 = -1$$
$$\gamma_{11} - \varphi = 0$$
$$\varphi - \eta_{11} = 0$$
$$\delta_j - \phi_j = 0 \quad j = 1, 2, 3, 4 \qquad \text{(iv)}$$

$$\beta_{jj'} = \beta_{j'j} \quad j \neq j' \qquad j, j' = 1, 2, 3, 4 \qquad \text{(v)}$$

where, the restriction condition (i) ensures that all the decision-making units lie on or within the technique frontier; (ii) and (iii) limit the shadow price of desirable and undesirable outputs to be non-negative and non-positive, respectively; (iv) imposes the assumption of homogeneity of degree 1 on the outputs, ensuring that the value of distance function will increase in the same proportion as the increase of outputs given inputs and techniques and, accordingly, indicating that the output variables are weakly disposable; (v) endows the translog function with symmetry. There are no restrictions imposing on the input variables here.

Based on the derivative characteristics of the output distance function and its dual characteristic to benefit function and cost function, Färe *et al.* (1993) and Chung (1996) provide the formula shown below to estimate the shadow price of undesirable output:

$$\hat{p}_b^i = p_y^i \frac{\partial \widehat{D}_o(\mathbf{x}^i, \mathbf{y}^i, \mathbf{b}^i; 1, -1) / \partial \mathbf{b}^i}{\partial \widehat{D}_o(\mathbf{x}^i, \mathbf{y}^i, \mathbf{b}^i; 1, -1) / \partial \mathbf{y}^i} = p_y^i \frac{\partial \widehat{D}_o(\cdot) / \partial \mathbf{b}^i}{\partial \widehat{D}_o(\cdot) / \partial \mathbf{y}^i} \qquad (4.3)$$

in which, p_b^i and p_y^i are the shadow price of carbon dioxide and industrial gross output. As denoted by Färe *et al.* (1993), if you want to estimate the absolute value of shadow price of undesirable output, the most direct method is to assume that the shadow price of desirable output equals its market price, that is, p_y^i being 1 RMB. In this chapter, this assumption will be utilized to calculate the absolute value of carbon dioxide by using both the parametric and nonparametric methods.

Algebraic transformation of expression (4.3) will be undertaken further as follows:

$$\frac{\partial \widehat{D}_o(\cdot)}{\partial \mathbf{b}^i} = \frac{\partial \widehat{D}_o(\cdot)}{\partial \ln[1 + \widehat{D}_o(\cdot)]} \cdot \frac{\partial \ln[1 + \widehat{D}_o(\cdot)]}{\partial \ln \mathbf{b}^i} \cdot \frac{\partial \ln \mathbf{b}^i}{\partial \mathbf{b}^i} \qquad (4.4)$$

$$\frac{\partial \widehat{D}_o(\cdot)}{\partial \mathbf{y}^i} = \frac{\partial \widehat{D}_o(\cdot)}{\partial \ln[1 + \widehat{D}_o(\cdot)]} \cdot \frac{\partial \ln[1 + \widehat{D}_o(\cdot)]}{\partial \ln \mathbf{y}^i} \cdot \frac{\partial \ln \mathbf{y}^i}{\partial \mathbf{y}^i} \qquad (4.5)$$

Then (4.3) could be rewritten as:

$$\hat{p}_b^i = p_y^i \cdot \frac{\partial \ln[1+\widehat{\overline{D}}_o(\cdot)]/\partial \ln \mathbf{b}^i}{\partial \ln[1+\widehat{\overline{D}}_o(\cdot)]/\partial \ln \mathbf{y}^i} \cdot \frac{\mathbf{y}^i}{\mathbf{b}^i}$$

(4.6)

It then can be derived as below according to the form of translog in expression (4.1):

$$\frac{\partial \ln[1+\widehat{\overline{D}}_o(\cdot)]}{\partial \ln \boldsymbol{b}^i} = \hat{\eta}_1 + \hat{\eta}_{11} \ln \boldsymbol{b}^i + \hat{\varphi} \ln \boldsymbol{y}^i + \sum_{j=1}^{4} \hat{\phi}_j \ln \boldsymbol{x}_j^i$$

(4.7)

$$\frac{\partial \ln[1+\widehat{\overline{D}}_o(\cdot)]}{\partial \ln \mathbf{y}^i} = \hat{\gamma}_1 + \hat{\gamma}_{11} \ln \mathbf{y}^i + \hat{\varphi} \ln \boldsymbol{b}^i + \sum_{j=1}^{4} \hat{\delta}_j \ln \mathbf{x}_j^i$$

(4.8)

Based on the coefficients estimated by linear programming (4.2), the absolute shadow price of carbon dioxide of *i*th unit can be calculated using the following formula:

$$\hat{p}_b^i = \frac{\hat{\eta}_1 + \hat{\eta}_{11} \ln \boldsymbol{b}^i + \hat{\varphi} \ln \boldsymbol{y}^i + \sum_{j=1}^{4} \hat{\phi}_j \ln \mathbf{x}_j^i}{\hat{\gamma}_1 + \hat{\gamma}_{11} \ln \mathbf{y}^i + \hat{\varphi} \ln \boldsymbol{b}^i + \sum_{j=1}^{4} \hat{\delta}_j \ln \mathbf{x}_j^i} \cdot \frac{\mathbf{y}^i}{\mathbf{b}^i}$$

(4.9)

4.3.2 Nonparametric measurement of shadow price

The studies that estimate output directional distance function and calculate the shadow price by using the nonparametric method have greatly increased in recent years. Following Lee *et al.* (2002) and Kaneko *et al.* (2010), the piecewise linear production technique and linear programming to estimate the output directional distance function in this chapter is specified as follows:

$$\overline{D}_o^i(\mathbf{x}_j^i, \mathbf{y}^i, \mathbf{b}^i; 1, -1) = Max_{\beta, \lambda} \ \beta$$

$$s.t. \ \ \mathbf{Y}\lambda \geq (1+\beta)\mathbf{y}^i; \ \mathbf{B}\lambda \leq (1-\beta)\mathbf{b}^i; \ \mathbf{X}\lambda \leq \mathbf{x}^i; \ \mathbf{i}^T\lambda \leq 1; \beta, \lambda \geq 0$$

(4.10)

where, $\mathbf{X}$, $\mathbf{Y}$ and $\mathbf{B}$ and are the matrix of input, desirable and undesirable output to represent all the units. $\mathbf{i}$ is the unit column vector. λ is the intensity vector, containing the weights to project the units into the production frontier. According to the conclusions of Tu and Xiao (2005), Li and Li (2008) and Zhang *et al.* (2009), we assume that Chinese industry is experiencing the non-increasing return to scale (NIRS) by specifying $\mathbf{i}^T\lambda \leq 1$.[3] After obtaining the value of distance function $\widehat{\overline{D}}_o(\cdot)$ by using expression (4.10), the shadow price could be derived on the production technique frontier, along the efficient path $\mathbf{y}^*, \mathbf{b}^*$ of the observations $(\mathbf{y}, \mathbf{b})$. The directional vector that makes the units more efficient is specified as:

$$\mathbf{y}^* = (1 + \widehat{\vec{D}}_o(\cdot))\mathbf{y}; \quad \mathbf{b}^* = (1 - \widehat{\vec{D}}_o(\cdot))\mathbf{b} \tag{4.11}$$

Maximizing the profit of each unit along the above efficient path, that is:

$$\max_{\mathbf{x},\mathbf{y},\mathbf{b}} \quad p_y \mathbf{y}^* + p_b \mathbf{b}^* - \mathbf{w}^T \mathbf{x}$$

$$s.t. \quad \vec{D}_o^*(\mathbf{x}, \mathbf{y}^*, \mathbf{b}^*) = 1 \tag{4.12}$$

in which, $\mathbf{w}$ is the price vector of input factors. The corresponding Lagrangian function is:

$$\max_{\mathbf{x},\mathbf{y},\mathbf{b},\mu} \quad p_y \mathbf{y}^* + p_b \mathbf{b}^* - \mathbf{w}^T \mathbf{x} + \mu[\vec{D}_o^*(\mathbf{x}, \mathbf{y}^*, \mathbf{b}^*) - 1] \tag{4.13}$$

where, μ is the Lagrangian multiplier. The two first-order conditions of the above profit maximizations relative to desirable and undesirable outputs are:

$$p_b + \mu \cdot \frac{\partial \vec{D}_o^*(\mathbf{x}, \mathbf{y}^*, \mathbf{b}^*)}{\partial \mathbf{b}^*} \cdot (1 - \widehat{\vec{D}}_o(\cdot)) = 0 \tag{4.14.1}$$

$$p_y + \mu \cdot \frac{\partial \vec{D}_o^*(\mathbf{x}, \mathbf{y}^*, \mathbf{b}^*)}{\partial \mathbf{y}^*} \cdot (1 + \widehat{\vec{D}}_o(\cdot)) = 0 \tag{4.14.2}$$

The relative shadow price of environmental pollution to industrial gross output could be derived according to the formula of both (4.14.1) and (4.14.2):

$$\frac{p_b}{p_y} = \frac{\partial \vec{D}_o^*(\mathbf{x}, \mathbf{y}^*, \mathbf{b}^*)/\partial(\mathbf{b}^*)}{\partial \vec{D}_o^*(\mathbf{x}, \mathbf{y}^*, \mathbf{b}^*)/\partial(\mathbf{b}^*)} \cdot \frac{(1 - \widehat{\vec{D}}_o(\cdot))}{(1 + \widehat{\vec{D}}_o(\cdot))} \tag{4.15}$$

Thus, to compute the shadow price of ith unit, it is necessary to first calculate the value of distance function $\vec{D}_o^*(\cdot)$ along the efficient path. The corresponding linear programming is specified as below:

$$\vec{D}_o^{i*}(\mathbf{x}_j^i, \mathbf{y}^{i*}, \mathbf{b}^{i*}; 1, -1) = \underset{\beta,\lambda}{Max} \ \beta$$

$$s.t. \ \mathbf{Y}\lambda \geq (1+\beta)\mathbf{y}^{i*}; \ \mathbf{B}\lambda \leq (1-\beta)\mathbf{b}^{i*}; \ \mathbf{X}\lambda \leq \mathbf{x}^i; \ \mathbf{i}^T\lambda \leq 1; \ \beta,\lambda \geq \mathbf{0} \tag{4.16}$$

If $\tau_{(i)}^i$ and $\tau_{(ii)}^i$ are utilized to represent the Lagrangian multipliers corresponding to the restrictions of desirable and undesirable outputs in Lagrangian function constructed from the linear programming (4.16), their ratio is just equal to the ratio of $\partial \overline{D}_o^{i*}(\cdot)/\partial y^{i*}$ 和 $\partial \overline{D}_o^{i*}(\cdot)/\partial b^{i*}$ in expression (4.15). Letting p_y^i equal 1 RMB, like the parametric method, the absolute shadow price of environmental pollution of ith unit could be computed by using the following formula:

$$\widehat{p}_b^i = p_y^i \cdot \frac{\partial \overline{D}_o^{i*}(\cdot)/\partial b^{i*}}{\partial \overline{D}_o^{i*}(\cdot)/\partial y^{i*}} \cdot \frac{1-\widehat{\overline{D}}_o^i(\cdot)}{1+\widehat{\overline{D}}_o^i(\cdot)} = \frac{\tau_{(ii)}^i}{\tau_{(i)}^i} \cdot \frac{1-\widehat{\overline{D}}_o^i(\cdot)}{1+\widehat{\overline{D}}_o^i(\cdot)} \tag{4.17}$$

The advantages of nonparametric method include no prior assumption of functional form and no sign restriction of shadow price of undesirable output. Thus, theoretically, the sign of shadow price computed by the nonparametric method could be positive or negative (Boyd *et al.*, 1996).

4.4 The measurement of CO₂ shadow price

Table 4.1 reports the estimated 28 coefficients, standard deviation, t statistics and the p value of parametric translog directional distance function by using linear programming (4.2). Except for β_{34}, all the remaining 27 coefficients are extremely significant in statistics, leading to the reliable calculation of shadow price by using such coefficients. Substituting the estimated coefficients in Table 4.2 into formula (4.9), we will obtain the shadow price of CO_2 for sun-industries in China.

Table 4.2 reports the averaging CO_2 shadow price of 38 sub-industries and aggregated industry over the sample period by using the parametric and nonparametric methods, respectively. Figures 4.1 and 4.2 depict the trend of estimated CO_2 shadow price for aggregated industry, light and heavy industry, also by using two methods. In this study, the CO_2 shadow price estimated using parametric method is all negative. The advantages of the nonparametric method include: no prior assumption for functional form; no restriction of positive or negative sign for shadow price of undesirable output. Due to no sign restriction being imposed on the shadow price, the estimated CO_2 shadow price using nonparametric method is almost positive. To reflect the feature of negative externalities of CO_2 emission, we also artificially endow the CO_2 shadow price estimated by the nonparametric method with negative sign, the same as that by the parametric method, in Table 4.2 and Figures 4.1 and 4.2. The heterogeneity of the CO_2 shadow price for 38 sub-industries could be clearly seen in Table 4.2. Because Table 4.2 only reports the simple average of the shadow price over the entire period, the time varying trend cannot be seen from this table. Instead, Figures 4.1 and 4.2 depict the time varying pattern of shadow price obviously. The heterogeneity of 38 sub-industries is condensed into that of light and heavy industry. As shown in Note 2 of Chapter 1, here we use the capital to labour ratio to classify 38 sub-industries into light and heavy industry. The weights used to compute the values in Table 4.2 and the former two figures in this chapter are the share of CO_2 emission for each sub-industry.

Seen from Table 4.2, Figures 4.1 and 4.2, the CO_2 shadow price estimated by the parametric and nonparametric methods has a similar measuring value and

Table 4.1 Estimation results of parametric method

Coefficients	Estimators	s.d.	t statistics	p value	Coefficients	Estimators	s.d.	t statistics	p value
α_0	0.00042	0.00004	9.96	0.0000	β_{34}	-0.00008	0.00033	-0.25	0.4027
β_1	0.00161	0.00013	12.54	0.0000	γ_1	-0.50049	0.00002	-22257	0.0000
β_2	0.00150	0.00014	10.65	0.0000	γ_{11}	0.00181	0.00015	12.26	0.0000
β_3	0.00182	0.00016	11.34	0.0000	η_1	0.49951	0.00002	22214	0.0000
β_4	0.00163	0.00014	11.30	0.0000	η_{11}	0.00181	0.00015	12.26	0.0000
β_{11}	0.00280	0.00020	14.15	0.0000	φ	0.00181	0.00015	12.26	0.0000
β_{22}	0.00251	0.00021	11.98	0.0000	δ_1	-0.00232	0.00009	-25.99	0.0000
β_{33}	0.00255	0.00023	11.29	0.0000	δ_2	-0.00200	0.00011	-17.59	0.0000
β_{44}	0.00292	0.00022	13.17	0.0000	δ_3	-0.00290	0.00013	-22.44	0.0000
β_{12}	0.00229	0.00018	12.46	0.0000	δ_4	-0.00221	0.00010	-21.55	0.0000
β_{13}	-0.00138	0.00035	-3.95	0.0000	ϕ_1	-0.00232	0.00009	-25.99	0.0000
β_{14}	0.00204	0.00019	10.62	0.0000	ϕ_2	-0.00200	0.00011	-17.59	0.0000
β_{23}	0.00081	0.00023	3.57	0.0002	ϕ_3	-0.00290	0.00013	-22.44	0.0000
β_{24}	0.00188	0.00017	10.83	0.0000	ϕ_4	-0.00221	0.00010	-21.55	0.0000

Table 4.2 Estimated shadow price of CO_2 for each industrial sector (unit: 10,000 RMB/ton CO_2)

Industrial Sectors	Parametric	Nonparametric	Industrial Sectors	Parametric	Nonparametric
Coal Mi.	-0.04	-0.02	Medicine	-1.24	-0.77
Petroleum Ext.	-0.06	-0.04	Fibres	-0.26	-0.15
Ferrous Mi.	-0.68	-0.65	Rubber	-0.95	-0.85
Non-Ferrous Mi.	-1.18	-1.19	Plastic	-3.87	-3.80
Nonmetal Mi.	-0.47	-0.36	Nonmetal Ma.	-0.12	-0.10
Wood Exp.	-0.41	-0.23	Ferrous Press	-0.10	-0.09
Food Proc.	-0.89	-0.74	Non-Ferrous Pr.	-0.47	-0.42
Food Ma.	-0.55	-0.40	Metal Products	-3.44	-2.92
Beverage	-0.72	-0.41	General Mac.	-3.62	-3.13
Tobacco	-2.99	-1.92	Special Mac.	-1.95	-1.30
Textile	-1.01	-0.89	Transport Eq.	-2.64	-1.95
Apparel	-4.49	-3.91	Electrical Eq.	-12.04	-8.79
Leather	-3.83	-3.86	Computer etc.	-34.74	-22.01
Wood Proc.	-1.21	-0.74	Measuring Inst.	-16.09	-16.21
Furniture	-5.49	-5.82	Electric Power	-0.01	-0.01
Paper	-0.24	-0.20	Gas Prod.	-0.03	-0.02
Printing	-5.18	-5.46	Water Prod.	-1.05	-0.69
Cultural Articles	-10.67	-11.03	Others	-1.33	-0.74
Petroleum Pro.	-0.01	-0.01			
Chemical	-0.14	-0.12	Industry as a whole	-3.27	-2.68

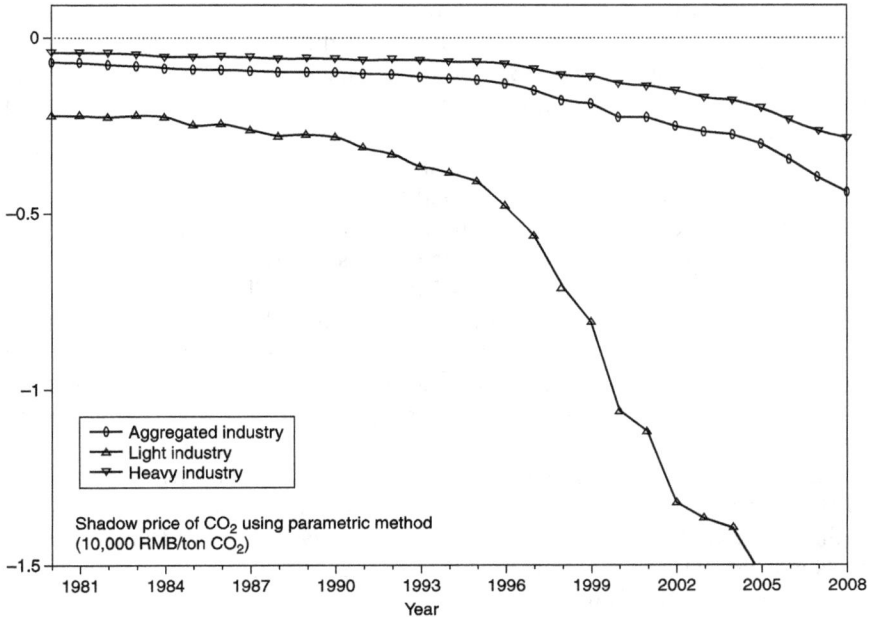

Figure 4.1 Estimated shadow price of CO_2 emission for aggregated industry, light and heavy industry using parametric method (1980–2008).

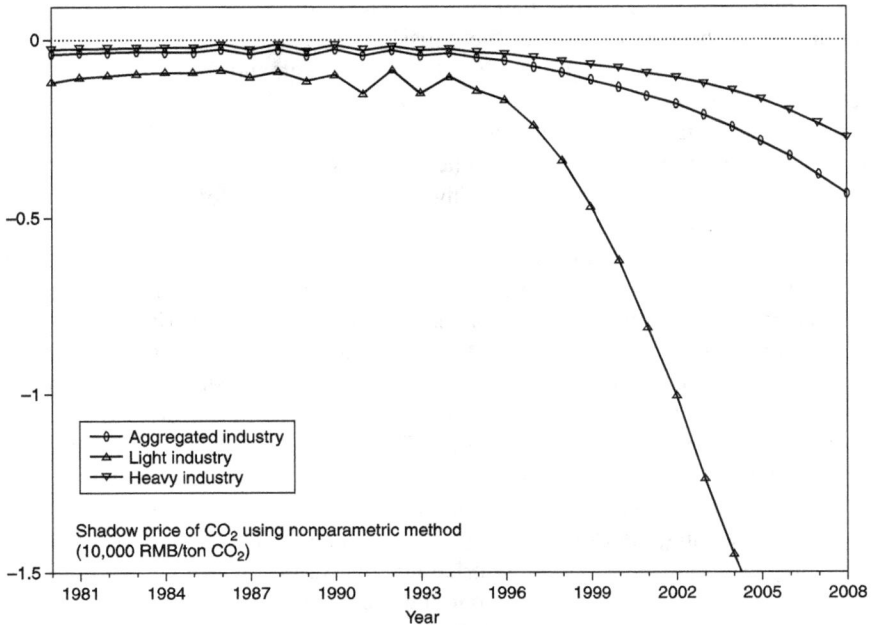

Figure 4.2 Estimated shadow price of CO_2 emission for aggregated industry, light and heavy industry using nonparametric method (1980–2008).

varying trend. Table 4.2 shows the obvious change of CO_2 shadow price for different sub-industries. The sectors that have the relatively small estimated absolute CO_2 shadow price, approximately hundreds of RMB per ton CO_2 emission, belong to heavy industry, including these sectors – Production and Supply of Electric Power and Heat Power, Processing of Petroleum, Coking, Processing of Nuclear Fuel, Production and Supply of Gas, Mining and Washing of Coal, and Extraction of Petroleum and Natural Gas. The sectors with a high CO_2 shadow price, more than 100 thousand RMB per ton emission, are mainly light and advanced technique industry such as Manufacture of Articles For Culture, Education and Sport Activities, Manufacture of Electrical Machinery and Equipment, Manufacture of Measuring Instruments and Machinery for Cultural Activity and Office Work, Manufacture of Measuring Instruments and Machinery for Cultural Activity and Office Work, and so on. It is not strange to obtain such results. The nature behind CO_2 abatement is the issue of energy utilization. Due to its high energy utilization efficiency, strong energy-saving techniques and lower energy consumption, it is difficult to save energy further in the advanced industry. That is to say, the cost of saving more energy and abating emission is extremely high. On the contrary, there exists more room to save energy and lower abating costs in heavy industry because of its lower energy efficiency and more wasteful use of resources. That is why we see different emission abating cost for different enterprises or sectors. This is consistent with the new results produced by Tu (2009) who measured the SO_2 shadow price for different regions in China. He argues that the value of the SO_2 shadow price is dependent on the emission levels and productivity in different regions. If there is high SO_2 emission and low productivity, the cost of abating the emission is relatively low; otherwise, the emission abating cost is high when productivity is high and the emission level is low. For example, the SO_2 shadow price in Beijing is 279.6 thousand RMB per ton of SO_2, the highest among all the provinces in China. Wang *et al.* (2005b) also find that the marginal abating cost in heavy industry (such as the electricity and coal sectors) is relatively low, indicating it is more flexible for heavy industry to reduce CO_2 emission.

The estimated CO_2 shadow price for aggregated industry is 26.8–32.7 thousand RMB per ton emission, reported in Table 4.2, which is lower than the estimated SO_2 shadow price in the literature. For instance, the average industrial SO_2 shadow price between 1999 and 2005, estimated by Tu (2009), is 82.6 thousand RMB per ton emission. Kaneko *et al.* (2010) estimate that the SO_2 shadow price in the Production and Supply of Electric Power and Heat Power sector is about 47.3 thousand RMB per ton emission in 2003. This is reasonable because the shadow price represents the interior evaluation of society on environmental pollution, and the social pricing of pollutants with higher negative externalities is also higher. Normally speaking, the CO_2 shadow price should be the lowest, the NO_2 shadow price highest, and the SO_2 shadow price inbetween. Of course, the value of the estimated shadow price is also related to the choice of the directional vector in the directional distance function. As denoted by Lee *et al.* (2002), the shadow price equals the negative slope on the distance functional frontier and its absolute value is dependent on the direction of the efficient path that projects into the frontier. Generally speaking, with the shift of efficient observations from the outer to the origin along the technique frontier, the estimated shadow price based on the most exterior Shephard distance function is the lowest (Coggins and Swinton, 1996); that in terms of the distance function on the middle position and with the fixed emission assumption becoming large (Turner, 1995; Boyd *et al.*, 2002); and that based on the standard directional distance function, close to the origin, is larger (Boyd *et al.*, 1996), as in this study. I have estimated the industrial CO_2 shadow price by using the emission-fixed directional distance function proposed by Boyd *et al.* (2002) and obtained the value to be 1.8 and 3 thousand RMB per ton emission, obviously lower than the estimation in this chapter.

The conclusion that CO_2 shadow price of light industrial sectors is higher than that of heavy industrial sectors, shown in Table 4.2, is also obviously seen in Figures 4.1 and 4.2 in which the difference between the two industries becomes larger and larger over time. Both figures also reveal that the shadow price of CO_2 emission is increasing over time, either for aggregated industry or light and heavy industry. Due to the larger weights of heavy industrial sectors, the varying pattern of the CO_2 shadow price of aggregated industry is dominated by that of heavy industry, with a relatively lower growth rate of shadow price than light industry. The rising shadow price over time could be explained as follows. Lee *et al.* (2002) denote that, as the units become more and more efficient, the abating room by reallocating the given resources becomes smaller and smaller, leading to a larger and larger shadow price of pollution emissions. Our findings are similar to some other studies. For example, Färe *et al.* (1993) also estimate a shadow price that has the same heterogeneous and time-varying patterns as ours. Zhang and Baranzini (2004) find that the abating cost rises as the concentration of carbon dioxide increases. There also exist some studies that provide us with the opposite or inconsistent conclusions. Kaneko *et al.* (2010) find that the SO_2 shadow price in the Production and Supply of Electric Power and Heat Power sector has decreased by half from 2003 to 2006, because of the governmental fiscal support on the SO_2 abatement. The SO_2 shadow price estimated by Tu (2009) fluctuates considerably between 1999 and 2005 and does not display a consistently increasing or decreasing trend.

As shown above, the two estimates of CO_2 shadow price using the parametric and nonparametric methods are similar. The nonparametric method has its advantages when compared to the parametric one. But, as denoted by Lee *et al.* (2002), in nonparametric estimation, the slope on the efficient observation is not necessarily unique and so the corresponding shadow price is not unique, so the maximized slope value is often used to calculate the shadow price in the empirical studies. From this viewpoint, the shadow price estimated by the parametric method seems to be more reliable than that by the nonparametric method. Also you will see from the analysis in the following section, in the analysis of public policy, the parametric method is more convenient than the nonparametric one. That is why the parametric shadow price estimation appeared earlier and is still utilized extensively nowadays.

4.5 The application of the estimated CO_2 shadow price

4.5.1 *Application in environmental public policy*

Hueting (1991) shows that the shadow price of environmental pollution is the cornerstone of environmental public policy and growth accounting. Because there is no market for environmental pollutants, it is not possible to directly observe the market price of the pollutants, and so the shadow price could be regarded as their environmental opportunity cost or real value. Therefore, it is very important to estimate the shadow price that will achieve the fulfillment of many tasks in the field of environmental policy and green accounting. For example, green GDP is one of the important indicators for the measurement of sustainability. If its value is positive, it shows that the market output is greater than the value of the environmental damages, indicating that the development is weakly sustainable; if negative, the development is weakly unsustainable. It is almost impossible to measure green GDP because there are many types of environmental pollutants and it is difficult to add them directly. If the shadow price of many pollutants can be estimated, we can calculate the total value of all the environmental pollutants and then adjust the market GDP to finally obtain the green GDP. Kuosmanen and Kortelainen (2007) utilize the estimated shadow price to undertake the cost-benefit analysis of environmental policies. With the shadow price, the environmental demand function can also now be constructed. Because the relative shadow price is the marginal technique substitute rate of desirable output relative to undesirable output, based on this, the short-term change of environmental quality resulting from the growth of desirable output can also be estimated. Two sectors (such as Manufacture of Communication Equipment, Computers and Other Electronic Equipment, and Production and Supply of Electric Power and Heat Power) are used as examples here. In 2008, their industrial gross outputs were about 7.5 and 0.7 trillion RMB, CO_2 emissions were 4.1 and 2858.5 million tons, and the estimated CO_2 shadow price using the parametric method were -1730.7 and -0.2 thousand RMB per ton, respectively. Assuming that the efficiency keeps constant and the two sectors will grow by 10 per cent, in terms of the estimated shadow price, we can calculate that the two sectors will increase the CO_2 emissions by 430.3 and 348,466.5 thousand tons, respectively.

There are two ways to save energy and abate emission by using market based means: one is the levy of an environmental tax and the other is the trade of pollution emission rights, in which the price mechanism is used. Therefore, the shadow price of environmental pollutants could be used as the reference price of the environmental tax rate and emission trade pricing. Let's discuss the environmental tax firstly. The two sectors of Manufacture of Communication Equipment, Computers and Other Electronic Equipment, and Production and Supply of Electric Power and Heat Power, shown above, are still selected as examples here to compute the optimal tax scale. Assume that the government plans to decrease the CO_2 emissions of the two sectors to only half of the original growth rate – that is, 2.4 per cent and 4.6 per cent, respectively, in 2008 relative to 2007, given the efficiency and other inputs. Substituting the new CO_2 emission level into the formula (4.1), we obtain the result that the new industrial gross outputs are 7.6 and 0.7 trillion RMB, respectively. Based on this new calculated CO_2 emission and gross output levels, and using the formula (4.9), the shadow price of the two sectors can be computed; that is, -1728.4 and -0.235 thousand RMB per ton of CO_2 emission is the optimal CO_2 tax rate that could reduce the two sectors' CO_2 emission to half of their original growth rate. Therefore, if the target level of CO_2 abatement is set first, the corresponding CO_2 tax rate that should be levied can be calculated by using the estimated coefficients of the parametric model. The estimated shadow price and then CO_2 tax rate in this chapter is larger than the estimates in other studies. For example, Wei and Glomsrod (2002) specify the carbon tax rate as 42–83 RMB per ton of CO_2 emission. Wang *et al.* (2009) estimate the carbon tax rate to be 40 RMB per ton of CO_2 emission according to the current CDM market value in China. My conclusion is that, although it is larger, the shadow price estimated in this chapter reflects the real cost of CO_2 abatement, and it is better to levy carbon tax in terms of this rate. Zhang and Baranzini (2004) argue that the carbon tax currently levied in the developed countries is too low to stabilize the CO_2 concentration in the atmosphere; if the carbon tax is the only policy for abating CO_2, its rate should become higher. As opposed to other environmental policies, carbon tax provides the pollutant emitters with the economic incentives to change their behaviour through the market mechanism. Thus, to fully reflect the institutional value of an environmental tax or a carbon tax, the carbon tax rate should be high enough to influence the emitters' behaviour, and the additional social cost must suffice to stimulate the emitters' awareness of environmental protection.

As for the carbon emission trade, because the responsibilities (specified by the Kyoto Protocol) and the capabilities of Annex 1 and non-Annex 1 countries are different in abating carbon emission, the developed countries can buy the carbon emission rights from the developing countries in order to make up for any shortfall on achieving the carbon emission quota. China has become the largest exporter of CDM, but does not have the pricing rights for the carbon emission trade. Thus, the deal price of the carbon trade is far from the international market price, leading to a big loss of carbon assets in China. Obviously, the transaction price of carbon emission should reflect the cost to abate the emission, and the CO_2 shadow price is just to measure the marginal abating cost of carbon. As shown in Table 4.2, the heavy industrial sectors in China have a relatively low shadow price and larger opportunity to abate emission, which delivers the capability to provide carbon emission rights to the developed countries. For example, in 2008, Baosteel sold carbon emission rights to UK and Switzerland companies. Of course, the advanced technique sectors in China have a high shadow price and big emission abating difficulties, and could also buy the emission rights from less developed countries. According to Färe *et al.* (1993), the unequal shadow prices of different sectors, shown in Table 4.2, imply that the resources allocation is not efficient. To change this, the shadow price could be used as the reference price of emission trade, leading to the equal evaluation of environmental pollutants among countries, regions, and even industries or sectors, by using the price mechanism. Specifically, the purchaser of the carbon emission right is willing to buy the emission quota at the lower price than its own shadow price, to increase desirable output and CO_2 emission, while the seller is willing to provide the emission quota at the higher price than its own shadow price, to reduce the desirable output and CO_2 emission simultaneously. As revealed by the parametric model, the absolute shadow price of pollutants is negatively relative to the desirable and undesirable output; thus, the shadow price of the seller will rise and that of the purchaser falls, and the transactions of carbon emission will continue into the finally equal shadow price among countries, regions and sectors. It can be seen from this that the carbon emission trade is not sustainable for the developing countries because the benefit from the sale of carbon emission rights in the short run will harm potential economic growth in the long run. Therefore, some appropriate policies should be implemented to overcome the sale of emission quotas. For example, the funds obtained by selling the emission rights should be used to develop energy-saving and emission-abating techniques, to increase the efficiency and technique levels of heavy industrial sectors.

4.5.2 Calculation of productivity indices

Fixed base total factor productivity (TFP) could be defined as:

$$TFP_{t,0} = \frac{Y_t/X_t}{Y_0/X_0} = \frac{Y_t/Y_0}{X_t/X_0} = \frac{Total\ output\ index_{t,0}}{Total\ input\ index_{t,0}} \tag{4.18}$$

in which, the subscript 0 indicates the base period and t the reporting period. Obviously, to calculate TFP index, it is necessary to calculate the total output index and total input index. Because the input and output in actual production are multiple, it needs the calculation of a composite statistical index. The traditional composite index has four types: Laspeyres index, Paasche index, Fisher index and Törnqvist index. The common point to calculate four such indices is to obtain the price information of all output and input used as the calculation weights. Due to the scarcity of the market price of pollutants, they are often not included in the calculation of the productivity index. If the shadow price of pollutants is estimated, it is possible to estimate the green productivity index by using the traditional composite indices when the pollutants are taken into account.

Laspeyres index uses the price at the base period as the weights and the Paasche index uses the price at the reporting period as the weights. Based on these, the productivity index can be calculated by using the formulae (4.19) and (4.20):

$$TFP_{t,0}^L = \frac{Y_t/Y_0}{X_t/X_0} = \frac{\sum_{i=1}^{2} p_0^i y_t^i / \sum_{i=1}^{2} p_0^i y_0^i}{\sum_{j=1}^{4} p_0^j x_t^j / \sum_{j=1}^{4} p_0^j x_0^j} = \frac{\sum_{i=1}^{2} w_0^i \dfrac{y_t^i}{y_0^i}}{\sum_{j=1}^{4} w_0^j \dfrac{x_t^j}{x_0^j}} \tag{4.19}$$

where, $w_0^i = p_0^i y_0^i / \sum_{i=1}^{2} p_0^i y_0^i$ and $w_0^j = p_0^j x_0^j / \sum_{j=1}^{4} p_0^j x_0^j$ are the value share of two output ($i = 1,2$) and two inputs ($j = 1,2,3,4$) at the base period. Thus, the formula

(4.19) is in fact to use the weighted arithmetic average index to calculate the total output index and total input index. Here, output Y includes industrial gross output and undesirable CO_2 emission, and input X consists of capital stock, labour, energy consumption and industrial intermediate input. When calculating the productivity index, the negative externalities of CO_2 is reflected by using the negative shadow price, and based on this, the calculated value share of CO_2 is its calculation weights, following Aiken and Pasurka Jr (2003).

$$TFP_{t,0}^{P} = \frac{Y_t/Y_0}{X_t/X_0} = \frac{\sum_{i=1}^{2} p_t^i y_t^i / \sum_{i=1}^{2} p_t^i y_0^i}{\sum_{j=1}^{4} p_t^j x_t^j / \sum_{j=1}^{4} p_t^j x_0^j} = \frac{\left(\sum_{i=1}^{2} w_t^i \left(\frac{y_t^i}{y_0^i}\right)^{-1}\right)^{-1}}{\left(\sum_{j=1}^{4} w_t^j \left(\frac{x_t^j}{x_0^j}\right)^{-1}\right)^{-1}} \tag{4.20}$$

where, $w_t^i = p_t^i y_t^i / \sum_{i=1}^{2} p_t^i y_t^i$; $w_t^j = p_t^j x_t^j / \sum_{j=1}^{4} p_t^j x_t^j$. The formula (4.20) is in fact to calculate the total output index and total input index, respectively, by using the weighted harmonic mean index.

Laspeyres index can reflect the quantity change of output and input precisely, but deviates from the actual case at the reporting period due to the use of the base-period price weights. Paasche index overcomes the drawbacks of Laspeyres index but involves the joint influence of quantity and price, in addition to the influence of quantity change, because the price at the reporting period is included in the calculation. Thus, as the index of quantity indicator, Laspeyres index is better than Paasche index. Fisher ideal index attempts to overcome the drawbacks of Laspeyres and Paasche indices by calculating their geometric mean, specified as follows:

$$TFP_{t,0}^{F} = (TFP_{t,0}^{L} \times TFP_{t,0}^{P})^{\frac{1}{2}} \tag{4.21}$$

Though it is the tradeoff of two indices, Fisher's ideal index does not have such an obvious economic meaning as Laspeyres and Paasche indices, and needs more files for calculation due to its inclusion of both indices.

Törnqvist index is defined as weighted geometric mean of individual quantity indices, the weights being the simple arithmetic mean of the weights used to calculate Laspeyres and Paasche indices. That is:

$$TFP_{t,0}^{T} = \frac{Y_t/Y_0}{X_t/X_0} = \frac{\prod_{i=1}^{2} \left(\frac{y_t^i}{y_0^i}\right)^{\frac{w_0^i+w_t^i}{2}}}{\prod_{j=1}^{4} \left(\frac{x_t^j}{x_0^j}\right)^{\frac{w_0^j+w_t^j}{2}}} \tag{4.22}$$

The calculation of the above index often adopts the following logarithmic form:

$$\ln(TFP_{t,0}^{T}) = \sum_{i=1}^{2} \frac{1}{2}(w_0^i + w_t^i)\,(\ln y_t^i - \ln y_0^i)$$
$$- \sum_{j=1}^{4} \frac{1}{2}(w_0^j + w_t^j)\,(\ln x_t^j - \ln x_0^j) \tag{4.23}$$

Caves *et al.* (1982) denotes that the translog production function based on the Solow neoclassical growth model corresponds to the discrete Törnqvist productivity index. Because the translog production function is regarded as the better second-order approximation of any functional form, Törnqvist productivity index is also regarded as the most appropriate productivity index. In fact, formula (4.23) is just the classical expression of Solow residuals. Because of such good properties, Törnqvist productivity index is currently extensively used.

The chain TFP index can be easily computed according to the fixed base TFP index, described above.[4] Figure 4.3 depicts the four types of chain green productivity index of aggregated industry based on the CO_2 shadow price estimated by using both the parametric and nonparametric methods; this is the weighted mean

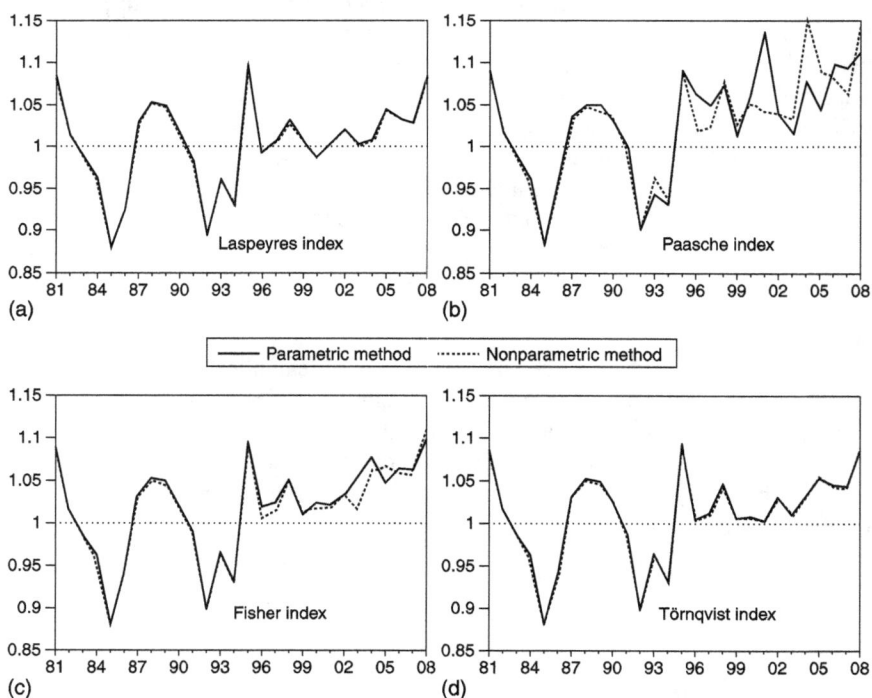

Figure 4.3 Four industrial green productivity indices calculated using two types of shadow price of CO_2 emission (1980–2008).

of sectoral productivity indices and the weights are the sectoral CO_2 emission shares, same as those in Figure 4.1 and 4.2.

The obvious feature shown in Figure 4.2 is that the calculated productivity indices based on the CO_2 shadow price estimated by using the parametric and nonparametric methods are very similar, especially close for those calculated using Laspeyres and Törnqvist indices. The Paasche index shown in sub-figure (b) is much higher than Laspeyres index in sub-figure (a) after the mid-1990s, which has maybe resulted from the joint influence of quantity and price change in Paasche index because the price at the reporting period is included into the calculation, as denoted previously. Fisher's ideal index, shown in sub-figure (c) falls between the former two, consistent with the calculation by using the geometric mean of the former two indices. As the quantity indicator, the Laspeyres index in sub-figure (a) is better than Paasche index, in theory. In fact, the Laspeyres index seems to be very close to the Törnqvist productivity index shown in sub-figure (d), the most appropriate productivity index denoted in the literature. Next, Törnqvist productivity index will be exemplified to analyse the varying pattern of green TFP growth in Chinese industry.

During the entire reform period, only in two sub-periods of 1983–1987 and 1991–1995 does the negative growth of green TFP appear in Chinese industry. The negative growth of green TFP in 1983–1987 could be explained by the blind development of energy-intensive small enterprise such as coal mines, with heavy environmental pollution and serious resources waste, resulting from the policy to encourage energy production to resolve the energy shortage issues at the beginning of the reform period. Until the end of the 1980s, central government had to rectify the resource market such as the coal market. The industrial reform in 1980s was incremental in nature and did not touch the ownership structures. The opportunistic behaviour of enterprises contractors, together with the presence of significant numbers of under-utilized employees, high asset-liability ratios, the government apportionment of charges and social burden, among other factors, finally caused the serious financial deterioration of state-owned enterprises (SOEs). Thus, it is not surprising to see the negative growth of TFP in the early 1990s.

After the mid-1990s, central government started substantial reform of the SOEs by grasping the large and letting go of the small enterprises. Accordingly, more than 100 thousand energy- and emission-intensive small enterprises were shut down, leading to the first decline of energy consumption and carbon dioxide emission during the long-run increasing trend. That is why we see the obvious growth of green TFP since 1996. Due to the reappearance of the heavy industrialization phenomenon in China since the beginning of this century, the economy has grown rapidly, with a more rapid increase of energy consumption and carbon emission, leading to relatively low TFP growth, shown in sub-figure (d). The central government realized the problems and proposed the scientific development outlook and the construction of a harmonious society. In the field of industry, the road to new industrialization was required. Such policies lead to the rise of green TFP growth, as shown in sub-figure (d) of Figure 4.3.

4.6 Summary and comments

The shadow price of environmental pollutants has a valuable application in the field of environmental public policy and green growth accounting. This chapter uses carbon dioxide as the proxy of environmental pollutants and estimates the carbon dioxide shadow price for 38 Chinese sub-industries between 1980 and 2008 by using the parametric and nonparametric environmental directional distance function. The estimates show that heavy industry has a relatively low carbon shadow price due to its low energy utilization efficiency and large scope for emission abating; on the contrary, light industry has a high shadow price because of its high energy efficiency and smaller opportunity to abate the emission. From the time horizontal, the shadow price of carbon dioxide increases over time because the cost of emission abatement rises with the decreasing abating scope. The estimates of shadow price change the circumstance where many environment related analyses cannot be undertaken due to the lack of market price information of environmental pollutants. Based on the estimated shadow price of carbon dioxide, this chapter discusses its application in the field of environmental tax and emission trade. The shadow price of carbon dioxide could be used as the reference price of a carbon tax rate and carbon emission rights trade. This chapter also calculates the green TFP growth of Chinese industry by using traditional productivity indices because the shadow price of carbon dioxide is available now. The green productivity index will be very helpful in enacting appropriate environmental regulations in China.

5 Energy and environmental policies and factors driven industrial growth

5.1 Introduction

After the overthrow of infant capitalism sixty years ago, China has experienced rapid economic growth averaging 8 per cent annually, and has performed spectacularly well since its market reforms in 1978 with an average growth rate of 9.8 per cent, held more steadily than before. It's well known that such achievements were accomplished by employing a heavy-industry-oriented development strategy, transforming resources from agriculture to industry, which resulted in a severe dual economic structure in China. For instance, the share of GDP in agriculture declined from over 50 per cent in 1952 to only 11.7 per cent in 2006, with a below average rate of growth (3.4 per cent); while industry maintained a high growth rate of 11.5 per cent annually and increased its GDP composition in a stable pattern from 17.6 per cent in 1952 to 43.3 per cent in 2006. Such high growth of industry depended seriously on heavy inputs of important resources. During the reform period, the averaged growth rate of industrial gross output, capital stock, energy consumption and CO_2 emission reached annually was 11.2 per cent, 9.2 per cent, 6 per cent and 6.3 per cent, respectively, but that of the labour force only 1.9 per cent. The averaged proportion of industrial GDP, 40.1 per cent, was achieved by consuming 67.9 per cent of national energy and emitting 83.1 per cent of national carbon dioxide for the period from 1978. Obviously, the high growth of Chinese industry has the typical characteristics of high investment, high energy consumption and high emission. This chapter plans to analyse the factors driven economic growth model in Chinese industry by using green growth accounting, in which, both the energy consumption and carbon dioxide emission will be treated as the input factors. To see the influence of energy and environment on industrial growth, it is necessary to review the evolution of energy and environmental policies in China.

5.2 Energy and environment policies implemented in post-reform China

As outlined, on average, industrial output accounts for around half of national output at the expense of most of the energy consumption and carbon emission. The elements of economic expansion emanating from the central planned economy, such as the planned allocation of resources, the distorted energy pricing system, insufficient financial resources for geological surveys and the exploration of energy sources and the compartmentalization of administrative organizations led to an inadequate supply of energy for a long time, which has been a bottleneck for economic growth in China. Therefore, it became necessary to reform the Chinese economy, including its energy and emission polices. Next, we will briefly review the evolution of energy and emission policies implemented in China, corresponding to each stage of industrial reform, which are introduced in detail in Chapter 6.

5.2.1 Encouraging phase (1978–1992)

This represented a trial phase of industrial reform in China, in which the main task was to solve the product shortage, including tight energy supply, resulting from the long-term command economy. Therefore, the early reform of energy industry was implemented by encouraging the production of energy. In response to the energy shortage, the Chinese government also formulated the policy to save energy in 1980.

Similar to the industrial reform to increase operating autonomy in the urban areas, energy intensive industries (such as coal mining, petrochemical and electric power enterprises) underwent the reform of the contract responsibility system one after the other. Among them, the coal industry was the sector that experienced the most thorough reforms; it was where competition was fiercest because the central government decentralized authority to local governments, relaxed the entry barriers of coal enterprises and encouraged townships, commune and brigades (*shedui*), individuals and foreign investors to mine coal. Driven by the chance to acquire wealth rapidly through mining, and supported by local government to obtain local interests, a surge of coal mining appeared across the state for a time, resulting in a dramatic increase of non-state-owned coal-mines and production of coal. In particular, town and village enterprises (TVEs) contributed most in this expansion process (there was a net increase of 150 million tons of coal produced by TVEs among the total increase of 250 million tons of coal in the period of the national sixth Five-Year Plan: 1981–1985). As in the case of small coal mines, a large number of other small enterprises producing coal power, steel and iron, machinery, fertilizer and cement also expanded at the same time which dispersed the production layout of nationwide industry and led to a reduction in the concentration of Chinese industry during this stage.

Overall, the industrial reform in this phase not only eased effectively the difficulties of energy supply but also greatly stimulated light industries, such as textile and food. This was key in transforming the development of Chinese industry from the catch-up strategy before reform to focusing on labour- and energy-intensive sectors that reflected comparative advantages.

5.2.2 Restricting stage (1992–2001)

Deng Xiaoping's southern tour gave an enormous impetus to the historically decisive reform of Chinese industry at this stage. Such reforms as the confirmation that the market economy would be a transition target in 1992, fully liberalizing product prices in the early 1990s and the privatization of state-owned enterprises (SOEs) in the late 1990s laid a solid foundation for World Trade Organization (WTO) accession in 2001. This also represented the stage of environmental protection from the formation of awareness in the mind to implementation in practice. The government started to restrict the development of energy enterprises, rather than the encouragement provided in the 1980s.

The blind development of small energy-intensive enterprises in the first stage improved the energy shortage initially, but the production of coal was surplus and out of control. Furthermore, the destructive exploitation, the lack of environmental protection measures and the relatively low production cost of small coal mines not only indirectly led to the financial loss of state-owned coal mines but also caused serious damage and the waste of coal resources and environmental pollution. Throughout the 1980s and earlier, industrial pollution prevention and control focused only on control of the point source of pollution and the treatment of end-of-pipe pollution. Though China has to clean up the coal market in 1989, it was not until the concept of sustainable development had become accepted worldwide in the early 1990s that the consciousness of environmental protection really grew in China and the measures of environmental conservation and policies to save energy were effectively implemented by the government. China was one of the countries involved in the composition of 'Our Common Future' in 1987 and signed the 'Rio Declaration' and 'United Nations Agenda 21' in 1992. In 1994, China took the lead in the first national release of 'China Agenda 21'. The official document 'Outlines of the Ninth Five-Year Plan and Long-term Target towards to 2010', authorized by the People's National Congress (PNC) in 1996, confirmed sustainable development as one of the basic national development strategy goals, and energy as the important area to achieve the sustainable development strategy; this once again emphasized the energy policy of 'saving and developing energy simultaneously' originally proposed in 1980. In 1998, the Energy-Save Law came into effect in China. Phased goals in every area of sustainable development were put forward in the national tenth Five-Year Plan in 2001. Industrial pollution prevention and treatment become the important area of environmental protection, the strategy of which from now on was transformed from end harnessing to control of source and production process, from control of the point source pollution to comprehensive governance of drainage and regional areas, from the treatment of certain enterprises to adjustment of the industrial structure and development of circulative economy, among many measures.

In the context of the substantial reform of SOEs – implementing 'grasp the large and let go of the small' (*zhuada fangxiao*) – and the conversion of policy from encouraging to restricting energy industry, from expanding energy supply to emphasizing financial profits and environmental conservation, between 1996 and 2000 the Chinese government closed down about 84 thousand small energy- and emission-intensive enterprises and curbed the disorderly production of energy. Thus, the degree of industrial concentration increased.

At the beginning of the 1990s, the well-known dual-track pricing system, in which the plan and the market coexisted, was successfully liberalized in the products market, but the factors market was still underdeveloped. In 1994, the dual-track price of coal was first phased out of the state plan and then fully liberalized, unlike that of product oil and electric power. The full formation of a coal market, together with the closure of many small coal mines and the furlough policy (*xiagang*) employed in SOEs, balanced the demand and supply of coal and reversed the overall loss of the coal industry. The reform of the petrochemical and electric power industry was limited in comparison to that of coal manufacturing. The petrochemical industry was strategically reconstructed in 1998 by building two groups, such as petroleum and petrochemical corporations. The electric power industry was reorganized by separating government functions from enterprise management and power plant from network. But both were still operated under the national monopoly.

Overall, the development strategy focusing on labour-intensive light industry was still implemented during this stage. Though the development of energy and emission-intensive heavy industry was restricted, the production of energy still increased steadily. Likewise, the policy of saving energy and abating emission improved noticeably. As known already, the consumption of primary energy and carbon emission began to decline or maintain a standstill for the first time between 1997 and 2001, after the long-term increase. The energy and emission intensity also had its largest reduction at this stage.

5.2.3 Reappearance of heavy industrialization and proposition of scientific development outlook (2001–present)

In this stage, the Chinese industry still followed the strategy of emphasizing labour-intensive light industry and sustainable development, characterized by energy-save and emission-abate. The sixteenth Party Congress in 2002 was the first to describe the new road to the next phase of industrialization. On this basis, its Third Plenum put forward the scientific outlook on development and the Fourth Plenum put forward the proposition for constructing a harmonious society.

In 2004, 'Outlines of China Medium and Long Term Energy Saving Plan 2004–2020' was enacted. Climate change was the popular topic in the energy policy area. In 2002, China authorized the Kyoto Protocol. In 2004, the official document 'The People's Republic of China Initial National Communication on Climate Change', including national greenhouse gas inventories, was submitted to the tenth conference of the parties to the United Nations Framework Convention on Climate Change (UNFCCC). The Chinese government released the 'China National Plan for Coping with Climate Change' in 2007, the first national green plan among developing countries. In summary, up to 2008, the Chinese legislature has drafted about 30 national laws related to resources, energy and environmental conservation and the State Council has issued more than 100 administrative regulations on sustainable development. Guided by these energy and emission policies, from 2001 to 2004, China's government published the product lists for outdated production capacity three times in succession, shut down 30 thousand resource-wasteful enterprises, and stopped the establishment of 1900 projects in serious pollution sectors such as cement, steel and iron, and electrolysis aluminum. In 2005, the government continued, closing 2600 energy- and emission-intensive enterprises.

However, the phenomenon of heavy industrialization appeared again after 2002. As described in Figure 1.1, the consumption of primary energy and carbon emission rose to unprecedented levels after 2001, and reached the peak in 2010. Industrial energy and emission intensities also reversed the long-term decreasing trend and displayed a temporary rise for several years. In the first two stages, the averaged elasticity of energy consumption was about 0.5, indicating that one per cent of GDP growth was accompanied by only 0.5 percent of growth of energy consumption; after the start of the new century, the averaged elasticity exceeded 1 which was also the sign of the reappearance of heavy industrialization. Heavy industrialization is also reflected in the structure of exporting products. From the late 1990s, Chinese exports changed from labour-intensive light products such as textile to capital- and energy-intensive products such as machinery, chemicals, cement, and so on. When the GDP per capita is low, the complexity degree of exporting products should be low too. But, the complexity degree of exporting products in China reached the level for those countries where the GDP per capita is three times that of China (Rodrik, 2006). Anderson (2007) further denotes that the export surplus in heavy industry from 2005 was one of the important reasons for the dramatic increase in the foreign trade surplus of China at the current time.

This phenomenon led to a big debate in China. Is the heavy industrialization inevitable and necessary in China today? Can high energy consumption and carbon emission drive industrial growth sustainably into the future or not? The appearance of heavy industrialization at this stage may be different from that under the catching-up strategy before the reforms, as light industry is still developing quickly, but it seems to deviate from the new industrialization route proposed at the beginning of this century. That merits special attention. It could be attributable to either short-term reasons (such as continuous and massive infrastructure investment, new entry of private capital in heavy industries due to the low price of natural resources, accelerated transfer of international manufacturing into China), or some long-term causes including the upgrade of consumption structure, the increase of urbanization level and the empirical law that heavy industry experiences faster development at the middle stage of industrialization.

Based on this, the eleventh national Five-Year Plan in 2006 put forward quantitative goals to save energy and abate emission, that is, decrease energy consumption per GDP by 20 per cent and total pollution emission by 10 per cent between 2006 and 2010. The concrete measures to promote the conversion to a low carbon economy were justified as follows. First, provide financial support for energy-save and emission-abate projects. The energy industry has no capability to accumulate the financial resource to develop by itself because it is a high capital intensive sector but has a low value of output due to low energy prices. According to the theory of public finance, this is the government's responsibility because the enterprises aiming at profit maximization are not willing to invest in it. Thus, the treasury bonds and financial investment of central government for Energy and

Environment projects reached 23.5 billion RMB in 2007 and 50 bn in 2008. Second, adjust foreign trade policy from 2006, such as cancelling tax rebates, increasing tariffs, cutting down quotas and listing some products forbidden for export because of their high energy consumption, pollution emission and serious resource waste. Other measures included restricting the development of energy-intensive projects from different channels such as land, credit and foreign investments, consolidating the levy of taxes and fees for environmental pollution, reducing tax for those enterprises attaining the standards of the lowest pollution emission and producing the environment-conservation products, liberalizing the price of resources such as product oil, and so on.

However, it's challenging for China to save energy and abate emission at the present time. China will continue its industrialization and urbanization, and the energy- and emission-intensive industries such as chemical products, cement and iron and steel still play a fundamental role in the economy, both at present and in the foreseeable future. Thus, China faces a dilemma between transforming the growth model and promoting industrialization, especially under the current international financial crisis. The challenge to reducing pollution emission comes also from the unbalanced composition of primary energy consumption in China. China is rich in coal reserves and there is a dominant share of coal consumption in China, although its dominance has declined in recent years. The CO_2 emitted per thermal unit of coal is 36 per cent and 61 per cent higher than that of oil and gas, respectively. The poor endowment of cleaner energy, the low level of energy using and exploiting technology and a lack of investment will inevitably cause the rigidity of structure of energy consumption and therefore a difficulty in reducing pollution emission for a long time. Therefore it is crucial to investigate the influence of energy and emission on the industrial growth in China, as that will play an important role in pushing the new type of industrialization forward in the short run.

5.3 Sustainable development and the role of energy and environment on economic growth: literature survey

The concept of sustainable development was first defined by the Brundtland Commission and has become accepted worldwide following the 1992 Earth Summit and the adoption of the United Nations' Agenda 21 (WCED, 1987; United Nations, 1993). Though a variety of practical definitions have emerged in distinct areas, the concept has evolved to encompass such major points of view as economic, technological, environmental and social issues, and many models and indicators are used to measure the sustainability of development. For example, the World Bank (2001) declared that the GDP indicator is still necessary to be used to evaluate the success of the sustainable strategy of each country. Lin (2004) asserts that technical progress is the most important indicator to measure the sustainability of economic growth. By adapting technological know-how from advanced countries at a lower cost, China is very likely to maintain around 8 per cent GDP growth rate for another twenty or thirty years. Brock and Taylor (2005) argue that structural upgrade is the factor to drive sustainability. Islama *et al.* (2003) evaluate the prospect of sustainability by using cost-benefit analysis. Most studies make use of the framework of neoclassical growth theory and the transformation of the growth model to judge the sustainability of economic development (Solow, 1957, 1993; Ofer, 1987; Krugman, 1994; Young, 1995, 2003; Irmen, 2005; Zheng *et al.*, 2007). They argue that the extensive growth model based on the perpetual expansion of inputs, the so-called level effect, is unsustainable but only an intensive growth model with the continuous and substantial improvement of total factor productivity (TFP), referred to as growth effect in the terminology of growth modelling, is sustainable in the long run. For example, Krugman (1994) and Young (1995) conclude that the growth of East Asian economies, such as Korea and Singapore in the 1960s or 1970s, is driven by inputs expansion and not sustainable, based on their estimated TFP. Of course, there are also some studies which criticize their conclusions (Chen, 1997; Hsieh, 2002; Zhang *et al.*, 2003).

However, the early researches related to the neoclassical growth model usually consider output to be determined by only conventional inputs of capital and labour, ignoring the important nexus between energy, environment and the sustainability. As described previously, the rapid growth of Chinese industry has been accomplished at the expense of high energy consumption and carbon emission, particularly, during the reform period. Under the high degree of economic globalization today, the great increase of energy prices and a sharp deterioration of environmental quality will certainly have a far-reaching impact on economy, so it's impossible to investigate the sustainable development of Chinese industry without consideration of the elements of energy and the environment. Therefore, a detailed analysis of the linkages between energy, environment and economy is very crucial. Since the world oil crisis in 1973, researchers' enthusiasm to apply modelling techniques in energy studies has increased tremendously. The enthusiasm was further enhanced by worldwide awareness and concerns about environmental issues such as global warming and climate change from the 1980s. As a result, a number of modelling techniques have been developed to address complex energy and environmental issues.

The literature on energy, environment and economies can be classified into two categories. One focuses on the analysis of activities or factors that affect the change of energy consumption and pollution emission and their efficiency (Zhang, 2003; Cole *et al.*, 2005; Wang *et al.*, 2005a; Steenhof, 2006; Fisher-Vanden *et al.*, 2004, 2006; Fan *et al.*, 2007a; Wang, 2007; Wei *et al.*, 2007; Gassebner *et al.*, 2008). They normally use the decomposition approach to decompose the energy or emission (efficiency) into some components such as the effect of structural change and technological innovations, and find the principal factors, such as those which drove the stagnancy or decline of energy use and pollution emission since the mid-1990s in China. Another approach is interested in how and to what extent is economic growth affected by energy consumption and/or pollution emission, by using the neoclassical growth equation. This chapter follows the latter and includes both energy consumption and CO_2 emission, along with traditional capital and labour, into the translog production function to undertake green growth accounting. Next, we briefly describe the role of energy and emission variables in the growth equation.

In recent years, many theoretical and empirical studies concluded that it is appropriate to incorporate energy as an input in the production process; see Kummel *et al.* (2002), Pokrovski (2003), Ayres *et al.* (2003), Murillo-Zamorano (2005), Thompson (2006), Kander and Schön (2007), Lee and Chang (2007, 2008), Lee *et al.* (2008), Kasahara and Rodrigue (2008) and others for details. During this process, energy is considered not only as an ordinary intermediate product that contributes to the value of produced products by adding its cost to the price, but also as a value-creating factor which has to be introduced in the list of production factors in line with conventional capital and labour. For example, the well-known KLEM model, that is, the decomposition of inputs into capital, labour, energy and intermediate materials, for analysis of productivity growth was first proposed and then applied to the post-war US economy by Jorgenson *et al.* (1987) and this was followed by many other researchers.

In related literature, there are usually two ways to treat waste emission in the production process. One way is to treat it as an input factor during the production process, but an unpaid input. The literature in which emission is incorporated into production function, together with capital and labour, include Tahvonen and Kuluvainen (1993), Nielsen *et al.* (1995), Palmer *et al.* (1995), Mohtadi (1996), Aghion and Howitt (1998), Dean (1999), Brock and Taylor (2005), Tzouvelekas *et al.* (2006), Xepapadeas *et al.* (2007); while the researches where both energy and emission are included as inputs in the growth equation include Carter (1974), Bovenberg and Smulders (1995), Lansink and Silva (2003), Ramanathan (2005), Considine and Larson (2006), Lu Xuedu *et al.* (2006), Oda *et al.* (2007), among others. As an input, emission may affect growth through two channels. If the natural environment (for example, its function as a waste sink) is considered to be an input into the production process, then emission represents the use of natural environmental capital, implying a positive effect of emission on growth. If environmental quality enters the production function as an input, then pollution exerts negative effects on growth by lowering the quality of the natural environment. Another method regards waste emission as an undesirable, or bad, output and incorporates it into analysis by using the Shephard or Directional output distance function; see Färe *et al.* (1993, 2001), Coggins and Swinton (1996), Chung *et al.* (1997), Boyd *et al.* (2002), Jeon and Sickles (2004), Picazo-Tadeo *et al.* (2005), Brummer *et al.* (2006), Kumar (2006), Sebastián and Gutiérrez (2007) for details.

In this study, we present a translog production function to estimate the productivity and undertake green growth accounting. Because the approach of the production function considers only one type of output, we analyse the gross output value as a normal output and do not treat emission as bad output (byproduct). That is, in this chapter, both energy and emission enter into the production function as the inputs, along with the traditional inputs, capital and labour.

5.4   Methodology

The production function of industrial sectors is specified as follows:

$$Y_{it} = f(K_{it}, L_{it}, E_{it}, C_{it}, t) \qquad (5.1)$$

where, the subscript i representing the industrial sector ($i = 1,2, \ldots, 38$); t being the time trend variable used to capture the technical progress ($t = 1,2, \ldots, 27$). Y is the gross industrial output value, and K, L, E, C are inputs of capital stock, labour, energy consumption and CO_2 emission, respectively.

Based on the definition of Solow (1957), differentiating equation (5.1) with respect to t and then dividing it by Y on both sides, we obtain the equation to calculate the growth rate of total factor productivity (TFP) as below:

$$\dot{TFP}_i = \dot{Y}_i - \alpha_{K_i} \dot{K}_i - \alpha_{L_i} \dot{L}_i - \alpha_{E_i} \dot{E}_i - \alpha_{C_i} \dot{C}_i \qquad (5.2)$$

where, the superior dots are used to denote the growth rate of each variable; $\alpha_K, \alpha_L, \alpha_E, \alpha_C$ represent the output elasticity of four inputs. Under the neoclassical assumption, such output elasticity of inputs is equal to respective cost share, and TFP growth approaches the rate of technical change. Equation (5.2) can be employed to undertake growth accounting analysis, too. The weighted average sectoral TFP obtained from equation (5.2), aggregates TFP growth of Chinese industry is thus obtained, which is a more reliable measure of aggregate TFP than that estimated according to the aggregate production function. Based on this disaggregate TFP, we're able to clearly find the engine and drag of aggregate TFP growth in Chinese industry.

To avoid the restrictions of the Cobb-Douglas (CD) production function such as constant return to scale and Hicks-neutral technical change, equation (5.1) will employ a more flexible translog form for the purpose of our study, which contains the CD function as a special case. The translog function not only imposes no priori constrains that CD function does, but also enables us to obtain varying output elasticity of inputs across sectors and over time. The concrete parametric industrial sector translog function is specified as follows:

$$\ln Y_{it} = \beta_1 + \beta_i + \beta_t t + \frac{1}{2} \beta_{tt} t^2 + \sum_{j=1}^{4} \beta_j \ln X_{itj}$$

$$+ \frac{1}{2} \sum_{j=1}^{4} \sum_{k=1}^{4} \beta_{jk} \ln X_{itj} \ln X_{itk} + \sum_{j=1}^{4} \beta_{tj} t \ln X_{itj} + \varepsilon_{it} \qquad (5.3)$$

In addition to the variables defined in equation (5.1), X represents four inputs in the form of matrix and subscripts $j,k = 1,2,3,4$ correspond to the types of inputs, that is, capital, labour, energy and emission in this study. β s are the regression coefficients required to be estimated in which β_i represents the individual effect of each sector if fixed effect model can be used for estimation. Note that the introduction of the interactive term $t \ln X$ aims at the capture of the biased technical change of each input.

In estimating any production function, it's necessary to consider the validity of the chosen estimation method. If the individual effect does not exist, the translog production model can be estimated by only using pooled OLS; otherwise, one should determine which method between fixed effect and random effect will be chosen to estimate the panel data. The former can be tested using the general F test and the latter by the Hausman Wald test. There may be additional problems of auto-correlation and heteroscedasticity, but these problems could be safely neglected since the time trend variable and individual effect will absorb the autocorrelated and heteroscedastic components in the disturbance term, and since the output and inputs were transformed into a logarithmic scale. Though multicollinearity is another potential problem existing in the specification of translog function, we do not modify them because the estimation consequences are relatively light; for instance, the unbiasedness and efficiency of estimator are still valid. The disturbance term ε_{it}, therefore, is assumed to follow the assumptions of classical linear regression model.

After estimation, the interested statistic could be calculated based on the estimated regression coefficients. The estimates of output elasticity of inputs are formulated as:

$$\alpha_j = \frac{\partial \ln Y_{it}}{\partial \ln X_{itj}} = \beta_j + \sum_{k=1}^{4} \beta_{jk} \ln X_{itk} + \beta_{tj} t \tag{5.4}$$

To justify the translog specification of equation (5.3) in our study, two hypothesis tests are necessary. Since much of the literature uses a CD production function, we first test the hypothesis that this specification of CD form is preferable to translog function, that is, $H_0 : \beta_{jk} = \beta_{tj} = 0$. Secondly, we perform the test on the existence of technical progress. The hypothesis of no technical change is specified as $H_0 : \beta_t = \beta_{tt} = \beta_{tj} = 0$. The hypotheses could be test by using the general F test if fixed effect model is employed in this study because it is in nature pooled least squares on the time demeaning of the original equation.

5.5 Empirical findings

As described previously, the output data used in this study is sub-industrial gross output value with the unit of 100 million RMB at 1990 price level, rather than value-added. This has an advantage in providing an explicit role for intermediate materials, say, energy in our model, in allocating economic growth at the sub-industrial level. For example, intermediate materials such as coal are indispensable to the production of electric power, particularly in China. By identifying these intermediate inputs explicitly, we can allocate the economic growth from investment in electric power between the production of coal and the production of electric power. The four input factors are sectoral capital stock, labour, energy consumption and carbon dioxide emission. To see the difference among sub-industries, we divide all the 38 sectors into low and high energy groups, low and high emission groups according to the ranking of energy consumption and carbon emission in 2004; see Table 5.1. That is, the former 19 sectors with a relatively low value of energy consumption and carbon emission belong to the low energy and emission group, and the latter 19 sectors with high value of energy and emission are divided into high energy and emission group.

The estimated parameters of the translog production function and its exact probability value for China sub-industries are reported in Table 5.2. The estimates

Table 5.1 Industrial sectors' ranking from the lowest to highest based on energy and emission in 2004

Number	Sectors	Ranking by		Number	Sectors	Ranking by	
		Energy	Emission			Energy	Emission
1	Coal Mi.	32	33	20	Chemical	37	34
2	Petroleum Ext.	30	32	21	Medicine	18	21
3	Ferrous Mi.	13	9	22	Fibres	24	28
4	Non-Ferrous Mi.	11	8	23	Rubber	16	18
5	Nonmetal Mi.	15	19	24	Plastic	20	14
6	Wood Exp.	2	6	25	Nonmetal Ma.	36	36
7	Food Proc.	26	26	26	Ferrous Press	38	35
8	Food Ma.	17	24	27	Non-Ferrous Pr.	33	30
9	Beverage	14	23	28	Metal Products	27	15
10	Tobacco	5	10	29	General Mac.	25	17
11	Textile	31	29	30	Special Mac.	21	20
12	Apparel	8	12	31	Transport Eq.	28	25
13	Leather	6	7	32	Electrical Eq.	19	11
14	Wood Proc.	10	16	33	Computer etc.	22	13
15	Furniture	1	3	34	Measuring Inst.	3	2
16	Paper	29	31	35	Electric Power	35	38
17	Printing	7	5	36	Gas Prod.	9	27
18	Cultural Articles	4	1	37	Water Prod.	12	4
19	Petroleum Pro.	34	37	38	Others	23	22

perform very well. Only two out of 21 main coefficients are insignificant at the 10 per cent level. The fraction of variance due to sector effect, rho, attains 94 per cent, which indicates that the variation of individual effect could explain most of the variation in industrial growth. The F statistic of 99.86 is in favour of the panel data model used in this study, rather than pooled OLS. The Wald statistic of the Hausman specification test is 283.26 which also justifies the fixed effect (within) regression model used in our analysis. Both the F value of 485.96 with degree of freedom (20,968) and the R square of 0.9094 within the group indicate that the translog production function here has a overall significant performance and a good fit.

Table 5.2 Estimates of fixed-effects (within) regression of translog production function

Regressors	Coef.	se	p-value	Regressors	Coef.	se	p-value
Constant	**5.9396**	0.6563	0.000	lnLlnC	**−0.0805**	0.0464	0.083
t	**0.1194**	0.0171	0.000	lnElnC	**−0.3189**	0.0677	0.000
(1/2)t2	**0.0059**	0.0007	0.000	(1/2)lnK2	**0.2496**	0.0776	0.001
lnK	**−0.4119**	0.1894	0.030	(1/2)lnL2	**0.6614**	0.0649	0.000
lnL	**−1.5114**	0.1976	0.000	(1/2)lnE2	0.1216	0.1102	0.270
lnE	−0.1755	0.2505	0.484	(1/2)lnC2	**0.5029**	0.0544	0.000
lnC	**0.7409**	0.1637	0.000	tlnK	**−0.0517**	0.0052	0.000
lnKlnL	**−0.0930**	0.0501	0.064	tlnL	**0.0306**	0.0042	0.000
lnKlnE	**0.5199**	0.0734	0.000	tlnE	**−0.0312**	0.0051	0.000
lnKlnC	**−0.4378**	0.0512	0.000	tlnC	**0.0336**	0.0042	0.000
lnLlnE	**−0.1211**	0.0661	0.067				
sigma_u		1.1360		sample size			1026
sigma_e		0.2762		number of groups			38
rho		0.9442		obs per group			27

Note: 1. R-sq: within=**0.9094; Overall Significance Test:** F(20,968)=**485.96**, Prob>F=0.0000.
2. Individual Effect Test that all u_i=0: F(37,968)=**99.86**, Prob>F=0.0000.
3. Hausman Specification Test: chi2(20)=**283.26**, Prob>chi2 =0.0000.

5.5.1 Growth accounting and production structure

Statistical tests concerning the production structure of China'S industrial sectors are undertaken as follows. The first tests the hypothesis that the sector production can be described by a Cobb-Douglas production function. This hypothesis is strongly rejected by F statistic of 38.7 at the 1 per cent level of significance, indicating that the translog form specified in this study is reasonable. The second test that there exists no technical progress is too rejected, according to the general F statistic of 100.4, which reveals that Chinese industrial sectors have achieved improvement of productivity since the reform. Table 5.3 reports the source of industrial growth for each of the 38 sectors where the contribution of an input is defined as the growth rate of that input multiplied by its estimated output elasticity, all values being simplu averaged over time to have output growth exactly equal to the sum of contributions of inputs and productivity growth.

Table 5.3 shows the importance of high-tech industries (such as Manufacture of Communication Equipment, Computers and Other Electronic Equipment, Manufacture of Measuring Instruments and Machinery, Transport Equipment, Manufacture of Medicines) and some light sectors (such as Manufacture of Chemical Fibres, Manufacture of Articles For Culture, Education and Sport Activities) which grew rapidly in both output and productivity. Slower growing sectors are almost all high energy-use and pollution-emit sectors such as Extraction of Petroleum and Natural Gas, Mining and Processing of Non-Ferrous Metal Ores and Nonmetal Ores, Exploiting of Wood and Bamboo, Production and Supply of Water, Manufacture of Raw Chemical Materials and Chemical Products, which

Table 5.3 Sources of Chinese industrial growth

Industrial Sectors	Output Growth	TFP Growth	Contribution of Inputs			
			Capital	Labour	Energy	Emission
Coal Mi.	0.080	0.055	0.022	−0.003	0.017	−0.012
Petroleum Ext.	0.013	−0.086	0.100	−0.023	0.055	−0.032
Ferrous Mi.	0.141	0.029	0.070	0.007	0.052	−0.016
Non-Ferrous Mi.	0.102	0.023	0.041	0.003	0.033	0.001
Nonmetal Mi.	0.086	0.042	0.008	0.015	0.025	−0.005
Wood Exp.	0.033	0.036	0.009	−0.010	−0.003	0.001
Food Proc.	0.100	0.047	0.043	0.000	0.025	−0.014
Food Ma.	0.104	0.081	0.014	−0.003	0.011	0.001
Beverage	0.127	0.085	0.029	−0.006	0.022	−0.003
Tobacco	0.116	0.074	0.025	−0.013	0.020	0.010
Textile	0.086	0.027	0.058	−0.013	0.031	−0.017
Apparel	0.128	0.138	−0.017	0.019	−0.012	0.000
Leather	0.131	0.121	−0.011	0.010	0.002	0.008
Wood Proc.	0.160	0.134	0.000	0.010	0.012	0.003
Furniture	0.138	0.137	−0.019	0.013	0.002	0.006
Paper	0.122	0.075	0.019	0.005	0.014	0.009
Printing	0.118	0.046	0.016	0.001	0.044	0.010
Cultural Articles	0.158	0.117	0.004	0.009	0.011	0.018
Petroleum Pro.	0.065	0.087	−0.042	−0.018	−0.024	0.062
Chemical	0.110	0.035	0.064	0.001	0.031	−0.022
Medicine	0.162	0.128	0.027	−0.006	0.013	0.000
Fibers	0.185	0.146	0.022	−0.015	0.026	0.005
Rubber	0.113	0.083	0.012	0.002	0.012	0.003
Plastic	0.154	0.089	0.039	0.003	0.040	−0.019
Nonmetal Ma.	0.109	0.068	0.037	−0.001	0.023	−0.018
Ferrous Press	0.112	0.033	0.063	−0.001	0.048	−0.030
Non-Ferrous Pr.	0.123	0.041	0.058	−0.005	0.069	−0.041
Metal Products	0.124	0.051	0.045	−0.005	0.038	−0.005
General Mac.	0.109	0.044	0.030	−0.015	0.027	0.022
Special Mac.	0.113	0.081	0.020	−0.011	0.024	−0.001
Transport Eq.	0.177	0.110	0.031	0.000	0.039	−0.003
Electrical Eq.	0.167	0.072	0.031	0.002	0.050	0.012
Computer etc.	0.307	0.122	0.060	0.021	0.125	−0.022
Measuring Inst.	0.169	0.115	0.019	0.001	0.027	0.009
Electric Power	0.133	0.086	0.008	−0.016	0.053	0.002
Gas Prod.	0.077	0.117	−0.030	−0.048	0.008	0.030
Water Prod.	0.071	−0.064	0.170	−0.027	0.098	−0.107
Others	0.145	0.069	0.047	0.015	0.010	0.003

show below-average output growth and low or even negative TFP growth. Heterogeneity can also be seen from this table, where the growth rate of output ranges from the lowest 1.3 per cent for Extraction of Petroleum and Natural Gas to the highest 30.7 per cent for Manufacture of Computer, among others, and estimated TFP growth also broadly falls within the interval [−8.6 per cent,

14.6 per cent]. Such sectoral characteristics observed from this table favour the necessity of the use of sectoral panel data in this study, rather than aggregate data.

From the perspective of growth sources, the first driving engine in both low energy and emission group is TFP growth and its contribution of labour is not high as expected; while the first engine in both high energy and emission group is not always TFP growth but sometimes capital and energy consumption, and its biggest drag is basically carbon emission. Concretely, the sectors driven firstly by capital and then energy rather than productivity are Extraction of Petroleum and Natural Gas, Production and Supply of Water, Mining and Processing of Ferrous and Non-Ferrous Metal Ores, Smelting and Pressing of Ferrous Metals, and Manufacture of Textile. In Manufacture of Raw Chemical Materials and Chemical Products, the first driving engine is capital, the second productivity and the third energy. The sectors where energy is the first driving force include Manufacture of Communication Equipment, Computers and Other Electronic Equipment, and Smelting and Pressing of Non-ferrous Metals, whereas the second and third engine for the former sectors are productivity and capital and those for the latter are capital and productivity. The few sectors listed above are those that are driven first not by productivity, in which Extraction of Petroleum and Natural Gas and Production and Supply of Water are the only two sectors with negative growth of TFP. The sectors with carbon emission being the second driving engine, inferior only to productivity but superior to the others, are just Processing of Petroleum and Coking, and Production and Supply of Gas. The sectors with labour being the second engine, only inferior to productivity, are labour-intensive Manufacture of Textile Wearing Apparel, Footwear and Caps, Manufacture of Leather, Fur, Feather and Related Products, and Manufacture of Furniture. The sectors with labour being the third engine include Extraction of Petroleum and Natural Gas, Mining and Processing of Nonmetal Ores, Processing of Petroleum and Coking, and Manufacture of Wood, Bamboo and Straw Products. Six sectors such as apparel, leather, furniture, wood proc., cultural articles and petroleum pro. have the lowest contribution of capital and the highest of productivity. There are no sectors where energy performs the worst. Such sectors as Extraction of Petroleum and Natural Gas, Mining and Processing of Non-Ferrous Metal Ores, and Production and Supply of Water have a higher contribution of capital and energy but lower output growth due to the relatively low even negative contribution of labour, emission and particularly productivity. The highest growing sector of Manufacture of Communication Equipment, Computers and Other Electronic Equipment experiences also very high productivity, strong labour absorbability, low capital demand, lowest energy and emission per value-added, which seems to be very instructive for China's new-style industrialization in the future.

Taking a quick look at Table 5.3, in most of the 38 industrial sectors, the growth effect of TFP, energy and capital is positive, but that of labour and carbon emission is negative. Thus, we can say that Chinese industry indeed achieved the improvement of TFP since the reform, and in most of the sectors, technical progress plays the biggest role in pushing the growth forward. Seen from the contribution of inputs, energy and capital are the main sources driving industrial growth in China, only inferior to technical progress or productivity, while labour and carbon emission have less or even negative impact on Chinese industrial growth. However, in some heavy industrial sectors with high energy consumption and carbon dioxide emission, the contribution of energy or capital to growth is higher than productivity, indicating that it's necessary to transform their growth pattern; of course, the effect of high emission on growth is still negative.

5.5.2 Productivity and technical change

Figure 5.1 displays the sectoral contributions to aggregate total factor productivity growth in Chinese industry, in which each sector's contribution is calculated as the product of sector productivity growth and its weights (gross output shares) averaged over 1980–2006. The sum across 38 sectors equals 6.36 per cent per year for the period studied and provides one estimate of aggregate TFP growth of China industry. This figure reveals the wide range of sector contributions to TFP growth, reflecting the variation in both sector productivity growth and its weights. For example, the Manufacture of Chemical Fibres experiences highest productivity growth of 14.6 per cent, while contributing only 0.17 percentage points to TFP due to its lower output share. By contrast, such sectors as Manufacture of General Purpose Machinery, Manufacture of Non-metallic Mineral Products, and Manufacture of Electrical Machinery and Equipment show below average productivity growth, 4–7 per cent, but make a top six contribution due to their larger size.

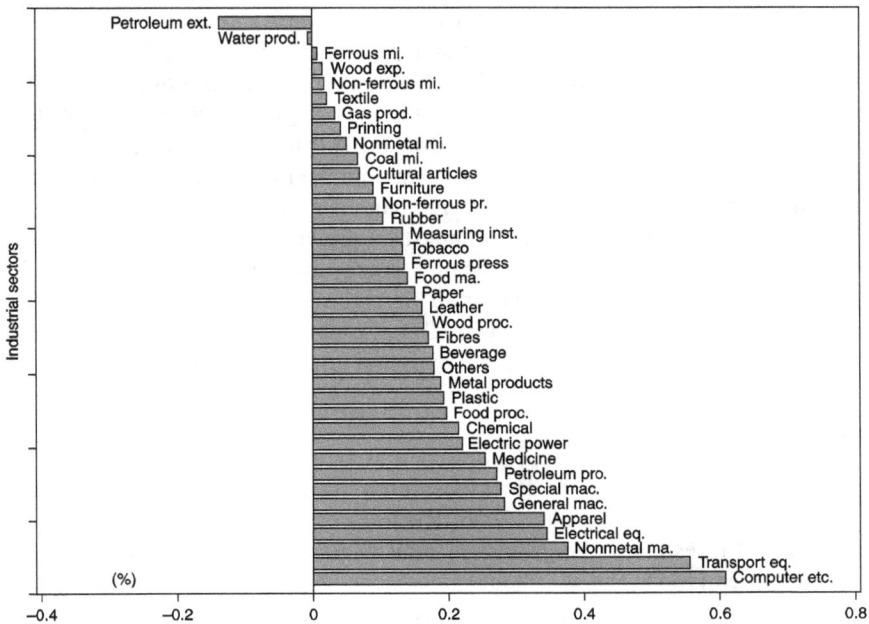

Figure 5.1 Sectoral contributions to aggregate industrial TFP growth in China.

An industrial sector with a negative productivity growth rate makes negative contributions to aggregate TFP growth. Obviously, the sectors of Extraction of Petroleum and Natural Gas and Production and Supply of water are the only two with negative productivity growth, being −8.6 per cent and −6.4 per cent, which become the greatest drag on TFP growth, lowering aggregate TFP growth by 0.1497 and 0.1273, respectively. Table 5.3 has illustrated that rapid growth of TFP is the important source that is driving the transformation of the growth pattern into sustainability in Chinese industry; furthermore, Figure 5.1 reveals that the sectors contributing highly to the aggregate TFP are almost all high-tech sectors such as Manufacture of Communication Equipment, Computers and other Electronic Equipment, Manufacture of Transport Equipment and so on. The sectors with lower contributions to aggregate TFP growth are almost all energy- and emission-intensive sectors such as Extraction of Petroleum and Natural Gas, Production and Supply of Water, Mining and Processing of Ferrous and Non-Ferrous Metal Ores, and Exploiting of Wood and Bamboo. Figure 5.1 again shows that many sectors have made important positive contributions to TFP growth, while others show negative or very low productivity growth that pulls down the aggregate TFP. To understand the full breadth and complexity of productivity growth, it is essential to examine each industrial sector individually.

Table 5.3 and Figure 5.1 reveal the heterogeneity of productivity growth and contributions of inputs among industrial sectors, ignoring the time trend. Figure 5.2 displays the varying pattern of accumulated TFP level (198082 = 1) over the entire period 1980–2006, compressing sector features into simple low and high energy and emission groups. Both sub-figures give the similar changing pattern that Chinese industry experienced substantial improvement of productivity and can be divided into four phases: slow growth of TFP in 1980s, rapid improvement of TFP before the mid-1990s, decreasing, even stagnating, growth at the turn of new millennium, and greatly increasing growth of TFP after 2003. Such a varying pattern of TFP growth in China is consistent with the findings documented in the literature on the studies of Chinese productivity, and in fact reflects the gradually transforming process of the Chinese industrial growth model. Although similar, however, the productivity level in the high energy and emission group is always less than that in the low level group. This confirms the conclusions found in Figure 5.1 that, to increase the sustainability of Chinese industry, it's necessary to improve technical levels in energy and emission intensive sectors.

5.6 Summary and comments

This chapter presents a sector decomposition of aggregate growth for the China industrial economy from 1980 to 2006. The translog sectoral production function is employed to estimate the growth of total factor productivity constrained by energy and emission. But we should remember here that, in this process, both energy consumption and carbon dioxide emission are treated as the input factors that are very likely to influence the TFP estimates.[1] In this chapter, growth accounting clearly displays the effect of energy consumption and carbon emission

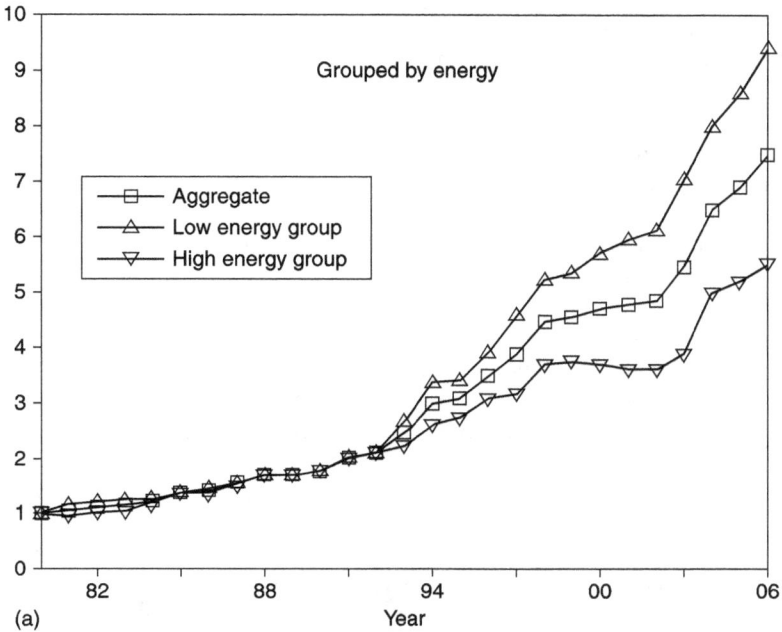

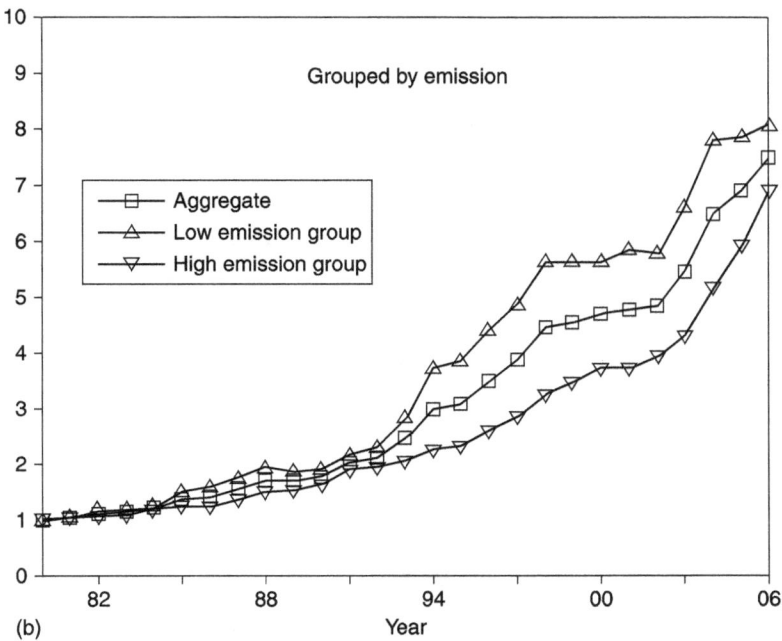

Figure 5.2 Averaged TFP level for aggregated industry and different industrial groups.

on industrial growth in China. The aggregate industrial TFP is also decomposed into contributions of 38 sectors to reveal the main contributors to productivity improvements in China. The principal conclusions of this chapter are obtained as follows.

1   For most Chinese industrial sectors, the total factor productivity improvements are substantial and TFP growth becomes the first engine that drives its output growth. But the main contributors to aggregate productivity are almost all high-tech sectors such as Manufacture of Communication Equipment, Computers and Other Electronic Equipment. The TFP level in energy- and emission-intensive heavy sectors is less than that in light industry; and some heavy sectors are still driven by the expansion of inputs and need to transform their growth model urgently.
2   Seen from the perspective of contributions of inputs, in addition to productivity being the important driving force, energy and capital are also the main engines that push industrial growth forward. The contributions of labour and emission to output growth are relatively low, even negative, holding back the growth of industrial output. To keep high growth of industrial output, especially under the current financial crisis, energy use and capital investment are obviously key inputs but should be used in a sustainable way.

6 Structural change, factors reallocation and industrial growth

6.1 Introduction

Chenery *et al.* (1986) argue that economic development is equivalent to an update of the economic structure. Therefore, structural change is the radical way to achieve sustainable development in the long run, when the economic development stage is finally reflected by the structural stage. Even for the US, during the period of the current financial crisis, most of the stimulus has been invested in such as way as to adjust the economic structure and develop the new energy and low-carbon techniques, in order to transform its old economic growth model. For China, it is more urgent to change the economic structure and transform its traditional growth model.

For a long time, China's economic growth has often been regarded as being traditionally extensive in nature, with high investment, high energy consumption, heavy emission, high growth and large fluctuations, which have resulted from the long-term imbalance of economic structures, such as regional and industrial structures. For example, after the overthrow of infant capitalism and three years of land reforms, more than half a century ago, the central government opted for a heavy-industry-oriented development strategy to catch up with the developed world. The strategy of utilizing China's comparative disadvantage has resulted in the persistence of a dual economy, leading to massive distortion in the factor market. The danger of the imminent collapse of China's economy pushed the central government to commence economic reforms from 1978. The evolution of China's market economy from its old system necessitated profound structural changes. In fact, there was a substantial fall in the share of the labour force in primary industries, from 83.5 per cent in 1952 to 39.6 per cent in 2008, and a steady increase in tertiary industry, broadly consistent with the general characteristics of the structural transformation process documented in the literature of transitioning economies. The composition of the labour force in secondary industry increased continuously, from a low of 7.4 per cent in 1952 to a peak of 27.2 per cent in 2008, which is different from the experience of industrialized economies which show a hump-shaped pattern. This indicates that China's industrialization is still in the early phase and has room to absorb more labour and further develop its labour-intensive sectors. Corresponding with this structural change, the share of industrial value-added increased from a low of 17.6 per cent in 1952 to a high of 44.1 per cent in 1978, under the catch-up strategy, and has remained stable around 40 per cent until today. The share in primary industry has decreased continuously to only 11.3 per cent in 2008, while the share in services has also grown sharply after the reform.

However, the radical structural imbalance still exists in China. This chapter endeavours to investigate the impact of structural change on the transformation of the economic growth model in China, especially under the constraints of high energy consumption and heavy environmental pollution. Many studies analyse the structural effect across three strata of industry or across regions in China, but they rarely discuss the structural adjustment and factor shifts across industrial sectors. Economists also believe that resources are restricted within sectors, and, as the industrial development literature suggests, it is necessary to reallocate the factors across sectors to boost industrial productivity and output growth. Chen and Huang (2003) describe the relationship between structural change and the industrialization levels. That is, the transformation of industrial growth from sharp structural adjustment and high growth to stable growth is the basic symbol that industrial growth has changed from quantitative expansion to qualitative increase, when industrialization is achieved successfully. During the process of industrialization, structural change within the industry will successively experience three stages of heavy industrialization, high-processing industrialization and technique intensification, reflecting the objective process from labour-intensive to capital- and technique-intensive industry. Now, China is on its way towards final industrialization and industry is the principal part of the Chinese economy. Therefore, industrial reform truly reflects China's entire transition experience. This chapter will also emphasize the structural change in China's industry and assesses its effect on industrial growth. The evolution of structural change in Chinese industry will be briefly introduced in the following section in order to provide readers with the background to the analysis in this chapter.

6.2 Review of industrial structural reform in China

China has changed its industrial development strategy from heavy-industry-oriented before the reform in 1978 to the parallel importance of light and heavy industry under the reform period; this has released great productive energy, leading to more stability and a higher than average growth rate of industrial GDP. Accordingly, substantial industrial structural change has occurred in China, as described here.

6.2.1 Trial phase (1978–1992)

The first period can be characterized by Deng Xiaoping's metaphor 'crossing a river by feeling for the stones' (*mozhe shitou guohe*), which reflected the exploratory nature of early reform.

Inspired by the successful reform in the countryside in the first six years of this period, the Chinese Communist Party (CCP) initiated industrial reforms in urban areas in 1984. The reform features at this phase only involved restructuring the operating rights of state-owned enterprises (SOEs) without touching on the issue of ownership rights. A contract responsibility system was mainly implemented in the large and medium-sized enterprises (LMEs) and a leasing system was used in small ones to transform SOEs' operating mechanisms, in order to increase autonomy and incentives. Government also hardened SOEs' budget constraints by changing their fund resources from finance to credit (*bogaidai*).

Concurrently, the rural non-farm industry, called township and village enterprises (TVEs), emerged and flourished. This development was timely, as the TVEs were able to absorb excess farm labourers who otherwise would have been compelled to leave their villages to search for employment. Two factors contributed to the success of TVEs. First, local governments were able to increase revenues due to the 'fiscal contract system' in place since 1980. Second, the hybrid ownership of TVEs allowed them more flexibility in operations than SOEs while also affording them greater protection from local authorities than private enterprises (PEs). Such protection was crucial, since private property rights were not clearly defined on paper. Meanwhile, special arrangements such as Foreign Direct Investment (FDI) – entirely owned and managed by foreign investors – and special economic zones to grant preferential treatment to attract FDI were boosting the birth of new foreign funded enterprises (FFEs). Therefore, the non-state enterprises (NSEs), including TVEs, PEs and FFEs (most of them belonging to light industry), developed very quickly at this stage; in 1993, their share of industrial gross output exceeded SOEs for the first time during the reform era.

In general, controversy and incomplete comprehension of sound economic theory followed the reform process during this period. The nature of this reform was gradual and incremental because a 'dual track system' was in place, whereby new entrants in non-state sectors were tolerated and occasionally encouraged, while SOEs remained untouched. Such trials of reform resulted in the rent-seeking power, corruption, income disparity, inflation, and so on, ultimately causing serious social unrest at the end of the 1990s. After setbacks and debates for several years, the target of establishing the socialist market economy was finally clarified and announced at the fourteenth Party Congress in 1992.

6.2.2 Decisive reform (1992–2001)

This is the historically decisive stage of industrial reform, characterized by substantial breakthroughs. Restructuring the ownership rights of SOEs was obviously the main achievement of industrial reform in this period.

Although reform of SOEs in the1980s succeeded temporarily, deeply rooted problems remained. Jefferson *et al.* (1998) also stated that the restructuring of SOEs without formal ownership conversion would meet with limited success. The lack of true private ownership structures was the major impediment to efficient operations. In fact, the short-term behaviour of contractors, together with the presence of many under-utilized employees, high asset-liability ratios resulting from the ten-year implementation of the policy *bogaidai* and other factors, finally caused the serious financial deterioration of SOEs. In 1995, roughly half of the SOEs were unprofitable and required large subsidies for continuing operations. The whole state-owned sector posted its first net loss in 1996. Not until then did the central government realize that state ownership was the root cause of all the ills of the SOEs and that only radical structural change of their ownership rights would improve performance.

In 1997 the fifteenth Party Congress reaffirmed the shareholding system and made a call for the ownership structure to change by employing the policy of 'grasp the large and let go of the small' (*zhuada fangxiao*). Since then, China's industry has seen an enormous wave of ownership restructuring. By grasping the large, the government retained direct control over some of the large SOEs in strategic industries. The state relinquished plenty of smaller SOEs in non-strategic industries to private ownership through a variety of means such as mergers, equity sales, auctions, and others. Thousands of SOEs that could not be sold were permitted to go bankrupt. Fan (2002) also reported that more than 70 per cent of small SOEs had been privatized or restructured in three years. By the end of 2000, the CCP announced that the modern enterprise system had been launched in as many as 84 per cent of SOEs. Approximately 70 per cent of SOEs were officially making profits and the net profits reach 230 billion RMB (Movshuk, 2004).

As SOEs reform proceeded, non-state industry continued to exhibit its strong vitality, continuously increasing its proportion of gross output and taxes to state, and enjoying higher profit ratios than the SOEs. The ownership reform also extended to collective-owned enterprises (COEs), the only partners of SOEs before 1978, including the celebrated TVEs. From the mid-1990s onward, TVEs were completely privatized through the creation of joint stock companies where the local governments are shareholders. Li and Rozelle (2000) report that the privatization of rural industry was deep and fundamental and the number of large and medium-sized COEs declined by 35 per cent. At the same time, China put an export promotion regime into practice which boosted the development of export-processing enterprises. The effect was a trade surplus starting in 1994, along with the status of being the world's top exporter. In a word, it was the rapid development of non-state industry that made the ownership reform of SOEs possible in the 1990s; the same reform that failed only a decade earlier. NSEs not only absorbed widespread layoffs from SOEs due to the furlough policy (*xiagang*) but also provided massive capital during the course of shareholding reform of SOEs.

Necessary reforms to support SOEs conversion were also undertaken in the financial system. Establishment of stock exchanges were part of this plan; the exchanges have played a critical role in helping SOEs transform into joint stock companies that allow non-state capital investment. The massive bad debts that resulted from the policy of *bogaidai* implemented ten years earlier were transformed into equity overnight under the new policy of changing debts into shares (*zhaizhuangu*). The staggering deterioration of the SOEs' financial performance also accelerated the birth of the tax-sharing system (*fenshuizhi*, TSS for short) and value-added tax (VAT) in 1994, which were considered to be positive for growth (Chen and Zhang, 2009). Another powerful reform at this stage was the fulfillment of the product market in China. The transition from planned prices to true market prices has been one of China's greatest challenges since the reform. As is already known, in the 1980s the well-known dual-track pricing system had to be implemented. In 1986, the central government decided to liberalize all commodity prices within five years, but this round of price reform ended with high inflation, panic buying and social turmoil at the end of the 1980s. When the new upsurge of economic reform came in 1992, the price of all industrial products, except for a few important materials such as oil, was successfully liberalized during a deflationary period. A unified domestic products market has existed since the early 1990s and the phenomenon of product shortages, so familiar in a planned economy, disappeared. Pragmatically speaking, ideological and political clarity have contributed to the phase of decisive reform of China industry since 1992.

6.2.3 Assessment and adjustment (2001–present)

The industrial reform in the 1980s and 1990s undoubtedly led to substantial structural change and rapid growth; however, it was not without expense such as wasteful investment, high energy consumption, heavy pollution, and so on., as shown in the last chapter. In general, social contradictions remained embedded, and sometimes even intensified when GDP per capita of this country in transition reached US$1000; such was the situation for China at the advent of the new millennium. It was natural for the controversy about the future reform direction to become heated. Therefore, re-assessment of the growth model and further adjustment of the economic structure became the main consideration for future reform by the leadership from the beginning of this century. The sixteenth Party Congress in 2002 first described the new route to the next stage of industrialization and proposed the sustainable development strategy. On this basis, its Third Plenum put forward the scientific outlook on development and the Fourth Plenum raised the proposition of constructing a harmonious society. Even so, the phenomenon of heavy industrialization reappeared at this stage, which seemed to deviate from the proposed new industrialization strategy. This could be attributable to the huge expansion of the housing and car industries, the rapid urbanization, accelerated exports of energy and emission intensive products after the access into the World Trade Organization (WTO), the continuous and massive infrastructure investment, and the new entry of private capital into heavy industries due to the low price of natural resources. Of course, there still existed the structural change of Chinese industry during this stage. By the end of 2008, the shares of gross industrial output and number of enterprises for NSEs were 71.7 per cent and 95.1 per cent, respectively, as opposed to 28.3 per cent and 4.9 per cent for SOEs.

Although the reform over the thirty years has resulted in great achievements, two major obstacles, the characteristics of every planned economy, are still in the way of future reform. One is the underlying weakness of SOEs and industry policy. Now, the SOEs have become the platform for the government to implement its so-called industry policy. In 2003, China established the State Assets

Commission (SAC) to administer the largest and best SOEs in the pillar indus-tries. As the product of the industry policy, these SOEs enjoyed dramatic expan-sion and concentration. The potential drawback is that weak corporate governance gives management the opportunity to steer the firm in terms of their own private interests, while shielding them from risk under the umbrella of government protection. The phenomenon that some SOEs are becoming privileged groups by means of monopoly, their dual role of being both players and referees of the SAC, and their deviation from their radical target of serving the public interest have all recently given rise to a growing discontent among people across society, who all own the SOEs.

Another obstacle is the persistent dual economic system that restricts the further free allocation of production factors such as resources, labour and capital. For example, the formation of a mechanism of market pricing in the resource market is far from complete. The state still controls the pricing rights of more than ten kinds of important resources, such as product oil and electric power. The low-price regulation of natural resources is unable to reflect its scarcity and stimulates the over-development of resource-intensive industries that partly led to the heavy industrialization during the third period. Such accumulation of allocative distortion in factor markets will inevitably cause the increasing deterioration of factor allocative efficiency, which will be reflected finally in the poor performance of growth and productivity in China's industry. That is why energy and environmental factors are included in to the analysis in this chapter.

6.3 Literature survey

6.3.1 The role of structural change in economic growth

Kuznets (1979) states that it is impossible to attain high growth per capita without substantial shifts of production factors among various sectors. By structural change, therefore, we mean that production factors are reallocated from less productive sectors to more productive ones.

The hypothesis that structural change is an important source of growth is initially developed in Lewis' classical models of a dual economy (Lewis, 1954), and is a central element in Maddison's growth-accounting literature (Maddison, 1987). The effect of structural change and factors reallocation in the theory of economic devel-opment is extensively used by Chenery *et al.* (1986) and Syrquin (1995), who show that it is an important factor explaining growth performances. The models of indus-trial development proposed by Lucas (1993) from the supply side and Verspagen (1993) from the demand side also stress the importance of structural change in productivity growth. Harberger (1998) vividly presents a 'mushroom-process' where continuous factors shifts into specific dynamic sectors drive growth and then productivity varies considerably across sectors. This vision of growth contrasts with a 'yeast-process' in which economy-wide growth tendencies predominate.

Many researchers have found that the effect of structural change and factors allocation on economic performance is significantly positive (Nelson and Pack,

1999; Berthelemy, 2001; Akkemik, 2005; Ngai and Pissarides, 2007). Some researchers find that the effect does not exist or is very small (Caselli, 2005). In the examination of the role of structural change in productivity growth in the manufacturing sector of the four Asian New Industrialized Economies (NIEs), Timmer and Szirmai (2000) refer to this positive effect of factor reallocation across sectors on industrial growth as the structural-bonus hypothesis. This terminology has been used extensively since then.

As shown in the Introduction section, many researchers have investigated the effect of structural change on China's economic performance across three strata of industry or across regions; for example, see Fan *et al.* (2003), Fleisher and Yang (2003), Wu and Yao (2003), Heckman (2005), Au and Henderson (2006), Bhaumik and Estrin (2007), Bosworth and Collins (2008), Gong and Lin (2008), to name a few. They do not discuss the factor shifts across industrial sectors that will be analysed in this chapter. Also, the analysis of structural effect described above normally focuses on the reallocation effect of traditional factors such as capital and labour. There are fewer studies that include the energy and environmental factors in the analysis of structural change. Jung *et al.* (2000) checked the relations between structural shift, climate change and sustainable development. Brock and Taylor (2005) argue that the structural update from resources-consuming industries will lead to sustainable development. Murillo-Zamorano (2005) investigates the importance of energy and environmental factors in economic development analysis in terms of the restricted and full production functions. Schäfer (2005) finds that the energy shifts among industries follows the regular pattern, that is, from household sector to industry and then to tertiary industry. In agricultural industry, the energy use is often lower. In this study, the energy or environmental factors will be included in the analysis.

6.3.2 *How to measure the structural effect in economic growth?*

Though scholars acknowledge the importance of structural change on economic growth, very few researchers have tried to quantify the structural effect. The traditional method to estimate the structural effect is the shift-share analysis proposed firstly by Fabricant (1942). It decomposes labour productivity into the growth effect within the industries and the structural effect among the industries, in which the structural effect is further divided into two parts: static shift effect and dynamic shift effect. The studies using the shift-share method include Liu *et al.* (1999), Fagerberg (2000), Timmer and Szirmai (2000), Kumar and Russell (2002), Peneder (2003), Tsutomu *et al.* (2004) and Akkemik (2005).

The share-shift analysis is convenient but has its own drawbacks; it is only able to analyse the shift effect of labour among sectors. If there exist other factors, such as capital, as the single factor productivity, the labour productivity is obviously not the overall measure of productivity. Based on this, some researchers propose to decompose the structural effect from the total factor productivity (TFP); see Massell (1961), Syrquin (1984), Timmer and Szirmai (2000), among others. That means, first estimate the aggregated TFP growth and sectoral TFP growth based on the production function with two factors of capital and labour. Massell (1961) argues that, relative to sectoral TFP, the aggregated TFP contains the additional effect due to the shift of the production factors into more efficient sectors that has nothing to do with the sectoral TFP. Thus, the difference between the aggregated TFP and weighted mean of sectoral TFP[1] captures the structural effect, namely, the factors reallocation effect between sectors, which could be decomposed further into the labour reallocation effect and capital reallocation effect.

The structural effect analysis by decomposing the TFP has its drawbacks, too. For example, it can only access the aggregated effect of structural change and is not able to measure the sectoral effect of structural adjustment. It also depends on too strict assumptions of constant return to scale and technical neutrality. To overcome such drawbacks, this study will firstly propose one approach to measure the structural effect by using the deterministic translog production function, which is the extension of the stochastic frontier function employed in Chen *et al.* (2011).[2] For the newly proposed method, there exist many advantages, such as measuring TFP and decomposing the structural effect simultaneously, its ability to obtain the sectoral structural effect, and no need of constant return to scale and technical neutrality assumption, which will make the measure of the structural effect more accurate. Another advantage of this method is its ability to include the energy and environmental factors in the analysis of the production function.[3] In this study, we only include the energy consumption as the input, together with capital and labour. To see the impact of environment on TFP growth and the structural effect, as carried out in Chapter 5, we will also divide all the 38 sectors into low and high emission groups according to the ranking of carbon dioxide emission in 2004. That is, the low emission group corresponds to the top half of sectors with the lower carbon emissions, and the high emission group to the lower half of sectors with the larger carbon emissions. In the influential factors analysis of structural effect, we also include the variable of carbon dioxide emission.

6.4 Methodology

The sample used in this study is the input and output panel data for 38 industrial sectors between 1980 and 2006, in order to quantify the sectoral structural effect. Due to the inclusion of energy consumption as the intermediate input, the output variable used here is the industrial gross output Y. The input factors X include capital stock, labour and energy consumption.

Industrial sectoral production function is specified as follows:

$$Y_{it} = f(X_{it}, t) e^{\varepsilon_{it}} \qquad (6.1)$$

where, $i = 1, 2, \ldots, 38$ represents 38 industrial sectors and $t = 1, 2, \ldots, 27$ is the time trend variable for 1980–2006. Stochastic disturbance term ε_{it} enters into the model in exponential form and is assumed to follow normal distribution of white noise.

Taking natural logarithm, differentiating with respect to t, and dividing by Y on both sides of equation (6.1), we obtain:

$$\frac{\partial \ln Y_{it}}{\partial t} = \frac{\partial \ln f(X_{it}, t)}{\partial t} + \sum_{j=1}^{3} \frac{\partial \ln f(X_{it}, t)}{\partial \ln X_{itj}} \frac{\partial \ln X_{itj}}{\partial t} \tag{6.2}$$

where, $j = 1, 2, 3$ corresponds to capital, labour and energy, $\partial \ln f(X_{it}, t)/\partial \ln X_j$ is the output elasticity of factor j, denoted by α_{itj}. If the superior dot is used to represent the growth rate of a variable, then $\dot{Y}_{it} = \partial \ln Y_{it}/\partial t$, and $\dot{X}_{itj} = \partial \ln X_{itj}/\partial t$. We define technical change (TC) to be $TC_{it} = \partial \ln f(x_{it}, t)/\partial t$, and rewrite equation (6.2) as:

$$\dot{Y}_{it} = TC_{it} + \sum_{j=1}^{3} \alpha_{itj} \dot{X}_{itj} \tag{6.3}$$

The growth rate of total factor productivity (TFP) is traditionally defined as:

$$T\dot{F}P_{it} = \dot{Y}_{it} - \sum_{j=1}^{3} s_{itj} \dot{X}_{itj} \tag{6.4}$$

where, $s_{itj} = w_{itj} X_{itj} / \sum_{j=1}^{3} w_{itj} X_{itj}$, w_{itj} is the price of factor j in i sector and at time point t. Thus, s_{itj} represents the actual input share of factor j to total cost in i sector and at t point and serves as the weight to build the total factor, the summation of it being 1.[4] Inserting equation (6.3) into (6.4), we obtain:

$$T\dot{F}P_{it} = TC_{it} + (RTS_{it} - 1) \sum_{j=1}^{3} \lambda_{itj} \dot{X}_{itj} + \sum_{j=1}^{3} (\lambda_{itj} - s_{itj}) \dot{X}_{itj} \tag{6.5}$$

where, $RTS_{it} = \sum_{j=1}^{3} \alpha_{itj}$ represents the return to scale of industrial sector, summation of output elasticity of all factors. Thus, $\lambda_{itj} = \alpha_{itj}/RTS_{it}$ indicates the optimal marginal output share of factor j, equal to its output elasticity under the assumption of constant return to scale. If the return to scale is varying, the second term on the right side of equation (6.5) can be used to describe the productivity improvement resulting from the evolution of scale economy of industrial sectors, the so-called scale economy (SE).

Under the assumptions of a pure market economy, such as perfect competition and profit maximization, the market price of factors can fully reflect its marginal product value, which is the precondition of efficient factor allocation. That is, $w_{itj} = pf_{itj}$, and then the equality of optimal output share and actual cost share ($\lambda_{itj} = s_{itj}$) holds, in which it is feasible to replace the cost share with its marginal output share when calculating TFP growth. In the case of a transition economy like China, however, such conditions and relationships usually fail to be satisfied, due to underdeveloped factor markets and inefficient factor reallocation. The distortion where the actual factor allocation is far from its optimal combination is pervasive. Like a coin with two sides, of course, the distortion also provides more space for Chinese industry to increase productivity by adjusting structure and then reallocating factors than in a mature economy. Thus, the third term on the right side of equation (6.5) normally makes sense in China and is used to capture the structural change effect (SCE), that is, the productivity change due to structural change and then factors reallocation. Thus:

$$
\begin{aligned}
FAEC_{it} &= \sum_{j=1}^{3}(\lambda_{itj} - s_{itj})\dot{X}_{itj} \\
&= (\lambda_{itK} - s_{itK})\dot{K}_{it} + (\lambda_{itL} - s_{itL})\dot{L}_{it} + (\lambda_{itE} - s_{itE})\dot{E}_{it}
\end{aligned}
\tag{6.6}
$$

That is, the structural effect could be divided into three parts of capital reallocative effect, labour reallocative effect and energy reallocative effect. Thus far, in terms of equation (6.5), the growth rate of TFP can be decomposed into three components of technical change, scale economy and structural change effect. That is:

$$
T\dot{F}P_{it} = TC_{it} + SE_{it} + SCE_{it}
\tag{6.7}
$$

In order to obtain the varying coefficients and statistics across sectors and over time, the more flexible translog form is specified for sectoral production function.[5] That is:

$$
\begin{aligned}
\ln Y_{it} &= \beta_0 + \beta_t t + \frac{1}{2}\beta_{tt}t^2 + \beta_K \ln K_{it} + \beta_L \ln L_{it} + \beta_E \ln E_{it} + \frac{1}{2}\beta_{KK}(\ln K_{it})^2 \\
&+ \frac{1}{2}\beta_{LL}(\ln L_{it})^2 + \frac{1}{2}\beta_{EE}(\ln E_{it})^2 + \beta_{KL}\ln K_{it}\ln L_{it} + \beta_{KE}\ln K_{it}\ln E_{it} \\
&+ \beta_{LE}\ln L_{it}\ln E_{it} + \beta_{tK}t\ln K_{it} + \beta_{tL}t\ln L_{it} + \beta_{tE}t\ln E_{it} + u_i + \varepsilon_{it}
\end{aligned}
\tag{6.8}
$$

As defined previously, K, L and E are capital stock, labour and energy. u_i represents the individual effect for the panel data model. According to the estimated coefficients of formula (6.8), we could calculate the necessary statistics such as the output elasticity of capital, labour and energy, TC, SE and SCE.

6.5 The energy and environment constraint structural change effect in Chinese industry and its determinants analysis

6.5.1 Structural change effect and industrial growth

The aggregated industrial growth accounting is reported in Table 6.1, which includes the growth rate and contribution of industrial gross output, capital stock,

Table 6.1 Measurement of structural effect, TFP growth and growth accounting in Chinese industry

Period	Outpt Growth	Capital	Labour	Energy	TFP Growth	TFP Decomposition		
						TC	SE	SCE
1981–1992	0.09	0.09	0.02	0.05	0.07	0.04	−0.02	0.05
	100	36	17	3	73	43	−26	56
1992–2001	0.13	0.08	−0.02	0.01	0.12	0.09	−0.01	0.04
	100	17	−12	0	93	71	−6	29
2001–2006	0.20	0.11	0.02	0.12	0.13	0.11	−0.01	0.04
	100	21	4	8	67	54	−6	18
1981–2006	0.13	0.09	0.01	0.05	0.10	0.07	−0.02	0.04
	100	24	3	4	77	56	−12	33

Note: For each period, the first row contains the averaging growth rate of each variable, the second row reports their respective contribution (%). The TFF growth in the table is calculated by adding three decomposed components (TC, SEC and FAEC) together, rather the Solow residual by substracting the capital, labour and energy factor contribution from output growth.

labour, energy, TFP growth and its three decomposed terms during three sub-periods defined in Section 6.2 and the entire reform period, averaged over 38 industrial sectors. Figure 6.1 displays the trend of estimated TFP growth and three decomposed components of TFP growth during the whole period at the level of aggregated industry and disaggregated low and high emission group.

Obviously, the massive industrial structural reform has led to increasing industrial productivity and output growth. The aggregated growth rate of industrial gross output and estimated TFP was 13 per cent and 10 per cent, respectively, over the entire reform period. At the first stage between 1981 and 1991, the gross output increased by 9 per cent annually. Though the contribution of TFP to output is relatively high, the TFP growth of 7 per cent is still less than the capital growth of 9 per cent. The fact that productivity increased more slowly than the rate of inputs indicates that Chinese industry was still experiencing extensive growth at this early stage, which is consistent with the experience from East Asia in the 1960s and 1970s, suggesting that inputs accumulation is more important than productivity gains in the economic take-off stage (Lucas, 1993; Young, 1995; Berthelemy, 2001).[6] The industrial growth pattern is transformed from extensive at the first stage into intensive in the following two stages, supported by the evidence that 12 per cent and 13 per cent growth of productivity has exceeded the highest factor growth, that is, 8 per cent of capital and 12 per cent of energy at the second and third stages, respectively. Of course, the contribution of productivity to output growth attained the highest level (93 per cent) in the 1990s but declined to 67 per cent after the turn of the new millennium, indicating that the current role of productivity in industrial transformation was still not stable in China.

Components of TFP growth include technical change (TC), scale economy (SE) and structural change effect (SCE). As demonstrated in Table 6.1, the contribution of SCE to gross output and TFP growth dominated at the initial stage of the

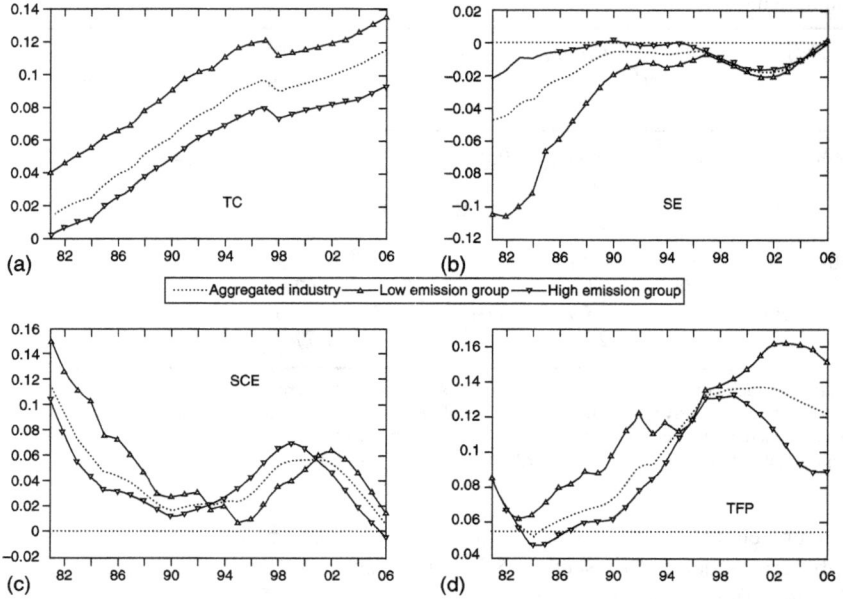

Figure 6.1 TFP growth and its decomposition of technical progress, scale effect and structural effect in Chinese industry.

reforms, accounting for 56 per cent of output growth, higher than 43 per cent of TC and 36 per cent of capital. At the second stage, the contribution of structural change to output has declined to 29 per cent, only ranked the second, inferior to 71 per cent of TC. At the third stage, the contribution of SCE has decreased to 18 per cent, ranked the third, lower than 54 per cent of TC and 21 per cent of capital. During the entire period, the structural change and then factor reallocation from less productive sectors to more productive ones increased output growth by 4 per cent annually and accounted for 33 per cent of output growth and 42.9 per cent of TFP growth, only inferior to 56 per cent of TC, and higher than 24 per cent of capital. Thus, we find a significant factors allocative effect, or so-called structural bonus in Chinese industrialization; this is main focus of our study. Seen also from Table 6.1, the negative growth and contribution of labour appears at the second stage, corresponding to the policy of furlough (*xiagang*). Since this century, though the output contribution of labour increases somewhat, it is far less than that in the early 1980s. Parallel to the SOEs reform by grasping the large and letting go of the small (*zhuada fangxiao*) and the shutdown of many small energy- and emission- intensive enterprises, the growth and contribution of energy consumption decreased to the lowest at the second stage. Since the start of this century, and with the reappearance of heavy industrialization, energy consumption has increased by a high 12 per cent and its contribution reached 8 per cent.

The scale economy has played a negative role in industrial output and productivity growth after the reform, but its negative role has become smaller and smaller.

Table 6.1 mainly reveals the time features (yeast effect, that is, an industry-wide effect) of industrial TFP and structural change effect, and the 38 sectoral growth accounting could reflect the heterogeneity, the so-called mushroom effect (sector-specific effect).[7] For example, slowly growing sectors are almost all heavy industries (such as the extraction of petroleum and natural gas, processing of petroleum and coking), which show low growth of output, TFP and TC. The sectors with the top five levels of growth of output, TFP and TC normally belong to light industry or advanced technique industry. The sectors with the relatively high structural change effect include both the natural monopoly industries, such as Production and Supply of Gas and Water, and sharp structural adjusting industries, say, Mining and Processing of Ferrous Metal Ores, Mining and Processing of Nonmetal Ores, Extraction of Petroleum and Natural Gas, and Processing of Petroleum, Coking, Nuclear Fuel. The different sectoral characteristics of productivity and its decomposition is condensed into the low and high emission group, as depicted in Figure 6.1,[8] in which the light industry with low carbon emission does enjoy higher TFP growth, TC and SCE than heavy industry. Therefore, there exists a mushroom effect during the process of Chinese industrialization. The opposite yeast effect is also found in these Figures, where both light and heavy industry experience similar trends to the aggregated industry. That is to say, in addition to heterogeneous factors, some general economy-wide factors, such as the common macroeconomic policy and external economic environment, play a role in industry, too. These factors tend to affect most sectors at the same time, rather than a limited number of sectors and, hence, improve the productivity in all industrial branches. As Nelson and Pack (1999) reveal for East Asian economies, the aggregated industrial productivity and its decomposition represented by the dotted line in the figure also seem to be driven by the expansion of the high emission group.

Corresponding with the different stages of industrial structural reform, the growth of TFP is low in the first stage, and relatively high in the second stage; but since the new millenium, industrial TFP has grown slowly, and especially slower for the high emission group (see Figure 6.1 (d). This is consistent with the evidence found by Li (1997) for Chinese industrial enterprises in the 1980s, Sun and Tong (2003), Yusuf *et al.* (2005), Jefferson and Su (2006) in the 1990s and Bai *et al.* (2009) in 1998–2005. As plotted in sub-figure (a), technical progress is the only factor contributing at an increasing rate, and its contribution is always ranked the first after 1992, though it falls somewhat over time. Mukherjee and Zhang (2007) refer to this as the paradigm of adaptive innovation, that the technology and know-how imported by China from abroad (through absorbing plenty of inflow of FDI and establishing many foreign funded enterprises) became the key to China's industrial success. Fisher-Vanden and Jefferson (2008) also argue that China's science and technology effort during the past 25 years has been moving away from a state-dominated system to one in which the locus of innovation has devolved to firms, research institutes and universities, with technology markets rapidly developing in China meanwhile. But the negative change of the scale economy, as shown in sub-figure (b), partly offsets the positive contribution of TC and SCE to TFP growth, due to its relatively small magnitude. Relatively speaking, the scale economy in the high emission group declined less than that in the low emission group and approached a similar level after the mid-1990s. This is maybe due to the presence of more small enterprises at the early stage of reform, that was not helpful in obtaining the effect of the scale economy. Overall, the return to scale (RTS) decreased for China's industry during the whole period. This phenomenon is also found by Tu and Xiao (2005) and could be explained by the restriction of free factor reallocation and limitation of optimal inputs combination during the industrial production process, such as scarce capital in light industry and skilled labour in heavy industry.

Figure 6.1 (d) presents the estimated structural effect, that is, the productivity change due to the factors allocation in Chinese industry. Obviously, a structural bonus does exist and is relatively large during the industrial reform period, which resulted from the structural reform in Chinese industry. At the beginning of the 1980s, the new reform policy and the overnight liberation of strict controls of labour from agriculture, at least for TVEs, released the vast potential energy of restricted production factors. This led to the most significant structural change effects that remedied the primarily low growth of technical progress and negative SE and pushed the early growth of TFP forward. Because radical structural reform was not undertaken at the early reform period, the structural change effect fell sharply, reached its lowest by the end of the 1980s, then remained at a standstill, even somehow falling in the early 1990s. Since 1992, such factors as the total liberalization of product markets, the massive conversion of SOEs into NSEs, and the export-oriented development strategy, among others. led to the second round surge of factor reallocation, and the aggregated industrial structural effect achieved its highest level in around 2001. From 2001 on, the disadvantages of the extremely underdeveloped factor markets and unreasonable industry policy began to be felt. The industrial restructuring -- like the inclination to emphasize higher value-added industries, the reappearance of heavy industrialization and industrial diversification -- led to the abandonment of the promotion of the traditional labour-intensive manufacturing sectors, such as textiles, and the encouragement of factors shift towards high-performing industries, like the electrical and electronic machinery sectors, towards certain high-profit industries such as mining and the manufacture of non-metallic mineral products, even towards services away from industry. Thus, the contradiction between structural adjustment and employment became acute. The high-tech and heavy industries were unable to absorb much labour. The labour-intensive industries that once hired massive workforces were experiencing reductions and facing the dilemma that they could not attract sufficient workers to stay for long due to the abnormal enterprise environment, or else they would lose labour cost competitiveness by increasing wages and improving the working environment. Regulations such as enforcing the new labour contract law in the labour market intensified the contradiction and caused the shutdown of many small enterprises and layoffs of many labourers, especially peasant workers.

All these factors acted as a rapidly decreasing structural change effect at the third stage that dragged back the increasing trend of TFP growth and its contribution to output in China's industry.

The evidence that the structural change effect follows a declining trend is also found in some of the literature. Dowrick and Gemmel (1991) show that the gain from a labour reallocation tends to decrease over time, as a country's level of development increases, and they argue that, in their sample period of the 1990s, the potential in many developing countries for such productivity gains from labour reallocation is still quite high, unlike more advanced countries. Berthelemy (2001) revealed that productivity gains achieved through the implementation of a successful structural adjustment policy could not be sustained beyond a point where the economy comes close to efficient macroeconomic management. And TFP gains through structural change are not likely to occur in the absence of appropriate adjusting policies that should keep factor distortions and wastes at the lowest possible level. Fan *et al.* (2003) state that the effect of structural change once predominant on past rapid growth will inevitably slow as the structure of the economy (for example, the shares of agriculture, industry, and services) reach a new balance.

6.5.2 *How does structural reform impact the structural bonus?*

The pattern of structural change exhibited in Figure 6.1 (d) could be observed roughly from Figure 6.2, which is based on the definition of SCE expressed by the third term on the right side of equation (6.5). Formula (6.6) shows that SCE could be further decomposed into capital reallocation effect, the labour reallocation effect and energy reallocation effect (or called structural adjustment effect of capital, labour and energy, respectively), which is dependent on factor distortion degree $\lambda - s$ and factor growth rate simultaneously. The calculating results show that the averaging share of labour marginal output, λ_L, is relatively low but increases over the sample period; while the share of capital and energy (λ_K, and λ_E) is large but keeps decreasing, indicating that there exists over-input of capital and energy and under-input of labour in Chinese industry.[9] The actual share of labour cost, s_L, rises a little at the first stage, keeps stable at the second stage and starts to fall since the beginning of this century; the actual share of capital cost, s_K, remains steady first and then rises; while the actual share of energy cost, s_E, is increasing gradually, then declines since 1998, and then starts to rise after 2002. The reduction of labour cost share and increase of capital and energy cost share correspond to the reappearance of heavy industrialization in recent years. The deviation of actual cost share of input factors from their optimal output share reveals the factor allocative distortion degree, as plotted in Figure 6.2 (b). Overall, the positive distortion degree of capital and energy is decreasing, while the negative distortion degree of labour is also decreasing. Because the optimal output share of energy begins to rise after 1996 during the decreasing trend, attains its peak in 2001 and then decreases sharply, the distortion degree of energy begins to rise at the same time, too. In Figure 6.2 (a), the averaging growth of capital and energy is higher than that of labour; corresponding with the policy of grasping the large and letting go of small, the growth of energy decreases after 1995, remains negative for the successive four years and changes to become positive from 2001; the growth of labour reverses to be negative after 1996 due to the the policy of furlough (*xiagang*), then starts to be positive till 2003;[10] parallel to the reappearance of heavy industrialization, the growth of capital and energy becomes higher at the same time. As shown in Figure 6.2, the different growth of factors in sub-figure (a) and the factor distortion degree in sub-figure (b) result together in the reallocative effect of three factors in sub-figure (c). Compared with the SCE in Figure 6.1 (c), we find that the overall structural change effect is dominated by the structural effect of capital and energy, rather than labour.[11] Specifically, the relatively high structural effect in the 1980s is mainly due to the positive reallocative effect of capital and energy. During this period, the labour reallocative effect is always negative and only ameliorates before 1986. The renewed increase of the overall structural change effect during the mid- and late 1990s is attributable to the successive rise of labour and energy reallocative effect, corresponding to the furlough policy and shutdown of many energy- and emission-intensive enterprises. As denoted previously, since the start of this century, the reappearance of heavy industrialization and closing down of many labour-intensive enterprises caused the rapid rise of energy and capital input cost and shrinking of labour input cost, that deviates from the optimal factor input direction. Thus, the structural adjustment effect of energy disappears quickly, the capital structural effect

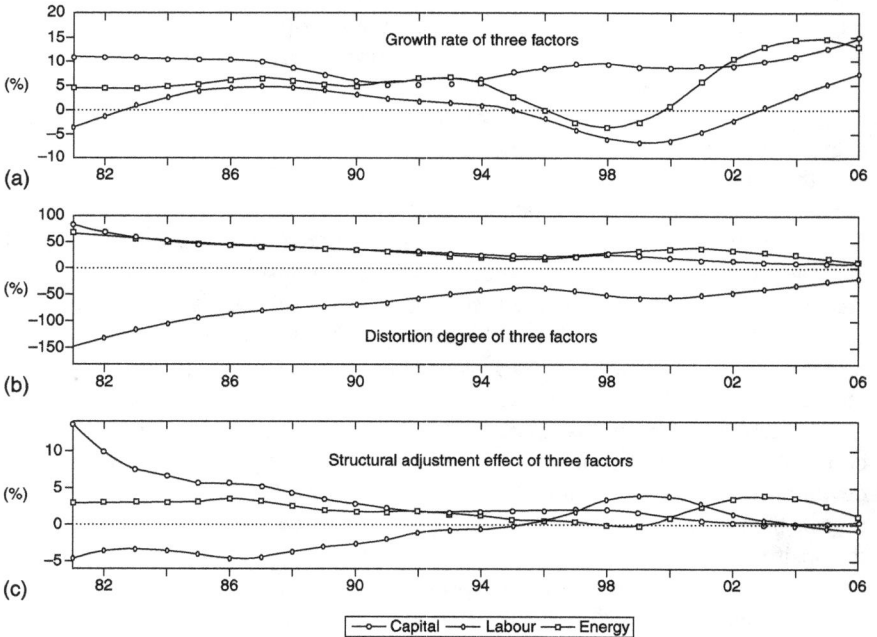

Figure 6.2 Distortion of factors allocation and structural adjustment effect for Chinese industry.

declines to almost zero, and the labour structural change effect even becomes negative; these together have pushed the rapid decline of the overall structural change effect since 2001.

Since structural change and then factor reallocation have played a substantial role in industrial productivity and output growth in China, we need a more in-depth study of the restructuring-growth nexus of SCE. Table 6.2 reports the regression analysis of the structural change effect on its determinants by using fix effect panel data models. Based on the availability of the data, the regression is undertaken over two periods, 1981–2006 (Model 1), and 1995–2006 (Model 2), respectively, The nexus variable, SCE, is the dependent variable taking the form of percentage. Among the explanatory variables, the structural variables are introduced first. The natural logarithm of capital to labour ratio represents the reform of investment and employment structure in the factor market. The second structural variable is energy structure, defined by the proportion of coal use to primary energy consumption (per cent). Both are used as the important structural variables to explain structural effect for the two models. They also serve as the variable of individual characteristics to reflect sectoral endowment of resources. In Model 2, in addition to the two structural variables described above, three structural variables are introduced further: ownership structure (the proportion of state-owned industrial value-added to total industrial value-added, per cent), scale structure (the value-added percentage of large and medium sized enterprises), and foreign funding structure (the share of foreign funded enterprises' value-added to total value-added, per cent), to capture how the structure reform of ownership, size and foreign investment affect the structural bonus. To obtain robust estimates, we control for several characteristic variables of individual sector. Since Chinese industry is often characterized by high growth, high investment, high energy consumption and low profit, The chosen control variables are the natural logarithm of industrial value-added per capita, the natural logarithm of carbon dioxide emissions, and the ratio of profit to cost (per cent). As shown in Table 6.2, most variables are statistically significant at least at the 10 per cent level (marked in a bold format). F statistics reveal that the two estimates are overall significant.

Following Kumar and Russell (2002), who check the effect of output per worker on productivity, we investigate the influence of output per capita on the structural change effect. The estimated coefficients are statistically negative for the two models, which indicates that the factor reallocation efficiency or the structural change effect declines over time with the growth of industry (if the value added per capital rises by 1 per cent, SCE will reduce by 7–8.8 per cent), similar to the findings of Dowrick and Gemmel (1991) and Berthelemy (2001). This evidence is in line with the theory of economic convergence that, as the level of industrial development rises, the adjusting space to push growth upward becomes smaller and smaller (Kumar, 2006). The large carbon dioxide emission reduces the structural change effect significantly – SCE will decline by 1.3–1.6 per cent with a rise of carbon emission by 1 per cent. The high emission group of course emits more carbon and leads to a lower structural change effect than the low emission group, which statistically confirms the mushroom effect of light and heavy

Table 6.2 Determinants analysis of structural effect (SCE)

Influential Factors	Model 1 (1981–2006)			Model 2 (1995–2006)		
	Coef.	s.e.	p value	Coef.	s.e.	p value
Constant	2.8856	17.8917	0.872	3.0288	1.3961	0.031
Sectoral Characteristics						
ln(industrial value-added per capita)	-6.9681	2.2833	0.002	-8.7710	2.0221	0.000
ln(carbon dioxide emission)	-1.3340	0.3711	0.000	-1.5688	0.6033	0.010
Profit ratio to cost (%)	0.0956	0.0930	0.304	0.0741	0.0607	0.223
Structural Variables						
ln(Capital to labour ratio)	-4.7210	2.8833	0.102	-8.5467	2.3200	0.000
Energy structure (%)	-0.3060	0.0991	0.002	-0.2011	0.0796	0.012
Ownership structure (%)				-0.1634	0.1077	0.130
Scale structure (%)				-0.1630	0.0604	0.007
Foreign funding structure (%)				0.1873	0.0594	0.002
Standard error of individual effect		11.5541			15.3838	
Standard error of disturbance term		8.3574			8.3786	
Rho		0.6565			0.7712	
Individual effect test	$F(37,945)=5.2$		0.0000	$F(37,410)=11.5$		0.0000
Overall significance test	$F(5,945)=3.42$		0.0046	$F(8,410)=4.50$		0.0000
Hausman test	$chi2(5)=12.18$		0.0325	$chi2(8)=30.88$		0.0001
Sample size	988			456		

industry revealed in Figure 6.1 (c). The positive coefficient of profit ratio to cost indicates that, as expected, the sector with a high profit rate does experience high factor reallocation efficiency, though not significantly.

After controlling for the sectoral characteristics, we find that a 1 per cent rise of the capital to labour ratio will cause a 4.7 per cent and 8.5 per cent decrease of structural change effect in Model 1 and 2, respectively. The capital to labour ratio is employed here to serve as the proxy of unbalanced investment and employment structure in China's industry, the typical features of underdeveloped factor markets. It is the rapidly changing rather than the constant growth of capital per capita that leads to increases in labour productivity unable to go hand in hand with increases in TFP, as revealed by the negative coefficient of industrial value-added per worker in the regression analysis. The ascent of industrial capital productivity since the late 1990s, after the long-term decline found in Fisher-Vanden and Jefferson (2008), seems not yet to have cured the investment hunger; on the contrary, the remaining over-investment trend today damages the allocative efficiency. Qin and Song (2009) ascribe this deterioration to imperfect capital markets, investment structural unbalance and the rigidity of structural change, among other things. They argue that policy-induced impulsive investment behaviour is still prevalent, soft loans are still available from the banking system, misallocation of financial resources is possible due to imperfect capital markets and investment structure is severely unbalanced, especially in view of the state sector. Gong and Lin (2008) assert that, in contrast with most OECD countries, the major financial resource for investment in China is credit. The easy and cheap credit provided by the government via its state banking system is certainly an important transitional feature of the Chinese economy. It reflects the strong intention of government to use its monetary policy to promote economic growth, in addition to usual demand management. Li and Xia (2008) declare that the state factor-allocation system in China still controls a vast amount of factor resources, such as capital in the forms of bank loans, subsidies and land. Chinese banks have been asked by the government to provide easy credit to the SOEs. In the absence of non-state financial institutions to allocate financial resources more efficiently, the financial sector under state monopoly tends to reinforce the already unequal distribution of financial resources in favor of SOEs. Fung *et al.* (2006) report that over half of capital investments were made by SOEs between 1998 and 2002, but the heavy investment by SOEs did not produce output proportional to their investment, as compared with that of non-state firms. Thus, it's not surprising to discover the negative influence of capital to labour ratio on the structural effect.

The coefficients of the energy structural variable in both models are all statistically negative. As expected, a 1 per cent rise of the coal share in primary energy consumption will reduce the structural change effect by 0.3 per cent and 0.2 per cent, respectively. Though it is difficult to change the coal dominated energy structure in the short-run, from the perspective of long-run it is necessary to decrease the share of coal consumption and increase the share of non-fossil energy and renewable energy. The coefficient of the ownership structural variable is negative but not too significant. A 1 per cent increase of the share of SOEs' industrial value-added reduces the structural change effect by 0.16 per cent. Thus, the reform of the ownership structure, converting state industry to non-state ones from the late 1990s, indeed ameliorates the factor reallocation efficiency, indicating that the reform of SOEs in China is also a reform of the government's regulatory practices from a grabbing-hand approach to a helping-hand approach; this is also found by Wan and Yuce (2007). This finding mirrors the studies of Li (1997), Sun and Tong (2003), Jefferson and Su (2006), Bai *et al.* (2009), among others. They discover that (labour) productivity has been improved by the ownership rights reform in China since the late 1990s. The survival of SOEs is, to a great extent, at the expense of state asset efficiencies, due to the agency problem; thus, ownership reform is vital to incentives and to economic performance. In order to ensure the long-term viability of high industrial growth, the restructuring of large SOEs will be the crux of the next wave of reform. Industrial concentration is the core of the theory of industrial organization, and the expansion of large and medium-sized enterprises (LMEs) helps to increase the degree of industrial concentration and should increase the structural effect. But it is not the case in this study. A 1 per cent rise of the share of LMEs' industrial value-added significantly reduces the factor allocative efficiency by 0.16 per cent, which means that industrial concentration won't remedy allocative limits but full competition among many medium and small-sized firms might. This is also consistent with the findings of decreasing return to scale in Chinese industry shown in the previous subsection. In addition to private enterprises, foreign funded enterprises have developed greatly after the reform. As revealed in Table 6.2, a 1 per cent increase of the share of foreign funded enterprises' industrial value-added significantly increases SCE by 0.19 per cent. As Yusuf *et al.* (2005) state, the combination of structural reform (such as state ownership, LMEs and foreign funded enterprises) enhances overall productivity due to structural change and factor reallocation, and this differentiates them among sectors, which statistically confirms the yeast and mushroom effect of the structural bonus found in our estimates in the previous subsection.

6.6 Summary and comments

This chapter investigates the impact of structural reform on the performance of Chinese industry using a panel data set of 38 two-digit industrial sectors over the entire reform period, especially under the constraints of energy and environment. We propose a deterministic translog production function and decomposition method to measure the changes in total factor productivity and its part due to structural change and then factor reallocation across industrial sectors. We also analyse the determinants of structural change effect, including the variable of carbon dioxide emission. We offer basic conclusions and corresponding remarks here.

1   Since the industrial development strategy was converted from being heavy-industry-oriented to recognize the parallel importance of light and heavy industry, reflecting the comparative advantages in 1978, China's industry has experienced substantial structural change, productivity growth and spectacular

output growth. The higher than input factor contribution of TFP growth, due to the technical change and structural change after 1992, indicates that the growth model of Chinese industry seems to have been transformed from being extensive to intensive. But this transformation is not stable or sustainable. With the reappearance of heavy industrialization, mainly because of the rapid reduction of the structural effect, the contribution of TFP to output decreases, implying that the momentum to drive the industrial transformation has become weak.

2   The factor inputs can affect industrial growth either directly through a volume effect, such as capital accumulation, or indirectly through an efficiency effect that promotes productivity by adjusting the economic structure and reallocating the factors from less productive sectors to more productive ones. The growth accounting in Table 6.1 reveals that, on average, the indirect structural change and factor reallocative effect plays a substantial role in industrial growth by pushing productivity upwards. Its contribution to output reaches 33 per cent, exceeding the sum of the three factors' contribution of 31 per cent. From the perspective of the volume effect distribution, capital stock has made the largest contribution to industrial output, followed by the energy consumption, and the labour force has the lowest contribution.

3   Timmer and Szirmai (2000) refer to the positive structural change effect as a structural bonus. We discover that the structural bonus does exist and matters in both the yeast and the mushroom processes of Chinese industrial transformation. But the structural change effect has rapidly decreased since the start of this century, making the force that drives industrial transformation smaller and smaller. The fixed effect panel model regression suggests that the reforms of investment structure, energy structure, ownership rights structure, size structure and foreign funded enterprises in China significantly contribute to a structural bonus at the former two stages, but also to the decrease of the structural change effect after 2001.

The factors that have driven the reduction of the factor reallocation effect since 2001 and produced the difference of allocative efficiency between the low and high emission group highlight the future reform directions for Chinese industry. The most urgent reform is to continue the development of factor markets, especially the energy or resources market. In order to relieve the factor distortion and establish long-run sustainable industrialization, it is necessary to reform the dual-track resource allocation system, reform the energy price system, balance the investment structure, provide non-state enterprises equal access to resources, and develop non-state financial institutions, as suggested by Fung *et al.* (2006), Gong and Lin (2008), Li and Xia (2008), Qin and Song (2009) and others. Those sectors that are exploiting China's comparative advantage successfully should be supported and promoted. Another challenging reform is to deepen the restructuring of state industry. Indeed, the speed of reforming its SOEs has distinguished China from other formerly centrally planned economies, and thus it has attracted much attention in the economics literature.

7 Undesirable output, environmental TFP and industrial transformation

7.1 Introduction

Continued depletion of fossil fuels and increasing environmental pollution indicate that the traditional economic development model is not sustainable and its low-carbon transformation is becoming more inevitable. For the case of China, however, not as usually imagined, it's not so easy to be optimistic that it can transform its development model from being extensive to intensive instead, because its economic growth depends heavily on extensive investment, high energy consumption and pollution emission, as stated previously, leading to its current position of being the largest CO_2 emitter and largest energy consumer in the world. An evaluation system based on growth is the root cause of all the ills, and nothing but the replacement of GDP with green development indicators will resolve this problem. Environmental or green total factor productivity (TFP) that allows for the negative externality of environmental pollution is the preferred choice, as this can reveal the actual situation of the development model transformation and be employed to guide the way to low-carbon development in China. This is what we wish to consider in this chapter.

7.2 How to treat energy and environmental variables when estimating environmental TFP

In fact, since the pioneering work by Solow, total factor productivity, as the engine to drive economic growth along with the quantitative inputs of factors, is increasingly being introduced into the analytical framework of neoclassical growth accounting, and the rising contribution share of TFP growth on output, as opposed to the declining one of all input factors, is becoming the key indicator used to evaluate whether the development model is transformed or not (Solow, 1957; Kim and Lau, 1994; Krugman, 1994; Young, 1995).[1] In the literature, however, the measure of productivity is only based on the traditional input factors of capital and labour, without the consideration of energy and environment factors that intimately relate to the concept of sustainable development; this causes a misleading estimate of TFP growth and corresponding sustainability analysis. Thus, greater inclusion of energy and environment in the measurement of productivity is extremely crucial. The growth of TFP should be restricted by the fact that some resources are increasingly rare and also heavy pollution emission. For example, the average proportion of industrial GDP to the overall total, 40.1 per cent, was achieved by consuming 67.9 per cent of national energy and emitting 83.1 per cent of national CO_2 during the reform period. After the turn of the millenium, heavy industrialization appeared again in China, leading to a unprecedented rise of energy consumption and CO_2 emission.[2] Of course, environmental regulatory policies are accordingly likely to improve productivity by saving and more efficiently using energy and abating emission, as stated by the Porter hypothesis (Porter, 1991). During the ownership right reform period, China closed tens of thousands of small inefficient mining operations and electricity power generators, causing the first stagnancy or decline of energy use and CO_2 emission between 1995 and 2002, and a more rapid decrease of energy and emission intensity during that period.

In what follows, we proceed to analyse two questions: What is the influence of energy and emission constraints on Chinese TFP growth and then sustainable studies? Has the energy-saving and emission-abating policy played a substantial role in low-carbon transformation of the industrial development model in China? To this end, of particular importance is how to include energy and emission into the growth accounting framework that generally only considers capital and labour factors. Many studies conclude that it is appropriate to incorporate energy as an intermediate input in the production process; see the well-known KLEM model by Jorgenson *et al.* (1987) as an example. The treatment of environmental pollution emission is more complex. For a long time, the emission variable has not been included in economic analysis due to its lack of market pricing and the difficulty of computing its production cost. With the increasing importance of environmental issues in sustainable analysis, many studies have begun to introduce emission into the production function as an input factor, though unpaid and unobservable, together with capital, labour and energy; see Mohtadi (1996) for detailed discussion. Some researchers find the output characteristics of the emission and regard the emission as a by-product, rather the input, of the production process; see Färe *et al.* (1994) for their radial data envelopment analysis (DEA) based on the Shephard distance function. However, at that time, the researchers had not considered the negative externality of the pollution by-product, still seeing it as a desirable output (or goods), the same as GDP. As Nanere *et al.* (2007) denote, the measure of productivity is very likely to be biased if it does not, or not correctly, allow for the environmental factors. Not until Chambers *et al.* (1996a) and Chung *et al.* (1997) put forward the directional distance function (DDF) and the environment regulatory activity analysis model (AAM) was the negative externality of pollution emission realized and characterized for the first time in the literature. Then the pollution variable was incorporated into the production function not only as a by-product but also as an undesirable output, or 'bads', along with the desirable output of GDP, making it possible to reasonably capture the restricting role of pollution emission in the production process and the impact of environmental regulation on the productivity growth, from the perspective of methodology. The approach was soon widely used; see Färe *et al.* (2001), Boyd *et al.* (2002), Jeon and Sickles (2004), Kumar (2006), among others.

In this chapter, the DDF based AAM is chosen as the target model, referred to as Model 4 thereafter, to address the energy and environmental issues proposed previously, the principle of which will be introduced in Section 7.4. The results of Model 4 will be compared with those estimated from three traditional models, in which the pollution emission will not be taken into consideration, will be treated as the input factor, and will be treated as the desirable output; these are named Model 1, 2 and 3, respectively. Actually, Models 1–4 hold the same distance function analytical framework, in which the energy consumption is all processed as input, the same as capital and labour, and the emission variable is treated in a different way following its methodology evolution process. In this study, the pollution variable used to analyse is CO_2 emission, and industry is also used as the example due to its higher energy use, heavier CO_2 emission and urgency of need of the development model transformation. As Jorgenson and Stiroh (2000) outlined, it is essential to disaggregate the industrial analysis to the sub-industrial level to find the true pattern behind the aggregation. Following this approach, this chapter avoids the limitations of an aggregation analysis by breaking down China's industry into 38 sub-industries for the period of 1980–2008. Specifically, based on the estimated values of the distance function, the sub-industrial TFP growth and its decomposition will be computed by using the Malmquist Productivity Index (MPI) for the first three models and by the Malmquist–Luenberger Productivity Index (MLPI) for Model 4. The productivity estimated by the target Model 4 should be more accurate, due to the reasonableness of DDF, and is called the environmental (adjusted) or green TFP, or actual TFP, from now on in this chapter. Hailu and Veeman (2000) once called it the environment sensitive TFP. Managi (2006) names it as the total TFP because he also considers both market output (i.e., agricultural production) and non-market output (i.e. environmental pollution) when estimating productivity.

7.3 Review of Chinese productivity estimates in the literature

There are plenty of research papers which study the topic of productivity; see Dorfman and Koop (2005) for a survey. Economic reform has led to substantial productivity and output growth in China since 1978. Therefore, many researchers are also interested in estimating the TFP growth in China and assessing its effect on the economy from different angles. This section focuses on a review of them.

The Chinese aggregate TFP is first considered and estimated in the studies. For example, Chow (1993, 2008) and Chow and Lin (2002) find that Chinese TFP growth was almost zero between 1952 and 1978 and 2.7 per cent annually after 1979. Perkins (1988) shows that the average TFP growth of China was 4.1 per cent, −1.4 per cent, 0.6 per cent and 3.8 per cent in 1953–1957, 1957–1965, 1965–1976 and 1976–1985, respectively. Perkins and Rawski (2008) estimate that annual TFP growth in China was 0.5 per cent in 1952–1978 and 3.8 per cent in 1978–2005, in which the TFP growth attained a peak of 6.7 per cent between 1990 and 1995 then decreased after 1995. The TFP growth rate estimated by Borensztein and Ostry (1996) was −0.7 per cent and 3.8 per cent in 1953–1978 and 1979–1994 in China. Collins and Bosworth (1996) concluded that the average annual TFP growth for China was 1.4 per cent and 3.3 per cent, respectively, during the periods of 1960–1973 and 1973–1994. Young (2003) reported that the TFP growth of non-farm industry averaged over 1978–1998 was 3 per cent using official data and 1.4 per cent using corrected data. Holz (2006) estimated that Chinese TFP growth in 1953–1978 and 1978–2005 was −0.6 per cent and 3.9 per cent. Zheng *et al.* (2009) first discussed the correction of Chinese capital, labour and GDP data in detail and, based on this, estimated that the annual average TFP growth was 3.2 per cent between 1978 and 1995 and declined to 0.8 per cent in 1995–2005 due to the speed-up of capital deepening. Some studies focus on the estimates of provincial or regional productivity in China; for example, see Lin (1992), Chen *et al.* (2008), Chen *et al.* (2009), Li (2009), Tuan *et al.* (2009), to name a few.

This chapter mainly concerns industrial productivity in China, and so does this section of the survey, as shown in Table 7.1. Since the initial work by Chen *et al.* (1988), there was a big surge of TFP estimates in Chinese industry as opposed to studies about other areas. Some studies estimated just Chinese aggregate industrial productivity (Woo, 1998; Bosworth and Collins, 2008); some study the industrial productivity of different ownership types such as SOEs and TVEs (Wu, 1995; Jefferson *et al.*, 2000; Zhang *et al.*, 2003); some estimated the TFP growth of light and heavy industry or industrial sectors (Zheng *et al.*, 2003; Chen *et al.*, 2011) and different industrial firms (Jefferson *et al.*, 2008; Liu and Wu, 2009); and some focused on the large and medium-sized enterprises (LMEs) in the industry (Tu and Xiao, 2005), and so on. As for the estimating methods, as illustrated in Table 7.1, most studies made use of Solow residuals or regression of Cobb-Douglas (CD) or the translog production function to estimate productivity; some used the parametric stochastic frontier production function; and some adopted the nonparametric deterministic frontier DEA framework and Malmquist productivity index (MPI), similar to the basic framework of distance function used in this chapter, to estimate the TFP growth. There is also a big variability of estimates of Chinese industrial productivity growth among different studies, as surveyed in Table 7.1, from a low of −1.1 per cent in Jefferson *et al.* (2000) to a high of 51.8 per cent in Liu and Wu (2009). Such measures of productivity will be partly compared with our estimates in Sections 7.5 and 7.6.

The above survey shows that the estimates of Chinese productivity in the studies are normally based on the traditional inputs of capital and labour, though sometimes the intermediate input is also included, without consideration of energy and environmental factors. As denoted in Section 7.2, the sustainable development analysis based on the productivity measures that do not, or not correctly, allow for the energy and emission role, is incomplete and even likely to be misleading because the restriction of resources and perniciousness of pollution will surely hinder the productivity improvement but energy-saving and pollution-reducing regulation (or policy) will achieve the opposite. This is particularly the case in China due to its extensive growth, using too much energy and emitting

Table 7.1 TFP estimates and its decomposition for Chinese industry in literature

Researchers	Data	Inputs	Output	Methodology	Sample Period	TFP Growth and Its Decomposition (%)	
Chen et al. (1988)	SOEs Industrial Data	K, L	GIOV	Solow Residual; C-D or Translog Production Function Regression	1953–1985 1978–1985	SOEs TFP 2.6 5.9	
Woo (1998)	Industrial Time Series	K, L	IVA		1979–1984 1985–1993	Industry TFP 3.3 0.7	
Jefferson et al. (2000)	Industrial Time Series	K, L, M	GIOV		1980–1996 1992–1996	SOEs TFP 1.7 –1.1	Aggregate TFP 2.8 1.5
Zhang et al. (2003)	TVEs Industrial Data	K, L	GIOV		1980–2000	TVEs TFP 6.7	Aggregate TFP 3.7
Bosworth and Collins (2008)	Industrial Time Series	K, L, Education	GDP		1978–1993 1993–2004	Industry TFP 3.1 6.2	Aggregate TFP 3.6 4
Jefferson et al. (2008)	Industrial Firms Data	K, L	IVA		2005/1998	SOEs TFP 15.6	
Wu (1995)	Industrial Panel Data	Fixed Assets, Working Capital, L	GIOV	Stochastic Frontier Production Function Regression	1986 1991	SOEs TFP 4.4 1.8	TVEs TFP 3.9 5.2
Tu and Xiao (2005)	LMEs Firms Panel Data	K, L	IVA		1996–2002	LMEs TFP 6.8	
Chen et al. (2011)	Industrial Sectoral Panel Data	K, L	IVA		1981–2008 1981–1992 1992–2001 2001–2008	Industry TFP 6.7 0.8 9.9 11.1	Technical Progress 10.2 3.4 11.5 18.1

	Data		Output	Method	Period	Heavy Industry TFP	Light Industry TFP
Zheng et al. (2003)	SOEs Firms Panel Data	K, L, M	GIOV	DEA, MPI	1980–1989	7	12
					1990–1994	8	6
Liu and Wu (2009)	Firms Data	K, L, M	GIOV		2001–2004	Industry TFP 51.8	Technical Progress 13.2

Note: K – Capital; L – Labour; M – Intermediate Input; GIOV – Gross Industrial Output Value; IVA – Industrial Value Added. All estimates shown in this table come or are calculated from the results in the corresponding literature.

heavy pollution. In spite of its extreme importance in the analysis of economic development, there are only a few papers regarding the role of energy and emission in Chinese sustainability studies. For example, Hu *et al.* (2008) re-rank the technical efficiency of Chinese provinces by taking account of environmental factors while implementing the DEA methodology. Wang *et al.* (2010) find that the environmental and market TFP change deviate from each other, indicating that environmental regulation is very difficult in China. This paper will analyse the impact of energy and CO_2 emission on industrial productivity change in China by using newly proposed directional distance function based activity analysis model.

7.4 The analytical framework of low-carbon development

As mentioned in Section 7.2, TFP is the important indicator used to evaluate the development model transformation. But it cannot be observed directly and must be estimated by the researchers. In the literature, the approach to estimating TFP may be roughly classified into three categories: the index approach, the Solow residual and the Frontier Production Function. The index approach is calculated according to the definition of TFP. The widely used index formulations to calculate the TFP include the Laspeyres, Paasche, Fisher and Törnqvist indices. However, the calculation of index formulations needs the price information of inputs and outputs and is incapable of including pollution variables, due to the absence of their market price. Since the estimating formulation of TFP in the form of the production function was first given by Solow in 1957, the approach of using the Solow residual became popular when measuring productivity. To that end, the output elasticity of inputs must be firstly set using the priori constant or by regressing the CD or translog production function. The approach of the frontier function used to estimate the TFP is good at the identification of inefficient production units and normally divided into two methods: the stochastic and the deterministic frontier function. The stochastic frontier function is often parametrically defined and has the weakness of priori assumption of the efficiency pattern and is incapable of modelling the production process with more than two kinds of output, such as the desirable and undesirable output to be investigated in this chapter. The deterministic frontier function may be further subdivided into parametric and nonparametric methods, the latter of which, often called the DEA approach, is more widely used in the literature and is regarded as the common analytical framework for the four models specified in this chapter. Zhou *et al.* (2008) survey the applications of the DEA approach in the energy and environmental area. Obviously, the DEA approach could overcome the above-denoted weakness of the index approach, Solow residual and stochastic frontier approach, and is particularly good at modelling the production process with multi-inputs and outputs, in which the DDF version can even differentiate between the desirable output and undesirable output, such as the environmental pollutions, as targeted in this chapter. In what follows, the principle of DDF, a standard analytical framework of low-carbon development, will be briefly introduced by using Figure 7.1, as opposed to the traditional Shephard distance function (SDF).

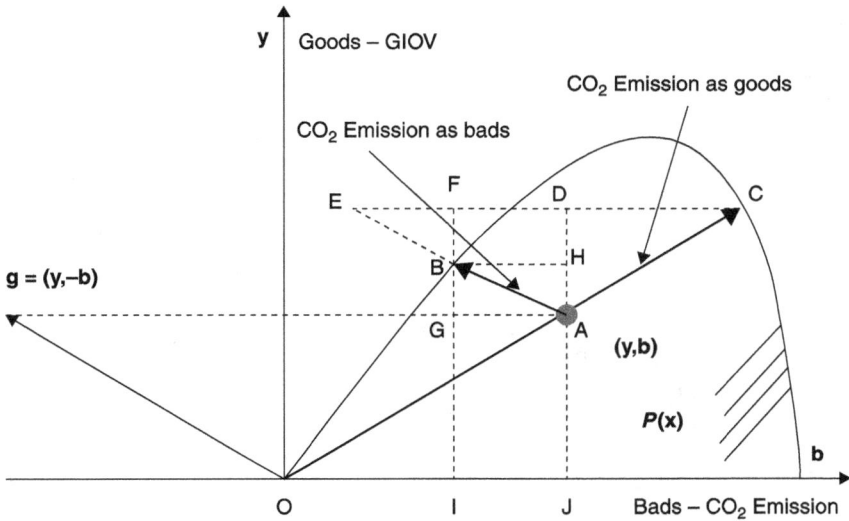

Figure 7.1 Principle of Shephard and directional distance functions.

Assume that there are n decision-making units (DMU) at t time point, and there are k types of input, l types of desirable output (or goods), and m undesirable output (or bads) for each DMU. For the ith DMU ($i = 1, 2, \ldots, n$), the column vectors $\mathbf{x}_i$, $\mathbf{y}_i$ and $\mathbf{b}_i$ represent the inputs, goods and bads, respectively. And $\mathbf{X}_{k \times n}$, $\mathbf{Y}_{l \times n}$ and $\mathbf{B}_{m \times n}$ are the input and output matrix for all the n DMUs. In this study, the DMU is the sub-industry; for each of them, $k = 3$, corresponding to capital, labour and energy, $l = 1$ being gross industrial output value (GIOV), and $m = 1$ carbon dioxide emission (CO_2).[3]

As illustrated in Figure 7.1, the technology is represented by the output set $P(\mathbf{x})$ to which the output vector of A point $(\mathbf{y}, \mathbf{b})$ belongs. The Shephard distance function radially scales the original vector from point A proportionally to point C to describe the simultaneous increase of desirable and undesirable output, in which the goods and bads are all assumed to be freely disposable without the consideration of negative influence of the bads. In contrast to this, due to the weak disposability assumption of the bads, say, pollution emission, in view of its negative externality that differs from the goods, DDF starts at point A and scales in the direction along AB, represented by the direction vector $\mathbf{g} = (\mathbf{y}, -\mathbf{b})$, to model the increase of desirable outputs and decrease of undesirable outputs simultaneously, which make this the most appropriate framework to capture the negative impact of pollution emission on productivity and output growth and to analyse the possibilities of low-carbon development. The strong or free disposability of bads in SDF indicates that the disposability of bads costs nothing and the environment imposes no restriction on the economy, which seems far from the reality. In fact, it's impossible for the producers to have no cost to reduce pollution and then voluntarily get rid of it, because the reduction seizes important inputs and then translates them into lost goods. The emission reduction can only be achieved by enforceable environmental regulation. The weak disposal assumption of the bads in DDF, to distinguish them from the goods, precisely reflects this story; thus, the DDF is also referred to as environment regulatory activity analysis model (AAM) by Chung *et al.* (1997) which can be used to evaluate the effect of environment governance on economic development. If we let $\mathbf{g} = (\mathbf{y}, \mathbf{b})$, DDF will reduce to SDF, indicating SDF is just the special case of DDF. SDF and DDF are just the Models 3 and 4 specified in this study. If we only consider one type of output, GIOV, and regard the bads of CO_2 emission as the fourth inputs, along with the other three inputs, SDF or Model 3 will reduce to Model 2. If we ignore the emission variable, Model 2 will further reduce to Model 1. In a word, the former three models are similar to those specified in most of the productivity literature, as surveyed in Section 7.3; and the DDF based Model 4 is the target model in our study. Mathematically, the DDF may model the environment regulatory activity that increases the goods and simultaneously decrease the bads, as:

$$\overline{D}_o^t(\mathbf{x}_i^t, \mathbf{y}_i^t, \mathbf{b}_i^t; \mathbf{g}_i^t) = \sup\{\beta : (\mathbf{y}_i^t, \mathbf{b}_i^t) + \beta \mathbf{g}_i^t \in P^t(\mathbf{x}_i^t)\} \tag{7.1}$$

Or, equivalently:

$$\begin{aligned}
\overline{D}_o^t(\mathbf{x}_i^t, \mathbf{y}_i^t, \mathbf{b}_i^t; \mathbf{y}_i^t, -\mathbf{b}_i^t) &= \sup\{\beta : (\mathbf{y}_i^t, \mathbf{b}_i^t) + \beta(\mathbf{y}_i^t, -\mathbf{b}_i^t) \in P^t(\mathbf{x}_i^t)\} \\
&= \sup\{\beta : ((1+\beta)\mathbf{y}_i^t, (1-\beta)\mathbf{b}_i^t) \in P^t(\mathbf{x}_i^t)\}
\end{aligned} \tag{7.2}$$

in which, β is the maximum feasible expansion of the goods and contraction of the bads when the expansion and contraction are identical proportions for a given level of inputs, which amounts to the value of DDF to be measured.

More specifically, for ith sub-industry, the version of DDF using the technology in t period and observation also in t period can be estimated by resolving the following linear programming:

$$\overline{D}_o^t(\mathbf{x}_i^t, \mathbf{y}_i^t, \mathbf{b}_i^t; \mathbf{y}_i^t, -\mathbf{b}_i^t) = Max_{\lambda, \beta}\ \beta$$

$$s.t.\ \ \mathbf{Y}\lambda \geq (1+\beta)\mathbf{y}_i; \mathbf{B}\lambda = (1-\beta)\mathbf{b}_i; \mathbf{X}\lambda \leq \mathbf{x}_i; \lambda \geq \mathbf{0} \tag{7.3}$$

The inequality for goods in Equation (7.3) makes it freely disposable, which means that the goods can be disposed of without the use of any inputs and then without the decrease of bads. The bads is modelled with equality that makes it weakly disposable. The inequality constraint of inputs illustrates also that the inputs are strongly disposable. The intensity variable $\lambda = (\lambda_1, \lambda_2, \ldots, \lambda_n)^T$ contains the weight assigned to each sub-industry when constructing the production frontier. The production function defined by the combination of input and output matrix $(\mathbf{X}, \mathbf{Y}, \mathbf{B})$ will be used as the benchmark to evaluate the efficiency of ith DMU, $(\mathbf{x}_i, \mathbf{y}_i, \mathbf{b}_i)$.

Based on the estimated values of SDF for the former three models in this study, TFP can be measured by computing the Malmquist productivity index (MPI), which is the geometric mean of Malmquist technical and efficiency change indices, MTCH and MECH, respectively. Following that, based on the values of DDF, the environment adjusted TFP can also be estimated by calculating the Malmquist–Luenberger Productivity Index (MLPI). To this end, four different types of DDF must be solved for each sub-industry: two use observations and technology at the time period t and $t+1$, $\overrightarrow{D}_o^t(\mathbf{x}_i^t, \mathbf{y}_i^t, \mathbf{b}_i^t; \mathbf{y}_i^t, -\mathbf{b}_i^t)$ and $\overrightarrow{D}_o^{t+1}(\mathbf{x}_i^{t+1}, \mathbf{y}_i^{t+1}, \mathbf{b}_i^{t+1}; \mathbf{y}_i^{t+1}, -\mathbf{b}_i^{t+1})$, the former illustrated in linear programming (7.3); and two use adjacent periods, for example, $\overrightarrow{D}_o^t(\mathbf{x}_i^{t+1}, \mathbf{y}_i^{t+1}, \mathbf{b}_i^{t+1}; \mathbf{y}_i^{t+1}, -\mathbf{b}_i^{t+1})$ calculated from t period technology with the $t+1$ period observation, and $\overrightarrow{D}_o^{t+1}(\mathbf{x}_i^t, \mathbf{y}_i^t, \mathbf{b}_i^t; \mathbf{y}_i^t, -\mathbf{b}_i^t)$ calculated from the $t+1$ period technology with the t period observation. Then MLPI is defined as below:

$$
MLPI^{t,t+1} = \left[\frac{1+\overrightarrow{D}_o^t(\mathbf{x}_i^t, \mathbf{y}_i^t, \mathbf{b}_i^t; \mathbf{y}_i^t, -\mathbf{b}_i^t)}{1+\overrightarrow{D}_o^t(\mathbf{x}_i^{t+1}, \mathbf{y}_i^{t+1}, \mathbf{b}_i^{t+1}; \mathbf{y}_i^{t+1}, -\mathbf{b}_i^{t+1})} \right.
$$
$$
\left. \times \frac{1+\overrightarrow{D}_o^{t+1}(\mathbf{x}_i^t, \mathbf{y}_i^t, \mathbf{b}_i^t; \mathbf{y}_i^t, -\mathbf{b}_i^t)}{1+\overrightarrow{D}_o^{t+1}(\mathbf{x}_i^{t+1}, \mathbf{y}_i^{t+1}, \mathbf{b}_i^{t+1}; \mathbf{y}_i^{t+1}, -\mathbf{b}_i^{t+1})} \right]^{1/2}
\tag{7.4}
$$

Different from the traditional index formulations and same as MPI, MLPI is the most widely used productivity index and is particularly attractive since it does not rely on prices – specifically the price of CO_2 which appeared in this study – in order to construct it. The MLPI can be decomposed as the product of two terms: the change of technical progress (MLTCH) and the change of production efficiency (MLECH); that is:

$$
MLPI^{t,t+1} = MLECH^{t,t+1} \cdot MLTCH^{t,t+1}
\tag{7.5}
$$

where,

$$
MLTCH^{t,t+1} = \left(\frac{1+\overrightarrow{D}_o^{t+1}(\mathbf{x}_i^{t+1}, \mathbf{y}_i^{t+1}, \mathbf{b}_i^{t+1}; \mathbf{y}_i^{t+1}, -\mathbf{b}_i^{t+1})}{1+\overrightarrow{D}_o^t(\mathbf{x}_i^{t+1}, \mathbf{y}_i^{t+1}, \mathbf{b}_i^{t+1}; \mathbf{y}_i^{t+1}, -\mathbf{b}_i^{t+1})} \cdot \frac{1+\overrightarrow{D}_o^{t+1}(\mathbf{x}_i^t, \mathbf{y}_i^t, \mathbf{b}_i^t; \mathbf{y}_i^t, -\mathbf{b}_i^t)}{1+\overrightarrow{D}_o^t(\mathbf{x}_i^t, \mathbf{y}_i^t, \mathbf{b}_i^t; \mathbf{y}_i^t, -\mathbf{b}_i^t)} \right)^{1/2}
\tag{7.6}
$$

$$
MLECH^{t,t+1} = \frac{1+\overrightarrow{D}_o^t(\mathbf{x}_i^t, \mathbf{y}_i^t, \mathbf{b}_i^t; \mathbf{y}_i^t, -\mathbf{b}_i^t)}{1+\overrightarrow{D}_o^{t+1}(\mathbf{x}_i^{t+1}, \mathbf{y}_i^{t+1}, \mathbf{b}_i^{t+1}; \mathbf{y}_i^{t+1}, -\mathbf{b}_i^{t+1})}
\tag{7.7}
$$

As illustrated by Färe *et al.* (2001), if there have been no changes in inputs and outputs over two time periods, then MLPI = 1. An improvement in productivity is signalled by MLPI > 1; a decrease in productivity by MLPI < 1. If technical change enables more goods production and less bads production, then MLTCH > 1;

whereas if MLTCH < 1, there has been a shift of the frontier in the direction of fewer goods and more bads. If MLECH > 1, it indicates that a DMU is closer to the frontier in period t + 1 than it was in period t, there existing the catch-up effect; if is less than unity, it indicates that a DMU is further away from the frontier in t + 1, and hence has become less efficient.

If for a given two-year period, the DMU satisfies the following three conditions simultaneously:

$$MLTCH_o^{t,t+1} > 1$$

$$\overline{D}_o^t(\mathbf{x}^{t+1}, \mathbf{y}^{t+1}, \mathbf{b}^{t+1}; \mathbf{y}^{t+1}, -\mathbf{b}^{t+1}) < 0 \qquad (7.8)$$

$$\overline{D}_o^{t+1}(\mathbf{x}^{t+1}, \mathbf{y}^{t+1}, \mathbf{b}^{t+1}; \mathbf{y}^{t+1}, -\mathbf{b}^{t+1}) = 0$$

then that DMU shifted the frontier for that two-year period and were innovators.

7.5 What's the energy and emission restricted actual productivity growth in China?

This chapter focuses on two-digit sub-industries datasets. The desirable output is gross industrial output value (GIOV) with the unit of 100 million RMB at 1990 price levels, rather than value-added, due to the inclusion of intermediate input, such as energy, into the analysis. CO_2 emission is exemplified as the representative of the undesirable output; the method to process it could be directly applied to other pollution. The capital stock cannot be obtained directly and is estimated by using a perpetual inventory approach, depreciated at constant 1990 prices in terms of the price indices of investment in fixed assets. The labour input is annual average employed workers (unit: 10 thousand workers) and energy input is total energy consumption with the unit of 10 thousand tons of coal equivalent (tce). Since the empirical work of Hoffmann (1958) and Chenery *et al.* (1986), the standard perception of industry transformation is a general shift in relative importance from light to heavy industry, in which the heavy industry is normally closer to high energy consumption than light. Therefore, we also divide all sub-industries into light and heavy industry according to the ranking of energy consumption in 2004. The classification of light and heavy industry is because 38 sub-industrial patterns of TFP change are too complicated to see clearly all at once, and sometimes the observation of the difference between the light and heavy industry instead is enough for the analysis. The descriptive statistics of the variables implies that there should be lower productivity growth for the heavy industry than light.

Based on the introduced sample data, the development speed of productivity, technical progress and production efficiency, weighted averages over all the sub-industries and the sample period geometrically, for the four models specified in Section 7.4 are estimated and reported in Table 7.2, in which, the weights are GIOV share for each sub-industry. Compared with the studies listed in Table 7.1, the variables included in this study are more informative in the sense that at least

Table 7.2 Average index of productivity, technical progress and efficiency for Chinese industry estimated by four models

Index	Not Allowing for Emission	Emission as Input	Emission as Goods	Emission as Bads
	(Model 1)	(Model 2)	(Model 3)	(Model 4)
MPI/MLPI	1.0545***	1.0626***	1.0387**	1.0229
MTCH/MLTCH	1.0522*	1.0592**	1.0346	1.0255
MECH/MLECH	1.0035	1.0044	1.0050	0.9981

Note: The index of the aggregated industry is the weighted mean of sub-industrial index (the weight is GIOV shares) and the geometric average over the entire sample period. Model 4 is the target model in this table. Null hypothesis of *t* test is that the estimated productivity, technical progress and efficiency by Models 1, 2, 3, respectively, are identical to those estimated by Model 4. ***, **, * indicate the level of significance to be 1%, 5% and 10% respectively.

the energy factor has been included as the input in the endeavour to deal appropriately with the CO_2 emission. As stated previously, Model 4 is the target approach due to being the most appropriate way to process the emission variable and the following analysis will be mainly based on the measures from it.

As shown in Table 7.2, the measurements of Model 4 tell us that, in the situation of processing the energy and emission variables properly, the actual average TFP, technical progress and efficiency growth for aggregate industry in China were 2.29 per cent, 2.55 per cent and −0.19 per cent, respectively. The results indicate that the improvement of actual industrial productivity in China during the reform period resulted mainly from technical progress rather efficiency, consistent with the findings in Wu (1995), Zheng *et al.* (2003), Tu and Xiao (2005), Liu and Wu (2009) and so on. The industrial TFP growth estimated by Model 4 in Table 7.2 is lower than almost all the measurements surveyed in Table 7.1, only greater than the estimate of 1.8 per cent by Woo (1998) based on the strictly adjusted data. In fact, in Table 7.2, the average growth of TFP, technical progress and efficiency estimated by the former three models is between the ranges of 3.87–6.26 per cent, 3.46–5.92 per cent and 0.35–0.5 per cent, respectively, consistently larger than the corresponding measures by Model 4 in which the change of production efficiency even begins to be negative in Model 4, from the positive results in Models 1–3. And the estimated productivities by Models 1–3 shown in Table 7.2 are similar to most of the estimates listed in Table 7.1, with the only exceptions being Chen *et al.* (1988) and Jefferson *et al.* (2000). Thus, we could preliminarily conclude that the traditional estimates of productivity, technical progress and efficiency that do not, or do not correctly, allow for environmental pollution variables often overestimate the actual performance. To confirm this conclusion, following Kumar (2006), this chapter tests whether the actual estimates by Model 4 are significantly lower than those estimated by Models 1–3 by using the Student's *t* test. As also shown in Table 7.2, the productivity estimates by the first three models are significantly greater than the estimate by Model 4; the technical progress measured by Models 1 and 2 is significantly larger than the corresponding estimate in Model 4; however, there is no statistically significant difference for the efficiency estimates among the four models. Now, we could safely say that the actual growth of TFP and technical progress after considering the negative externality of environmental pollutions (i.e., regarding the CO_2 emission as the undesirable output in Model 4) is lower than the conventional estimates which ignore or deal incorrectly with pollution emission, though their efficiency estimates are not significantly different. This is one of the main findings in this chapter that is similar to many other studies. For example, Jeon and Sickles (2004) report that the estimate of TFP increases for Japan, South Korea, Taiwan, Singapore and Hong Kong if the emission factor is neglected, but increases only for Japan if the emission is considered. Nanere *et al.* (2007) conclude that, without taking account of the external costs or benefits of production, productivity estimates can both over- and under-estimate productivity. Watanabe and Tanaka (2007) estimate Chinese industrial efficiency at the provincial level from 1994 to 2002 using two methods, that is, considering only desirable output, or considering both desirable and undesirable outputs simultaneously, and reveal that efficiency levels are biased only if desirable output is considered. The decrease of actual productivity measures after taking energy and environmental restrictions into account discovered in this paper seems to be similar to the finding by Young (1995) that the growth of most Asian economies after the Second World War is not driven by productivity improvement. Table 7.3 will give further analysis on this.

Table 7.3 reports the aggregated industrial growth accounting during three sub-periods and entire reform period;[4] in each period, the first row includes the averaged growth rate of GIOV, CO_2 emission, capital stock, labour, energy

Table 7.3 Average development speed of outputs, inputs and productivity estimated by four models during different periods

Periods	GIOV	CO_2 Emission	Capital Stock	Labour	Energy Consumption	MPI1	MPI2	MPI3	MLPI
1981–1992	1.09 100	1.05	1.11	1.05	1.06	1.0123 13	1.0146 16	1.0060 7	1.0068 7
1992–2001	1.12 100	0.98	1.09	1.00	1.03	1.0459 39	1.0578 49	1.0264 23	1.0315 27
2001–2008	1.21 100	1.05	1.09	1.07	1.11	1.1258 60	1.1380 66	1.1001 48	1.0356 17
1981–2008	1.13 100	1.03	1.10	1.04	1.06	1.0545 41	1.0626 47	1.0387 29	1.0229 17

Note: MPI1/MPI2/MPI3/MLPI are productivity indexes estimated by model 1–4. During each period, the first row reports the average development speed of each variable (also weighted average of all sub-industries and geometric mean over this period); the second row shows the contribution share (unit: %) of four kinds of estimated productivity on GIOV, each of which equals 100 minus the share of all inputs on GIOV.

consumption and TFP from the four models, averaged over 38 sub-industries, and the second row shows the contribution ratio of four kinds of TFP growth on GIOV growth.[5] Similar to the findings in Table 7.2, for each period, the estimates of TFP growth and its contribution for the former three models are greater than the corresponding estimates for Model 4, with only the measuring exceptions of Model 3 in the first and second sub-periods. Based only on the estimates of Model 1 and 2, we could reach a similar conclusion to Tu and Xiao (2005), Liu and Zhang (2008) and Chen *et al.* (2011)[6] due to the similar approach to how to deal with the environmental factor; that is, from the third sub-period, the output contribution of productivity has exceeded that of all the input factors, indicating that the industrial development model has been transformed from factors-driven to sustainably driven by productivity. This conclusion cannot be obtained based on the estimates of Model 3 and, in particular, the target Model 4 in which the contribution of MLPI on output decreases to very low value of 17 per cent at the third phase.[7] The fact that the contribution of real productivity on output is much lower than that of all the inputs, revealed by Model 4, indicates that the current development model of Chinese industry is still extensive and the transformation of it into an intensive one is extremely difficult and challenging; this matches the conclusions of Krugman (1994) and Young (1995). Seen also from Table 7.3, at the second phase, industrial CO_2 emission abates by an annual rate of 2 per cent, energy consumption grows by the lowest annual rate of 3 per cent, and labour decreases by 0.3 per cent per year (0.997 in the table if saving with 3 decimals); that corresponds with the ownership rights reform by grasping the large and letting go of the small (*zhuada fangxiao*), the environmental regulation by saving energy and abating emission, and the furlough policy (*xiagang*) from the mid-1990s. The productivity contribution estimated by Models 1–3 increases over three sub-periods, but that by Model 4 decreases at the third sub-period, obviously related to the above-mentioned appearance of heavy industrialization at this phase. Table 7.3 also shows that CO_2 emission and energy consumption rise sharply by an annual average of 5 per cent and 11 per cent at the third phase that is able to be captured by the DDF based Model 4; that is, Model 4 has fully taken the negative impact of energy and emission restriction on productivity into account, leading to a very low estimate of productivity contribution at this stage. If we compare the productivity growth estimated by the four models at each stage in this study with the corresponding estimate surveyed in Table 7.1, the actual productivity growth at the third stage estimated by Model 4 in this study is still the lowest among all estimates.

Unlike Tables 7.2 and 7.3 which report the measures of aggregated industry at different stage, Table 7.4 shows the growth accounting for 38 sub-industries averaged over the whole sample period, in which the geometrically averaged development speed for inputs and outputs is reported, MLPI/MLECH/MLTCH are measured by the target Model 4, and the productivity contribution in the last column is calculated by the same way as Table 7.3, following Wu (2008). Obviously, there is a great heterogeneity among the sub-industries. For example, the growth rate of GIOV varies from 1 per cent for Petroleum Extraction to

Table 7.4 Sub-industrial green growth accounting analysis based on Model 4 (1981–2008)

Sub-industries	GIOV	Capital Stock	Labour	Energy	MLPI	MLTCH	MLECH	Productivity Contribution (%)
Coal Mi.	1.08	1.06	1.01	1.04	1.0002	1.0004	0.9998	0.2
Petroleum Ext.	1.01	1.11	1.05	1.04	0.9991	1.0012	0.9979	−8.3
Ferrous Mi.	1.14	1.08	1.06	1.07	1.0072	1.0082	0.9990	5.0
Non-Ferrous Mi.	1.10	1.05	1.01	1.04	1.0070	1.0102	0.9968	7.0
Nonmetal Mi.	1.09	1.04	1.00	1.06	1.0023	1.0069	0.9954	2.7
Wood Exp.	1.03	1.02	0.98	0.99	1.0010	1.0039	0.9972	3.9
Food Proc.	1.10	1.11	1.03	1.04	1.0024	1.0097	0.9928	2.3
Food Ma.	1.10	1.10	1.03	1.05	1.0016	1.0053	0.9963	1.6
Beverage	1.12	1.11	1.03	1.06	1.0049	1.0086	0.9963	4.0
Tobacco	1.10	1.13	1.01	1.04	1.0471	1.0471	1.0000	47.7
Textile	1.09	1.09	1.02	1.04	1.0051	1.0116	0.9936	5.7
Apparel	1.12	1.12	1.05	1.10	0.9989	1.0219	0.9775	−0.9
Leather	1.12	1.10	1.06	1.06	1.0200	1.0256	0.9946	16.3
Wood Proc.	1.11	1.11	1.05	1.02	1.0083	1.0160	0.925	7.3
Furniture	1.13	1.10	1.05	1.04	1.0317	1.0357	0.9962	23.9
Paper	1.14	1.11	1.03	1.05	1.0031	1.0039	0.9992	2.2
Printing	1.12	1.11	1.03	1.08	1.0262	1.0299	0.9964	21.8
Cultural Articles	1.15	1.11	1.06	1.06	1.0473	1.0489	0.9985	30.7
Petroleum Pro.	1.06	1.11	1.04	1.09	0.9859	1.0103	0.9758	−24.1
Chemical	1.11	1.08	1.03	1.05	1.0012	1.0022	0.9990	1.0
Medicine	1.16	1.12	1.04	1.04	1.0053	1.0077	0.9976	3.3
Fibres	1.17	1.09	1.03	1.06	1.0675	1.0486	1.0181	39.6
Rubber	1.12	1.10	1.04	1.06	1.0008	1.0044	0.9964	0.7
Plastic	1.15	1.12	1.05	1.09	1.0226	1.0287	0.9940	14.6
Nonmetal Ma.	1.12	1.08	1.01	1.06	1.0009	1.0015	0.9994	0.8
Ferrous Press	1.12	1.08	1.02	1.07	1.0075	0.9966	1.0109	6.3
Non-Ferrous Pr.	1.13	1.09	1.05	1.11	1.0024	1.0076	0.9948	1.9
Metal Products	1.13	1.08	1.02	1.07	1.0229	1.0269	0.9961	17.6
General Mac.	1.11	1.05	1.02	1.03	1.0182	1.0213	0.9970	16.0
Special Mac.	1.12	1.05	1.01	1.03	1.0138	1.0177	0.9962	11.8
Transport Eq.	1.18	1.08	1.03	1.05	1.0204	1.0198	1.0005	11.3
Electrical Eq.	1.17	1.10	1.05	1.07	1.0458	1.0445	1.0012	27.1
Computer etc.	1.28	1.13	1.08	1.11	1.1256	1.1010	1.0223	44.5
Measuring Inst.	1.16	1.07	1.03	1.03	1.0576	1.0512	1.0061	36.0
Electric Power	1.12	1.11	1.04	1.09	0.9999	1.0001	0.9998	−0.1
Gas Prod.	1.07	1.10	1.03	1.04	0.9974	0.9985	0.9989	−3.9
Water Prod.	1.07	1.11	1.04	1.06	1.0057	1.0163	0.9896	8.2
Others	1.14	1.10	1.00	1.04	1.0317	1.0318	0.9999	22.5

Note: The contribution in last column is defined by the share of productivity change on GIOV growth. All the other values in this table are the geometric average speed of development over 1981–2008, for each variable.

28 per cent for the Manufacture of Communication Equipment, and Computers and Other Electronic Equipment (Computers etc. for short). The measured values of actual MLPI and MLECH differ from a low of 0.9859 and 0.9758 for Processing of Petroleum, Coking, Processing of Nuclear Fuel to a peak of 1.1256 and 1.0223

for Computers etc. And the value of MLTCH lies between the 0.9966 of Smelting and Pressing of Ferrous Metals and 1.101 of Computers etc. Slowly growing sub-industries are almost all in heavy industry such as Extraction of Petroleum and Natural Gas, Mining and Washing of Coal, Manufacture of Non-metallic Mineral Products, Mining and Processing of Nonmetal Ores, Manufacture of Raw Chemical Materials and Chemical Products, and Production and Supply of Gas, which show below-average output growth and very low or even negative growth of productivity and technical progress. It reflects not only the drawbacks of high energy consumption and heavy pollution emission but the necessity and urgency of achieving low-carbon development by reforming the traditional heavy industry. The sub-industries which grow rapidly in output, productivity and technical progress are in light industry such as Manufacture of Chemical Fibres, Manufacture of Furniture, and in high-tech industry like Manufacture of Communication Equipment, Computers and Other Electronic Equipment (all ranked the first), Manufacture of Transport Equipment, Manufacture of Measuring Instruments and Machinery for Cultural Activity and Office Work, and Manufacture of Electrical Machinery and Equipment, which shows the important role of high-tech industry and light industry in the update of industrial development.

With few exceptions, the sub-industrial growth of actual productivity and technical progress is positive but that of production efficiency is negative, which exhibits the same results as Table 7.2, showing that the improvement of industrial productivity results from technical progress instead of efficiency. The sub-industries with negative growth of productivity are Processing of Petroleum, Coking, Processing of Nuclear Fuel, Extraction of Petroleum and Natural Gas, Production and Supply of Gas, and Production and Supply of Electric Power and Heat Power. There are only two sub-industries of Smelting and Pressing of Ferrous Metals and Production and Supply of Gas with negative growth of technical progress. Except for Smelting and Pressing of Ferrous Metals, sub-industries with positive efficiency change (such as Computers etc., Manufacture of Transport Equipment, Manufacture of Electrical Machinery and Equipment, Manufacture of Measuring Instruments and Machinery for Cultural Activity and Office Work, Manufacture of Tobacco and Manufacture of Chemical Fibres) also have a positive growth of actual productivity and technical progress, which shows the importance of light and high-tech industry in industrial transformation again. Though the same as the findings in Table 7.3, that there are no sub-industries with the productivity contribution above 50 per cent – that is, all the sub-industries are driven by factors expansion and then grow extensively – the sub-industries with a productivity contribution of more than 10 per cent are all in light industry (except for the Manufacture of Plastics), in which those with productivity contribution close to 50 per cent are Manufacture of Tobacco and Computers etc. The minus growth of productivity in energy and capital driven heavy industrial sectors also makes negative contributions to its output, in which Processing of Petroleum, Coking, Processing of Nuclear Fuel has the lowest productivity contribution (−24 per cent). The difference of actual productivity between light and heavy industry is consistent with the implication of the descriptive statistics of the variables. From the viewpoint of input factors, the labour growth is not so high but the energy and capital growth is very high, which play a very important role in factors-driven extensive industrial growth.

7.6 Can environmental regulation lead to the transformation of low-carbon development in China?

It's not easy to explain the pattern of TFP change but it will be worthwhile to base study on the DDF based actual productivity measures, as opposed to the traditional one, from two points of view. On the one hand, as carried out in Section 7.5, the damage caused by environmental pollution that restricts the improvement of industrial productivity and hinders economic transformation in China can be detected by comparing actual productivity taking the negative externality of pollution into account and the traditional measures that do not, or do not correctly, deal with the environmental factors; on the other hand, the actual or environmental TFP is able to reflect the impact of environment regulatory policy on the economy and the development of low-carbon transformation because all the results from the energy-saving and emission-abating policy are expressed by the actual change of energy and emission data that is introduced into the production process through the DDF based AAM. Porter (1991) hypothesizes that a properly designed and strictly enforced environment regulatory policy can trigger innovation which then partially or fully offsets the costs of complying with the policy (the so called innovation offsets), leading to the win-win development possibilities that environmental quality and productivity increase simultaneously. Based on this, in this section the actual TFP estimated by the environmental DDF will be also understood as the comprehensive economic effect of all the environment regulatory policies such as developing new energy and low-carbon technology, saving energy and abating emission, and so on. The factors driven extensive growth is bound to be restricted by the law of diminishing marginal returns, and only a green revolution could break through the restriction of energy and environment and promote productivity growth by reallocating resources such as capital, labour and energy into the sectors with higher energy efficiency and emission abating ability, which is the only way to transform the industrial development model and realize the industrialization in China. To interpret the change of actual productivity from the perspective of environmental regulation, Figure 7.2 depicts the average development speed of MLPI, MLECH and MLTCH estimated by Model 4 for aggregated, light and heavy industry, respectively, over the whole reform period.[8] The classification of all sub-industries into light and heavy industry in Figure 7.2 enables us to see clearly not only the productivity heterogeneity of different types of sub-industries but also its time varying pattern. Seen from Figure 7.2, the productivity index and its decomposition for aggregated industry is dominated by heavy industry due to the bigger weights of heavy industry.

Look at the time varying pattern of actual productivity and its decomposition first. Both light and heavy industry experience similar trends of productivity, technical progress and efficiency to the aggregated industry, because some general

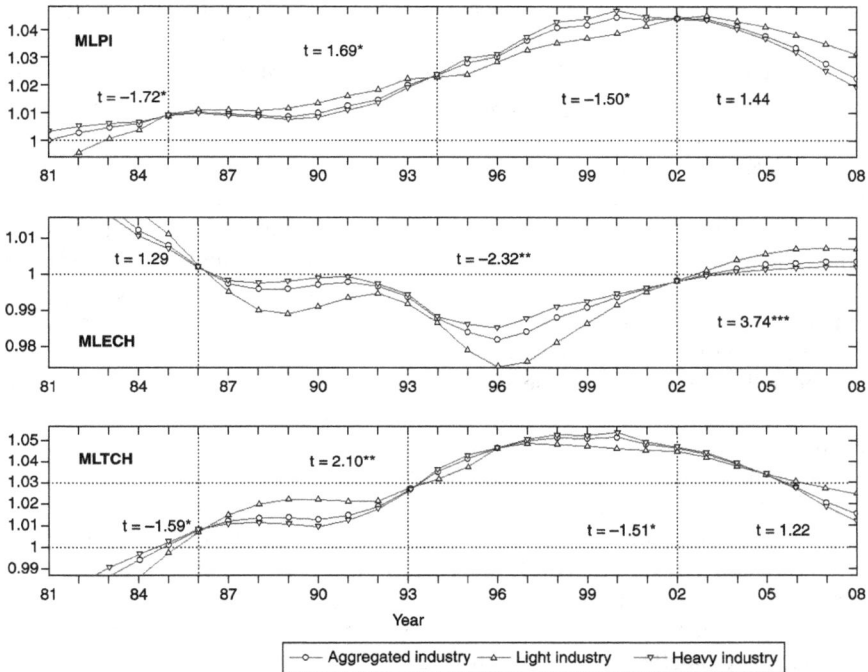

Figure 7.2 Average MLPI/MLECH/MLTCH of light, heavy and aggregated industry estimated by Model 4.

economy-wide factors such as the common macroeconomic policy and environmental protection tend to affect most sub-industries at the same time, rather than a limited number of sectors and, hence, influence productivity and its decomposition in all industrial branches. Specifically, during the beginning of reform, the production efficiency embodying the catching-up effect is relatively high but the technical change reflecting the innovative ability is minus; then technical progress changes to increase continuously and attains the peak of 5.2 per cent at the advent of the new millennium, while the catching-up effect by absorbing the frontier technology disappears and production efficiency falls; since the beginning of this century, production efficiency is improved but technical progress decreases. Influenced more by the technical change rather than efficiency, actual TFP growth in Chinese industry is very low at the beginning of the 1980s; then it experiences a long and steady increase; after reaching the highest rate of 4.4 per cent at the turn of this century, it begins to decline, consistent with most of the findings in the existing literature (Perkins and Rawski, 2008; Wu, 2008; Zheng *et al.*, 2009). The continuous and rapid increase of the actual TFP growth and its output contribution since 1990, especially after the mid-1990s, indicates that the energy-saving and emission-abating policy implemented during this period resulted in positive economic achievements and Chinese industry is experiencing a substantial low-carbon transformation. This could be attributed to the establishment of environmental protection awareness in the Chinese government, accompanied by the proposition of global sustainable development, since 1990, and the ownership rights reform at the second half of the 1990s. Due to the implementation of development of the new energy and renewable energy, the share of coal in energy consumption has fallen from 76.2 per cent in 1990 to 66 per cent in 2000, while the share of oil, gas, water and nuclear electricity, wind and solar energy in energy consumption has risen from 23.8 per cent in 1990 to 34 per cent in 2000. Since the beginning of this century, when the enforcement of environmental regulation has been somewhat relaxed and the phenomenon of heavy industrialization appeared again, the actual TFP growth decreased for the first time during the entire reform period, implying that the process of low-carbon transformation of industrial development model has been hindered in China. The difference between the environment adjusted TFP change in the 1990s and after the advent of this century could also be illustrated by the different changing pattern of energy consumption and CO_2 emission and their intensity shown in Figures 1.1, 3.3 and 3.7.

As for the changing pattern of technical progress and efficiency, the new reform policy and the overnight liberalization of strict controls of factors (e.g., labour migrating from successful agricultural reform at least to TVEs, with TVEs, commune and brigades (*shedui*), individuals and foreign investors also encouraged to extract coal) at the beginning of 1980s released the vast potential energy of restricted production factors and then significantly improved production efficiency, remedying the primarily negative growth of technical progress and pushing the early growth of TFP forward. However, the blind development of non-state small coal mines, together with the fact that the SOEs' reform in 1980s only restructured operating rights without touching on the issue of ownership rights and the significant numbers of under-utilized employees, caused financial losses in state-owned coal mines and then the first net loss for the whole state-owned sector in 1996, leading to the corresponding negative change of production efficiency for the industry as a whole (also the lowest in the year of 1996) (Chen *et al.* 2011). From 1997 on, central government began to reform the ownership structure by employing the policy of 'grasp the large and let go of the small' (*zhuada fangxiao*) and closed down a large number of small energy- and emission-intensive enterprises. The transition from plan prices to true market prices has been one of China's greatest challenges since the reform. In 1992, the well-known dual-track pricing system implemented in the 1980s was discarded and a unified domestic products market formed. In 1994, the dual-track price of coal was also phased out and later fully liberalized, as opposed to the prices of other factors such as oil and electric power, and a coal market appeared. The market began to work to curb disorderly production and increase the degree of industrial concentration, which improved production efficiency gradually and made its change positive from 2003, partially offsetting the synchronous decrease of technical progress over time. As plotted in Figure 7.2, though negative at the beginning of reform and decreasing since the turn of this century, industrial technical progress is still significant in China. Fisher-Vanden *et al.* (2004, 2006) find that R&D expenditure, ownership reform and industrial structural adjustment from the high to low emission groups are the main forces driving the increase of energy productivity and technology innovation. Though the significant technical progress from the mid-1980s to the beginning of this century is counteracted partly by the negative growth of efficiency, it will not change the situation that the environmental TFP growth is driven by industrial technical progress instead of efficiency, due to its lower absolute value.

What's the difference in the industrial productivity change between light and heavy industry and how can we explain this based on environmental protection policy? Though Table 7.4 concludes that the productivity growth of light industry is greater than that of heavy industry, also resembling the intuition or findings in some studies (e.g., Wu, 1995; Jefferson *et al.*, 2000), it's not necessarily the conclusion if analysed over different sub-periods, as seen in Figure 7.2. In fact, Figure 7.2 classifies the productivity trends into different sub-periods according to the different performances between light and heavy industry, in which the classification criterion is similar to, but different from, that used in Table 7.3. Figure 7.2 also reports the Student's *t* value to test if the difference of measurements between light and heavy industry is statistically significant.[9] In the beginning of reform, that is, the period of the sixth Five-Year Plan (1981–1985), the light industry, represented mainly by TVEs, started to develop and hold higher production efficiency than heavy industry, but this advantage is not significant and its technical progress is still significantly slower than that of heavy industry represented by SOEs, causing a significantly larger growth of productivity in heavy industry than that in light industry. In the following period of the seventh and eighth Five-Year Plans (1986–1995), the change of the industrial development strategy from heavy-industry-oriented before the reform to the parallel importance of light and heavy industry started to work. Therefore, the non-state enterprises (NSEs), including TVEs, private enterprises (PEs) and foreign funded enterprises (FFEs) (most of them belonging to light industry), developed very fast at this stage; in 1993, their GIOV share exceeds SOEs for the first time during the reform era. This may explain why the technical progress and productivity growth of light industry is significantly greater than those of heavy industry, as shown in Figure 7.2. To solve the energy shortage, during this period the government still encouraged the energy production. Akin to other SOEs' reform, heavy sub-industries (such as coal mining, petrochemical and electric power enterprises) also aimed at increasing operating autonomy by employing the contract responsibility system, in which the coal industry experienced the most thorough marketization reform. That may be one of the reasons why the improvement of production efficiency in heavy industry was faster than that in light industry at this stage.

Though the policy of encouraging energy production in the 1980s eased the energy shortage, the destructive exploitation of small coal mines caused serious waste of coal resources and heavy environmental pollution; therefore, China began to restrict the development of the energy industry in the 1990s. Based on the ninth, tenth and eleventh Five-Year Plans of state environmental protection approved by the State Council in 1996, 2001, and 2007, respectively, and the White Paper of China Environmental Protection (1996–2005) released by the

State Council Information Office on 5 June 2006, during the period of the ninth Five-Year Plan (1996–2000), the central government closed 8.4 thousand energy- and emission-intensive small industrial enterprises for the first time, resulting in the decrease of emission levels of 12 type pollutions such as SO_2, COD and others by 10–15 per cent when compared with that at the end of the period of the eighth Five-Year Plan. Exemplified by the change of CO_2 emission reported in Table 7.5, relative to the positive CO_2 growth of almost all the sub-industries between 1981 and 1995, CO_2 emission of 32 sub-industries among all the 38 samples decreases by 5–73 per cent at the end of the course of the ninth Five-Year Plan. Over the same period, however, the average growth of GIOV attains annual 12.7 per cent, much greater than the 7.6 per cent averaged over the period from the sixth to eighth Five-Year Plan. Thus, we see in Figure 1.1 the stagnancy or decline of energy use and carbon emission and the more rapid decrease of their intensity (see Figures 3.3 and 3.7) at this phase in China. The most specific findings from Figure 7.2 are that, in this period, technical progress in heavy industry obviously exceeds that in light industry, the improvement of production efficiency in heavy industry is still faster than that in light industry, leading to the corresponding higher growth of actual TFP in heavy industry than that light industry. Such findings are statistically supported by Student's t test shown also in Figure 7.2. The findings imply that, though the actual productivity growth was hindered by the constraints of energy and emission as a whole, from the mid-1990s to the beginning of this century, the policy of environmental regulation worked effectively and caused substantial productivity growth for both light and heavy industry and especially higher growth of actual productivity, technical progress and efficiency in heavy industry compared with light, which is different from the conclusion that productivity growth in light industry is consistently greater than that in heavy, found in most of the literature.[10] This is another main finding in this chapter, indicating that the low-carbon transformation of industrial development model is on the way in China.

Why does environment regulatory policy lead to faster productivity growth for heavy industry compared with light industry? The small enterprises closed down included not only the coal, electricity power, metallurgy, non-ferrous metals, petrochemical, and construction material in heavy industry but also pulp mills, tanneries, breweries, sugar refineries in light industry. We find by calculation that, during the course of the ninth Five-Year Plan (1996–2000), if CO_2 emission abates by 1 per cent, for light and heavy industry, their labour productivity will rise 0.5 and 22 thousand RMB per capita, energy productivity rise 0.2 and 6.7 thousand RMB per tce, and carbon productivity increase 0.4 and 22.4 thousand RMB per ton of CO_2, respectively. Though the single factor productivity (SFP) and total factor productivity are different concepts, the fact that the SFP of heavy industry increases more rapidly than that of light industry implies that, at this stage, the same abating magnitude of CO_2 emission contributes to improving micro efficiency more in heavy industry and finally, from the macro viewpoint, exhibits more rapid growth of actual TFP in heavy industry than that in light. Unfortunately, the effective environment regulatory policy implemented in the

Table 7.5 Change of sub-industrial CO_2 emission at end of Five-Year Plan relative to the end of last period (%)

Sub-industries	6th–8th Five-Yr Plan (1981–1995)	9th Five-Yr Plan (1996–2000)	10th Five-Yr Plan (2001–2005)	11th Five-Yr Plan (2006–2008, partial)	Sub-industries	6th–8th Five-Yr Plan (1981–1995)	9th Five-Yr Plan (1996–2000)	10th Five-Yr Plan (2001–2005)	11th Five-Yr Plan (2006–2008, Partial)
Coal Mi.	87	−19	95	43	Chemical	105	−13	47	16
Petroleum Ext.	226	71	−48	−1	Medicine	252	−45	16	0
Ferrous Mi.	58	−30	70	17	Fibres	273	9	7	7
Non-Ferrous Mi.	81	−54	9	0	Rubber	118	−56	48	5
Nonmetal Mi.	216	−9	42	11	Plastic	290	−57	69	2
Wood Exp.	41	−51	−27	−5	Nonmetal Ma.	193	−26	70	4
Food Proc.	187	−23	−14	20	Ferrous Press	113	−14	73	24
Food Ma.	136	−50	37	12	Non-Ferrous Pr.	220	−12	91	28
Beverage	197	−41	19	6	Metal Products	86	−53	27	−1
Tobacco	243	−38	−6	−17	General Mac.	−3	−59	7	3
Textile	75	−49	62	16	Special Mac.	21	−52	51	6
Apparel	236	−5	73	17	Transport Eq.	22	−23	18	7
Leather	171	−73	33	0	Electrical Eq.	87	−48	−17	6
Wood Proc.	101	−42	67	5	Computer etc.	1	−32	84	10
Furniture	65	−38	−32	1	Measuring Inst.	69	−59	−29	3
Paper	166	−19	76	13	Electric Power	242	26	92	37
Printing	24	−46	−17	3	Gas Prod.	54	40	28	17
Cultural Articles	156	−52	5	0	Water Prod.	370	15	−28	0
Petroleum Pro.	111	23	89	31	Others	54	−69	73	−15

period of the ninth Five-Year Plan was not to be maintained continuously. Based on the same file sources provided above, in the period of the tenth Five-Year Plan (2001–2005), although central government continued to close down about 3.3 thousand small enterprises because of heavy pollution, the rapid urbanization and the update of consumption structure driven by the sharp expansion of the car and housing industry pushed the reappearance of heavy industrialization, leading to the unachievement of the expected abating goal for the main pollutions. For example, industrial SO_2 emission even rises by 34.5 per cent at the end of this period instead of abating. As shown in Table 7.5, the number of sub-industries with a decline in CO_2 emission falls from 32 at the previous stage to only 9 in this period. Reflected in Figure 7.2, though production efficiency begins to grow positively, the growth of efficiency in heavy industry is significantly less than that in light industry; the situation is the same for technical progress and productivity. So it's not surprising to see the first decline of productivity growth and a setback of low-carbon transformation in Chinese industry during the period.[11] Table 7.3 also reveals that, after the turn of this century, industrial productivity growth is moderate and its output contribution decreases greatly as opposed to the previous stage. Of course, the reversion of the productivity index and its decomposition between light and heavy industry at this stage is not statistically supported by the Student's *t* test; therefore, it may be expected that actual productivity will continue to grow and the process of low-carbon transformation will not be broken if well-designed environment regulatory policy is enforced strictly and continuously.

Following Färe *et al.* (2001), this chapter also identifies the innovative sub-industries that shift the frontier over the adjacent two-year period. Table 7.6 reports the identification results based on Models 3 and 4. According to the target approach of Model 4, at the early stage of reform the innovative sub-industries are Manufacture of Textile Wearing Apparel and Manufacture of Electrical Machinery and Equipment; from the two-year period 1992–1993, Manufacture of Communication Equipment, Computers and Other Electronic Equipment began to push technical progress continuously. Known from Table 7.4, even considering the restriction factor of energy and environment, Computer etc. has the highest growth of output, productivity, technical progress and efficiency, the second highest output contribution (44.48 per cent), and the lowest energy and carbon emission intensity among all sub-industries, indicating that the transformation of Chinese industrial development model must be driven by both the green and information industries. Except for a few years, Manufacture of Tobacco with the largest output contribution (47.67 per cent) has been pushing the frontier technology of Chinese industry forward. Comparing the identification results between Models 3 and 4, if we do not consider the harmful influence of pollution emission, Processing of Petroleum and Coking and Production and Supply of Electric Power and Heat Power are wrongly identified as the efficient units, while the efficient units such as Manufacture of Tobacco and Manufacture of Textile Wearing Apparel cannot be recognized, emphasizing again the important role of environmental factors in evaluating technical progress and identifying the innovative units.

Table 7.6 Innovative industries identified by Models 3 and 4

Year	Emission as Goods	Emission as Bads	Year	Emission as Goods	Emission as Bads
1980–1981	[ ]	[ ]	1994–1995	10, 33, 38	10, 33, 38
1981–1982	[ ]	[ ]	1995–1996	10, 33	10, 33
1982–1983	[ ]	[ ]	1996–1997	10, 33	10, 33
1883–1984	[ ]	[ ]	1997–1998	10, 19, 33	10, 33
1984–1985	32	32	1998–1999	10, 19, 33	10, 33
1985–1986	10, 32	10, 32	1999–2000	10, 19, 33	33
1986–1987	10, 32	10, 32	2000–2001	19, 33, 35	33
1987–1988	10	10	2001–2002	19, 33, 35	33
1988–1989	10	10, 12	2002–2003	19, 33, 35	33
1989–1990	10, 12	10, 12	2003–2004	10, 19, 33, 35	10, 33
1990–1991	10, 12	10, 12	2004–2005	10, 19, 33, 35	10, 33
1991–1992	12	10, 12	2005–2006	10, 19, 33, 35	10, 33
1992–1993	12, 33	12, 33	2006–2007	10, 19, 33, 35	10, 33
1993–1994	33	12, 33	2007–2008	10, 19, 33, 35	10, 33

Note: The sub-industries with code 10, 12, 19, 32, 33, 35, 38 are Manufacture of Tobacco, Manufacture of Textile Wearing Apparel, Processing of Petroleum and Coking, Manufacture of Electrical Machinery and Equipment, Manufacture of Communication Equipment and Computers, Production and Supply of Electric Power and Heat Power and others. [] indicates no innovative units exist.

7.7 Summary and comments

By using the directional distance function that appropriately takes the negative externality of pollution emission into account, this chapter estimates the actual productivity and its decomposition for 38 sub-industries between 1980 and 2008. The first finding is that the actual productivity and technical progress estimated by DDF is much lower than those that do not, or do not correctly, consider the environmental variables, indicating that the negative impact of pollution emission on productivity growth is serious, and that development model transformation driving the increase of productivity contribution is still at the early stage. The second finding is that, though the actual productivity growth is greatly hindered by the constraints of energy and environment, the environmental regulatory policy is still pushing the improvement of actual productivity. In particular, from the mid-1990s to the beginning of this century, the government closed down more than 10 thousand small energy- and emission-intensive enterprises, leading to the first reduction of CO_2 emission and the highest growth of actual productivity for the industry as a whole, and a higher growth of productivity, technical progress and efficiency in heavy industry than that in light industry. This implies that the low-carbon transformation of industrial development model has always been making progress, and performed best from 1994 to 2002.

However, the effective environment regulatory policy has not been enforced continuously and, after 2002, heavy industrialization reappeared in China, resulting in the deterioration of the index of heavy industry as opposed to that of light industry and the first decline of industrial productivity over the entire reform period that deviates from the process of low-carbon transformation. But this is not confirmed by the statistical support. In fact, the Chinese central government thinks highly of the improvement of environmental productivity and the low-carbon transformation. According to HSBC's report (Robins *et al.*, 2009), with sizeable financial reserves and a tradition of long-term planning, in November 2008 China launched its 4 trillion RMB (584 billion US$) package. Almost 40 per cent of this is allocated to green themes, most notably rail, grids and water infrastructure, along with dedicated spending on environmental improvement. In November 2009, China decided to reduce the carbon dioxide emission per GDP (i.e. CO_2 intensity) by 40–45 per cent from 2005 to 2020. This is the first time that the Chinese government proposed the quantitative abatement target of CO_2 emission. The United Nations Environment Programme (UNEP) reports that Chinese investment on the area of renewable energy first became the largest, exceeding the US, in 2009, reaching more than US$ 20 billion. It can be expected that the environmental regulation will continue to promote the improvement of actual productivity and then the transformation of industrial development model, driven by the low-carbon plan, will finally arrive in China.

8 Evaluation of regional low-carbon economic transformation

Multiple emissions

8.1 Introduction

During the international economic crisis, the transformation of the original economic growth model has become the fundamental issue for many countries to achieve economic recovery and sustainable development in the long-run. When considering the current depletion of resources, rise of energy prices, deterioration of environmental pollution and the abnormal changes of global climate, then low-carbon economic transformation becomes the important method to realize economic transformation and sustainable development. The traditional high-carbon growth model, based on a very high level of utilization of fossil fuel, is not sustainable, and the development model in the future should be in terms of low-carbon and adopt the new energy and low-carbon techniques. Since the 1990s, China has started to emphasize energy saving and environmental conservation, but the rapid economic growth is still at the cost of resources waste and environmental pollution, due to the fact that the evaluation criterion using simply GDP is employed by local governments. After the rapid economic growth and the reappearance of heavy industrialization in the first decade of this century, low-carbon transformation and sustainable development is inevitable in China. However, there are very few theoretical and empirical studies on the topic of low-carbon economic transformation in the literature. This chapter aims to consider this topic. To this end, the appropriate economic theory on low-carbon transformation should be studied first. Drawing on this, it is necessary to construct the indicator to dynamically evaluate low-carbon transformation. Then empirical studies on the provincial or regional low-carbon transformation in China will be undertaken, in order to achieve the relative policy suggestions.

8.2 Literature survey and methodology choice

The notion of low-carbon comes originally from the concept of sustainable development that resulted from the energy crises and environmental disasters since the 1970s and was officially accepted by the United Nations Environmental and Development Committee in 1987. The definitions of sustainable development are many, but they all cannot ignore the dimension of the environment. The terminology of low-carbon itself appears firstly in the UK Energy White Paper (legislative proposal) 'Our Energy Future – Create a Low Carbon Economy' in 2003. However, the definition of the low-carbon economy is still not conclusive. In the literature, the studies often focus on the description of a low-carbon economy, the introduction of the low-carbon concept and some case studies. This is not enough. Because the low-carbon transformation has become the important way to recover the economy from the financial crisis, the urgent work is to investigate how to transform the economic growth model from one which is high-carbon based to one which is low-carbon in nature and provides local governments with the appropriate and convenient indicators to evaluate the low-carbon transformation process. This is what we attempt to do in this chapter.

To this end, the first task is to choose the appropriate theoretical framework base from which low-carbon development is analysed. In my opinion, the theory used to analyse low-carbon transformation should be consistent with that to analyse the general economic transformation. As surveyed in Section 7.2, modern economic growth theory and the growth accounting approach is the appropriate framework to analyse economic development. Using this, the total factor productivity (TFP), first estimated by Solow, can be adapted to quantify the growth quality, and the contribution of TFP to output, as compared with the factors quantitative contribution to output, can be employed as the indicator to evaluate whether the economic development is sustainable or not.The analysis of low-carbon transformation in this chapter will also follow this general economic framework and choose the same evaluation indicator but, differently, will include low-carbon factors, such as energy and environment, into the analysis. Precisely speaking, the low-carbon transformation is in fact economic transformation under the constraints of energy and environment. Similar to the analysis in Chapter 7, how to include energy and environment in the production framework is the main task we have to consider. The ideal directional distance function (DDF), also referred to as the environmental activity analysis model (AAM), will also be used as the basic analytical framework in this chapter, in which energy consumption is treated as the input factor and the environmental pollutants are regarded as the undesirable outputs, or called 'bads'. However, the standard DDF-AAM approach assumes that the increase of desirable output and the decrease of undesirable output follow the same proportion, which is too strict in reality. Thus, different from the methodology in Chapter 7, this chapter will relax the assumption of the standard DDF-AAM approach, to consider the slacks of input and output variables. That is to say, this chapter will choose the non-radial, non-angle slacks-based measure (SBM) DDF approach, also called SBM-DDF-AAM, to be the economic theory used to evaluate the economic transformation.[1] This approach is also studied by Tone (2001), Zhou *et al.* (2006), Fukuyama and Weber (2009), Färe and Grosskopf (2010), and others. Previously, the analytical target was often Chinese industry. However, this chapter will analyse the provincial low-carbon transformation. Also, multiple environmental pollutants such as CO_2, SO_2, COD, waste gas and waste water will be introduced into our analysis, rather than only the CO_2 emission considered previously. This is also the advantage of the DEA or

DDF framework in which multiple outputs and multiple inputs can be analysed. Under the framework of the parametric production function, only one type of outputs such as GDP can be analysed, like the examples in Chapters 5 and 6.

The preferable SBM-DDF-AAM approach convinces us that the correspondingly estimated environmental TFP growth quantifies the quality of economic growth more precisely and its contribution to output growth will become the more reliable indicator to evaluate the low-carbon economic transformation, as opposed to the previous studies. If the quality contribution exceeds the quantitative one, the economic transformation happens; if the indicator – that is, the contribution of environmental TFP growth to total output growth – is below 0.5, it indicates that the economic growth is still high-carbon in nature. Therefore, the evaluation indicator constructed in this chapter is in terms of the appropriate economic theory, which is able to not only reflect the degree of low-carbon development of each province over time, but also dynamically assess whether the transformation takes place or not. This differs from many indicators used in the studies to measure low-carbon or green economic development. For example, in 2010, Beijing Normal University and the National Bureau of Statistics of China constructed a green development indicator for 30 provinces in China which was averaged by 3 first-class indicators, 9 second-class indicators and 54 third-class indicators. The construction of a sustainable development indicator system by the United Nations Committee of Sustainable Development (UNCSD) is similar to this. These indicators are normally statistical in nature, drawn up by normalizing and averaging many sub-indicators which have no economic meaning and are not able to dynamically evaluate the transformation process of low-carbon economic development. There exist many studies working on the economic indicators which are not related to the low-carbon economy. For example, Fan *et al.* (2003) construct the relative indicator of the China marketization process. The construction method of these indicators could be followed by the low-carbon transformation indicator designed in this paper.

8.3 Methodology

8.3.1 The introduction of the SBM-DDF-AAM approach

Assume there are n decision-making units (DMUs) at t time point ($i = 1, 2, \ldots, n$). For each DMU, there are k inputs, l desirable outputs and m undesirable outputs. The symbols $\mathbf{x}$, $\mathbf{y}$, $\mathbf{b}$, $\mathbf{X}$, $\mathbf{Y}$ and $\mathbf{B}$ represent the column vector and matrix of inputs, desirable outputs and undesirable outputs, respectively. The production set P could be defined as:

$$P = \{(\mathbf{x},\mathbf{y},\mathbf{b}) \mid \mathbf{x} \geq \mathbf{X}\lambda, \mathbf{y} \leq \mathbf{Y}\lambda, \mathbf{b} \geq \mathbf{B}\lambda, \lambda \geq 0\} \tag{8.1}$$

The SBM-DDF-AAM based environmental efficiency score for ith DMU in terms of t-time observations and t-time technique could be obtained by resolving the following fraction programming:

$$S_{NLP}^t(\mathbf{x}_i', \mathbf{y}_i', \mathbf{b}_i') = \min \frac{1 - (1/k)\sum_{k=1}^{k}(s_k^{x,-}/x_{k,i}^t)}{1 + [1/(l+m)]\left[\sum_{l=1}^{l}(s_l^{y,+}/y_{l,i}^t) + \sum_{m=1}^{m}(s_m^{b,-}/b_{m,i}^t)\right]}$$

$$s.t. \ \mathbf{x}_i = \mathbf{X}\lambda + \mathbf{s}_i^{x,-}; \quad \mathbf{y}_i = \mathbf{Y}\lambda - \mathbf{s}_i^{y,+}; \quad \mathbf{b}_i = \mathbf{B}\lambda + \mathbf{s}_i^{b,-}; \mathbf{s}_i^{x,-} \ge 0;$$
$$\mathbf{s}_i^{y,+} \ge 0; \quad \mathbf{s}_i^{b,-} \ge 0; \quad \mathbf{i}'\lambda = 1; \quad \lambda \ge 0 \tag{8.2}$$

where, $\mathbf{s}_i^{x,-}$, $\mathbf{s}_i^{y,+}$ and $\mathbf{s}_i^{b,-}$ represent the over-inputs, under-desirable outputs and over-undesirable outputs, respectively, referred to as the slack variables. λ is the intensity vector, the summation of its elements being 1 indicating the assumption of varying return to scale (VRS). The value of this score lies between 0 and 1, and the larger the value, the higher the efficiency of the unit is; if $S_{NLP} = 1$, it reveals that the DMU is efficient that is located on the production technique frontier.

By using the Charnes-Cooper transformation, the above nonlinear programming (8.2) could be transfer into the equivalent linear programming as shown below:

$$S_{LP}^t(\mathbf{x}_i', \mathbf{y}_i', \mathbf{b}_i') = \min \ \tau - (1/k)\sum_{k=1}^{k}(S_k^{x,-}/x_{k,i}^t)$$

$$s.t. \ 1 = \tau + [1/(l+m)]\left[\sum_{l=1}^{l}(S_l^{y,+}/y_{l,i}^t) + \sum_{m=1}^{m}(S_m^{b,-}/b_{m,i}^t)\right]$$
$$\tau\mathbf{x}_i = \mathbf{X}\Lambda + \mathbf{S}_i^{x,-}; \quad \tau\mathbf{y}_i = \mathbf{Y}\Lambda - \mathbf{S}_i^{y,+}; \quad \tau\mathbf{b}_i = \mathbf{B}\Lambda + \mathbf{S}_i^{b,-}; \tag{8.3}$$
$$\mathbf{S}_i^{x,-}, \mathbf{S}_i^{y,+}, \mathbf{S}_i^{b,-} \ge 0; \quad \mathbf{i}'\Lambda = \tau; \quad \Lambda \ge 0; \quad \tau > 0$$

If the optimal solutions of linear programming (8.3) are symbolized by $(S_{LP}^*, \mathbf{S}^{x,-,*}, \mathbf{S}^{y,+,*}, \mathbf{S}^{b,-,*}, \tau^*, \Lambda^*)$, the optimal solutions of nonlinear programming (8.2) could be expressed accordingly as follows:

$$S_{NLP}^* = S_{LP}^*, \quad \lambda^* = \Lambda^*/\tau^*$$
$$\mathbf{s}^{x,-,*} = \mathbf{S}^{x,-,*}/\tau^*, \quad \mathbf{s}^{y,+,*} = \mathbf{S}^{y,+,*}/\tau^*, \ \mathbf{s}^{b,-,*} = \mathbf{S}^{b,-,*}/\tau^* \tag{8.4}$$

To calculate the environmental TFP based on the efficiency scores, four types of efficiency scores need computing: $S_{NLP}^t(\mathbf{x}_i^t, \mathbf{y}_i^t, \mathbf{b}_i^t)$, $S_{NLP}^{t+1}(\mathbf{x}_i^{t+1}, \mathbf{y}_i^{t+1}, \mathbf{b}_i^{t+1})$, $S_{NLP}^{t+1}((\mathbf{x}_i^t, \mathbf{y}_i^t, \mathbf{b}_i^t)$, and $S_{NLP}^t(\mathbf{x}_i^{t+1}, \mathbf{y}_i^{t+1}, \mathbf{b}_i^{t+1})$. This chapter chooses the Luenberger index to calculate the environmental TFP. Either the Malmquist index or the Malmquist-Luenberger index are needed to choose the calculating angle, that is, based on input or output. The Luenberger index proposed by Chambers *et al.* (1996b) does not need the choice of angle when calculating the productivity, which could be seen as the generalization of the Malmquist index and the Malmquist-Luenberger index. Based on four types of efficiency scores, the formulae to calculate the Luenberger technical change (LTCH), the Luenberger efficiency change (LECH) and the further Luenberger TFP (LTFP) are defined as below:

$$LTCH^{t,t+1} = \frac{1}{2}\{[1/S^{t+1}(\mathbf{x}_i^{t+1},\mathbf{y}_i^{t+1},\mathbf{b}_i^{t+1}) - 1/S^t(\mathbf{x}_i^{t+1},\mathbf{y}_i^{t+1},\mathbf{b}_i^{t+1})]$$

$$+[1/S^{t+1}(\mathbf{x}_i^t,\mathbf{y}_i^t,\mathbf{b}_i^t) - 1/S^t(\mathbf{x}_i^t,\mathbf{y}_i^t,\mathbf{b}_i^t)]\} \tag{8.5}$$

$$LECH^{t,t+1} = 1/S^t(\mathbf{x}_i^t,\mathbf{y}_i^t,\mathbf{b}_i^t) - 1/S^{t+1}(\mathbf{x}_i^{t+1},\mathbf{y}_i^{t+1},\mathbf{b}_i^{t+1}) \tag{8.6}$$

$$LTFP^{t,t+1} = LTCH^{t,t+1} + LECH^{t,t+1} \tag{8.7}$$

If several DMUs are found efficient, that is, their scores being 1, we could follow Tone (2002) and Du *et al.* (2010) to re-rank them by calculating their super-efficiency. If such reasons as history, endowment and institution cause the greatest technique difference among DMUs, it is possible for such DMUs to face different production frontiers. In this case, we could follow the meta-frontier method of Battese and Rao (2002) and clustering evaluation method of Tone (2010) to undertake the heterogeneous analysis of environmental efficiency scores for different DMUs. This is also what we plan to do in this chapter.

8.3.2 Variables in this study and construction of evaluation indicator

To evaluate the provincial low-carbon transformation process in China, we build the input and output panel data for 31 provinces over the reform period. Specifically, the panel database includes the variables of two desirable outputs (regional GDP, GDP for short; and regional gross output value, GOV), five undesirable outputs (carbon dioxide emission, CO_2; wasteful water, WW; wasteful gas, WG; sulphur dioxide, SO_2; chemical oxygen demand, COD), and seven inputs (energy consumption, E; electricity consumption, EL; investment in the treatment of industrial pollutions, IPI; investment in the treatment of environmental pollutions, EPI; area of afforestation, F; labour force, L; capital stock, K). Due to the availability of the variables, the sample size for each variable is different. That is to say, their starting time point is different, but the end of the sample period is normally in 2009. The descriptive statistics of such variables are reported in Table 1.2. In this chapter, the classification of all 31 provinces into two regions, that is, a low energy and emission region and a high energy and emission region, is also the same as that in Table 1.2. We do not discuss it again here. To forecast the low-carbon transformation in 2010, we firstly forecast the value of all the variables in 2010 by using a linear interpolation method. Generally speaking, if the intermediate inputs (such as energy consumption and electricity consumption) are included in the analysis of production function, the introduced output variable should be the gross output value rather the value-added such as GDP. But the regional gross output value constructed in the paper is closed in 2004. That is to say, this chapter has to use the regional GDP as the desirable output variable where the whole sample size is between 1980 and 2009. Thus, first, it is necessary to investigate the impact of two output variables on the estimation of efficiency scores and then environmental TFP growth. To do this, we construct a simple model in which only one undesirable output, CO_2, is included, and the inputs contain energy consumption, labour and capital stock.

To obtain the composite indicator to precisely evaluate the low-carbon transformation process, this chapter builds ten models according to the relationships among variables and their different sample size. The common variables contained in the ten models are regional GDP, CO_2 emission, energy (or electricity) consumption, labour and capital stock. The former five models contain more variables of wasteful water and wasteful gas; while the latter five models adopt their proxy variables of COD and SO_2. The sample size for the ten models is not the same. In order to achieve a robust indicator, this chapter firstly simple averages five indicators from the former five models over their relative sample period, and five indicators from the latter five models, respectively, to obtain two group indicators, and then simply averages the two group indicators over their corresponding period to finally obtain the composite indicator. The ten models are specified as follows:

Model	1	(1985 – 2010) :	*GDP*	*CO₂*	*WW*	*WG*	*E*	*L*	*K*			
Model	2	(1990 – 2010) :	*GDP*	*CO₂*	*WW*	*WG*	*E*	*L*	*K*	*EL*		
Model	3	(1998 – 2010) :	*GDP*	*CO₂*	*WW*	*WG*	*E*	*L*	*K*	*IPI*	*Group Indicator 1*	
Model	4	(2004 – 2010) :	*GDP*	*CO₂*	*WW*	*WG*	*E*	*L*	*K*	*EPI*		
Model	5	(2004 – 2010) :	*GDP*	*CO₂*	*WW*	*WG*	*E*	*L*	*K*	*EPI*	*F*	*Composite Indicator*
Model	6	(2000 – 2010) :	*GDP*	*CO₂*	*COD*	*SO₂*	*E*	*L*	*K*			
Model	7	(2000 – 2010) :	*GDP*	*CO₂*	*COD*	*SO₂*	*E*	*L*	*K*	*EL*		
Model	8	(2000 – 2010) :	*GDP*	*CO₂*	*COD*	*SO₂*	*E*	*L*	*K*	*IPI*	*Group Indicator 2*	
Model	9	(2004 – 2010) :	*GDP*	*CO₂*	*COD*	*SO₂*	*E*	*L*	*K*	*EPI*		
Model	10	(2004 – 2010) :	*GDP*	*CO₂*	*COD*	*SO₂*	*E*	*L*	*K*	*EPI*	*F*	

The ten models will produce ten evaluation indicators. The values of the produced ten indicators are dependent on the concrete variables and their units used to calculate them, which are bound to influence the final calculated composite indicator due to the simple averaging method. To overcome the biased influence of possible abnormal indicators on the composite indicator, this chapter follows Fan *et al.* (2003) to firstly normalize the produced ten evaluation indicators. That ensures that the ten indicators will play a even role in forming the two group indicators and the final composite indicators. The regional marketization indicators constructed by Fan *et al.* (2003) are totally relative ones. Unlike this, the evaluation indicator constructed in our study is not only relative but also absolute in nature.[2] According to the relationships between the value of the indicator and the turning point of low-carbon transformation, the approaches to normalize the produced indicators are classified as two types. If the value of the indicator is higher than the threshold of 0.5, the formula to normalize the indicators is as follows:

$$\text{Normalized indicator of } i\text{th DMU} = \left(1 + \frac{V_i - 0.5}{V_{max} - 0.5}\right) \times 0.5$$

where, the larger the value of the normalized indicator, the higher is the degree of the low-carbon transformation. If the value of the indicator is below 0.5, the indicator could be normalized by using the following formula:

$$\text{Normalized indicator of } i\text{th DMU} = \left(1 - \frac{0.5 - V_i}{0.5 - V_{min}}\right) \times 0.5$$

where, the larger the value of the normalized indicator, the closer the DMU approaches the threshold of the low-carbon transformation. In the above formulae, V_i represents the value of the produced indicator of ith province; V_{max} and V_{min} are the maximum and minimum indicators, respectively, among the 31 provinces during the sample period.

The composite indicator calculated from two group indicators and ten indicators effectively removes the possible influence of specification bias of some models on the final estimates, and is the combination of all the estimates from different approaches. The threshold value of 0.5 could be used to evaluate whether the low-carbon transformation happens or not; the value of the indicator could also be adopted to capture the degree of evolution of low-carbon transformation. The dynamic evaluation function of the composite indicator, produced endogenously from the low-carbon economic theory, is far better than the traditional calculated statistical development indicator, to the best of our knowledge, and is not addressed in other literature.

8.4 The evaluation and forecasting of provincial low-carbon transformation in China

8.4.1 Efficiency score and environmental TFP analysis

Figure 8.1 plots the weighted averaging TFP growth based on regional GDP and gross output, respectively. The weights are provincial GDP share. As stated previously, the validation model only contains one undesirable output of CO_2 and three inputs of energy, labour and capital. The sample of regional gross output and GDP is closed in 2004 and 2009. The figure reveals that the two lines have similar fluctuations and values, and no one line is consistently higher or lower than another line, which implies that the introduction of either regional GDP or gross output value as the desirable output will not greatly influence the results of the estimated environmental TFP and evaluation indicator of low-carbon transformation. Thus, in the following analysis in this study, we only use the series of regional GDP as the desirable output. As for the estimated environmental TFP growth, from the 1980s to the beginning of the 1990s, the environmental GFP growth in China is not low but fluctuates a great deal. The successful agricultural reform pushed the

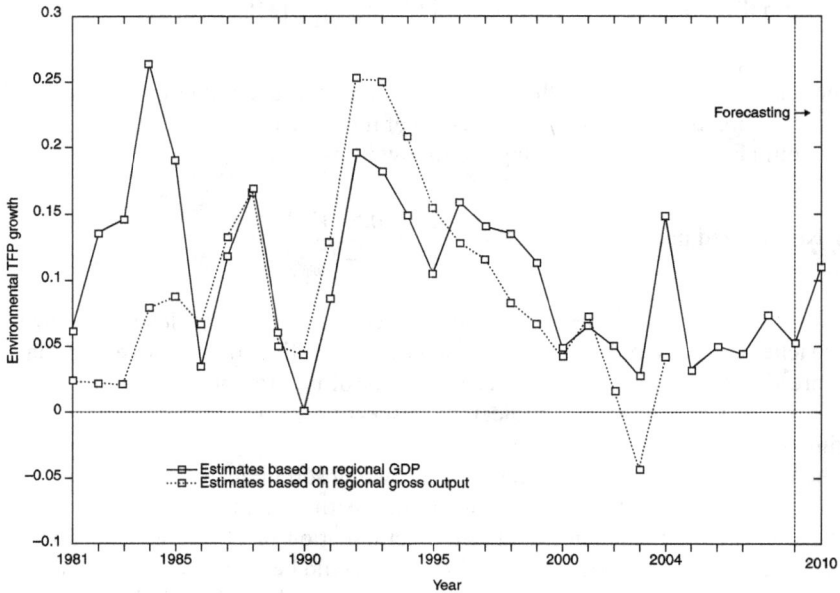

Figure 8.1 Averaging TFP growth estimated using regional GDP and gross output in China (1981–2010).

TFP growth in the early 1980s upwards, which reached the first peak in 1984 and 1985. This is consistent with the finding by Perkins (1988), Lin (1992) and Chow (1993). The TFP grows with fluctuations to a peak in the early and mid-1990s and changes to fall continuously to the lowest level in 2003, which corresponds with the reappearance of heavy industrialization,; this is similar to the findings by Wang *et al.* (2009a) and Zheng *et al.* (2009). After 2005, environmental TFP gradually rises again and, in 2010, the forecast for environmental GFP growth reaches 11 per cent. According to the decomposition of the Luenberger TFP index, the change of environmental TFP growth is dominated by technical progress and, most of the time, the environmental production efficiency is fluctuating around the zero horizontal axis.

The evaluation indicator and environmental TFP are calculated from the efficiency scores, therefore it is also important to analyse the estimated environmental efficiency scores. Table 8.1 reports the provincial efficiency scores estimated using the SBM-DDF-AAM approach. The environmental efficiency scores, calculated using two series of regional GDP and gross output value, are compared first. The efficient provinces, calculated in terms of regional GDP, include Beijing, Tianjin, Shanghai, Jiangsu, Guangdong and Tibet. In addition to the six provinces, the efficient provinces found from the regional gross output value are Zhejiang and Fujian. Though the efficiency scores estimated from the regional GDP are higher than those from regional gross output for most of the provinces, the ranking

Table 8.1 Provincial environmental efficiency, super efficiency and clustering efficiency scores in 2009, estimated by SBM-DDF-AAM method in China

Provinces	Environmental Efficiency (Based on regional gross output)	Super Efficiency (Based on regional GDP)	Clustering Region	Clustering Efficiency	Provinces	Environmental Efficiency (Based on regional gross output)	Super Efficiency (Based on regional GDP)	Clustering Region	Clustering Efficiency
Beijing	1	1.196	1	1	Henan	0.336		1	0.428
Tianjin	1	1.144	1	1	Hubei	0.361		1	0.484
Hebei	0.336		2		Hunan	0.338		1	0.490
Shanxi	0.290		2	0.775	Guangdong	1	1.165	1	1
Inner Mongolia	0.257		2	1	Guangxi	0.330		2	0.458
Liaoning	0.379		2		Hainan	0.597		1	0.736
Jilin	0.326		2		Chongqing	0.414		2	0.523
Heilongjiang	0.397		1	0.602	Sichuan	0.323		2	0.435
Shanghai	1	1.401	1	1	Guizhou	0.230		2	0.292
Jiangsu	1	1.350	1	1	Yunnan	0.263		2	0.364
Zhejiang	1		1	0.764	Tibet	1	1.002	1	1
Anhui	0.358		1	0.464	Shaanxi	0.336		2	0.467
Fujian	1		1	0.669	Gansu	0.304		2	0.365
Jiangxi	0.444		1	0.545	Qinghai	0.427		2	0.462
Shandong	0.614		1	0.898	Ningxia	0.335		2	0.365
					Xinjiang	0.293		2	0.403

of the provinces in the two series is similar. For example, only one province is different among the top ten provinces in the two series, and two provinces are different among the last eleven provinces in the two series. The province ranked the last in both series is Guizhou, with an efficiency score below 0.3. In two series, six provinces have the same ranking. The similarity indicates that the inclusion of regional GDP or gross output value as the desirable output does not have an obvious impact on the calculated environmental efficiency scores, as found by the environmental TFP in Figure 8.1. Therefore, we only use the regional GDP in the following analysis of this chapter.

Super-efficiency could be calculated to further rank the efficient provinces found by the SBM-DDF-AAM approach. As shown in Table 8.1, based on the recalculation of the super-efficiency scores, the ranking of six efficient provinces from the highest to lowest values is: Shanghai, Jiangsu, Beijing, Guangdong, Tianjin and Tibet. Due to different resource endowment and development history, different regions may face different technique frontiers. Looking at, say, the resource-intensive region and the resource-scarce region, the heavy industry and light industry probably have big technical differences. Thus, it is necessary to undertake a heterogeneity analysis for different provinces. Following the meta-frontier of Battese and Rao (2002) and clustering analysis of Tone (2010), this chapter also estimates the efficiency scores for different groups. As shown in the fifth and eleventh columns of Figure 8.1, 31 provinces are divided into two regions, where 1 indicates the low energy and emission region and 2 the high energy and emission region. The classification is the same as that in Table 1.2. Now, we construct two technique frontiers for the two regions. According to the technique frontier of the low energy and emission region, the efficient six provinces are the same as the six efficient provinces found by the common technique frontier of 31 provinces. That is to say, the common technique frontier for 31 provinces is determined by the technique of the low energy and emission group. According to the frontier of the high energy and emission region, we find 9 efficient provinces and more that are not efficient according to the common frontier of 31 provinces. They are Hebei, Inner Mongonia, Liaoning, Jilin, Guangxi, Chongqing, Sichuan, Gansu and Qinghai, which could be referred to as locally efficient but globally inefficient units. We do not discuss them further here. In the rest of this chapter, the analysis is still based on the environmental efficiency scores determined by the common technique frontier of all the provinces and estimated by the SBM-DDF-AAM approach.

8.4.2 The evaluation and forecasting of provincial low-carbon transformation in China

Figure 8.2 depicts the averaged composite indicator to evaluate and forecast the low-carbon transformation process for low and high energy and emission regions and China as a whole. The classification of the two regions is still same as that in Table 1.2 and the weights to calculate the averaged indicator are the provincial GDP share.

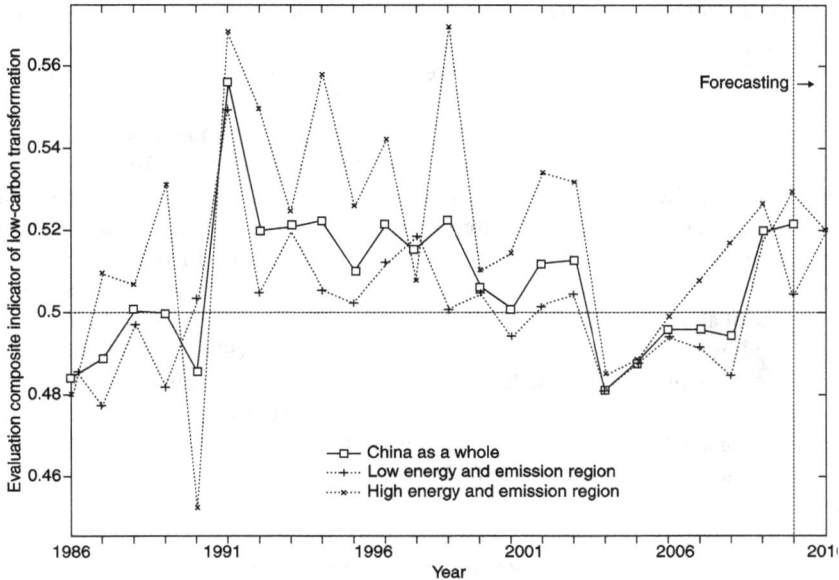

Figure 8.2 Averaging trend of evaluation composite indicator of low-carbon transformation for different regions and China (1986–2010).

As plotted in Figure 8.2, the evaluation composite indicator varies a great deal but experiences a regular pattern, based on which China's low-carbon transformation process can be divided into four stages. Between 1986 and 1990, the evaluation indicator is below or close to the threshold of 0.5, indicating that China experiences the traditionally high-carbon economic growth at this stage. In the 1980s, the central government encouraged the energy production to resolve the energy shortage resulting from the long term command economy. The energy intensive industry (such as coal mining, petrochemical and electric power enterprises) has undertaken the reform of contract responsibility system. Among them, the coal industry is the one that has experienced the most thorough reforms, and where competition is fiercest. As a result, the low-carbon transformation in high energy and emission region for a while had a g performance. However the production of coal eventually went out of control. The destructive exploitation and the lack of environmental protection measures caused serious damage, a waste of coal resources and environmental pollution. Therefore, at this stage, the low-carbon transformation did not go well in China, leading directly to the lowest value of the evaluation indicator of the high energy and emission region, in 1990. Because of this, the central government began to restrict energy development and started to rectify the coal market. In 1994, the dual-track price of coal was first phased out of the state plan and fully liberalized. The full formation of a coal market, together with shutdowns of many small energy-intensive enterprises during the period of SOEs reform and many other efficient environmental regulations, resulted in the excellent performance of low-carbon transformation in China from 1991 to 2002, in which the high energy and emission region contributed more to this transformation. The efficient environmental policies implemented at this stage were a strategy change: they included industrial pollution prevention and treatment from end harnessing to control of source and production processes; from control of point source pollution to comprehensive governance of drainage and regional area; from treatment of certain enterprises to adjustment of industrial structure and development of the circulative economy, among others. In 1994, China took the lead in the first national release of 'China Agenda 21'. The official document 'Outlines of the Ninth Five-Year Plan and Long-term Target towards to 2010' authorized by the People's National Congress (PNC) in 1996 confirmed sustainable development as one of the basic national development strategies, and energy as the important area to achieve the sustainable development strategy; it once again emphasized the energy policy of 'saving and developing energy simultaneously' originally proposed in 1980. In 1998, the Energy-Save Law came into effect in China.

Since the start of this century, however, the phenomenon of heavy industrialization appeared again, the consumption of primary energy and carbon emission rose to unprecedented levels, and the energy and emission intensity also reversed the long-term decreasing trend and displayed a temporary rise for several years. This worsened the low-carbon transformation performance of China between 2003 and 2007. Fortunately, as also shown in Figure 8.2, China was transformed into low-carbon economic development again in 2008 and 2009, with the values of the composite indicator attaining about 0.52. According to the forecast value of the evaluation indicator, China remains on the trend of low-carbon transformation in 2010. This could be attributed to the following environmental regulations. The seventeenth Party Congress in 2007 put forward the scientific outlook on development. The central government released the 'China National Plan for Coping with Climate Change' in 2007, the first national green plan among developing countries. In 2006 and 2009, China officially announced the binding target to reduce energy and carbon emission intensity. In the stimulus to recover the economy from the financial crisis at the end of 2008, green investment accounted for 38 per cent, attaining 221 billion US$. The finding that the high energy and emission region has the better low-carbon transformation than the low energy and emission region for most of the time points shown in Figure 8.2 seems to be different from what we would instinctively expect. One of the explanations is that the high energy and emission region has focused on environmental conservation more and has the more scope to save energy and abate emission, leading to more efficient results of low-carbon development, as compared with the low energy and emission region. Base on the model to produce the environmental TFP in Figure 8.1, we compute several testing results. The estimated environmental technical change, efficiency change and TFP growth are very close for both low and high energy and emission regions. They are about 10 per cent, −0.3 per cent and 10 per cent, respectively, and the difference between two regions is only reflected at the third digit after the decimal point. But the average GDP growth is 11.7 per cent and 10.6 per cent for the low and high energy and emission regions, the difference reaching one percentage. The evaluation indicator calculated in this chapter is just the ratio of environmental TFP growth to GDP growth, thus, it is not surprising that the more rapid growth of GDP in the low energy and emission region leads to its lower value of evaluation indicator, as compared with the high energy and emission region. Therefore, the low energy and emission region will achieve a better performance of low-carbon transformation if it could emphasize more the growth of environmental TFP at the same time as it has rapid economic growth.

Table 8.2 reports two group indicators, the composite indicators from the seventh to eleventh Five-Year Plan and the forecasting indicator in 2010 for 31 provinces in China. The values of the two estimated group indicators are different. For example, only four provinces have the equal group indicators; for eight provinces, the difference between the two group indicators is greater than 0.04. Ten provinces have the a group indicator 1 with a larger value than group indicator 2, and seventeen provinces have a larger value of group indicator 2 than group indicator 1. Thirteen provinces are located on different sides of the threshold value according to the group indicators 1 and 2. In terms of group indicator 1, twelve provinces have a value greater than 0.5; while, based on group indicator 2, nineteen provinces have a value over 0.5. The big difference between the two group indicators and implied by the 10 indicators justifies the construction of the composite indicator to evaluate low-carbon transformation. The evaluation indicator based on a single model is very likely to derive a biased conclusion. Consistent with the patterns depicted in Figure 8.2, during the seventh Five-Year Plan (exactly corresponding to 1986–1990), only twelve provinces have evaluation indicators greater than 0.5, the lowest among five Five-Year Plan periods and exhibiting extensive growth characteristics; similar to the better performance of low-carbon transformation in 1990s, the number of provinces that realize the low-carbon transformation during the period of the eighth and ninth Five-Year Plan reaches 20 and 16, respectively; the reappearance of heavy industrialization hinders low-carbon transformation again from the beginning of this century, and, accordingly, the number of provinces that realize low-carbon transformation reduces to the second lowest total of 14 in the sample period; with the proposition of the scientific development outlook and policy support on the low-carbon economy, 22 provinces have achieved low-carbon transformation in the period of the eleventh Five-Year Plan, and in 2010, the number of provinces that experienced sustainable development has attained 24.

As stated previously, the evaluation indicator is the ratio of environmental TFP growth to regional GDP growth, thus, the ranking of provinces according to green TFP growth is not necessarily consistent with the ranking according to evaluation indicators. As a result, the findings from Figure 8.2 are not consistent with those from Table 8.2. Some provinces perform well now in low-carbon transformation even if they belong to the low energy and emission region and have high efficiency scores, such as Beijing and Shanghai. Such provinces as Jiangsu and

Table 8.2 Provincial evaluation indicators of low-carbon transformation in China

Indicators	Group Indicator Indicator 1 Indicator 2	Averaged Evaluation Composite Indicator					Forecasting of Composite Indicator
Five-Year Plan	10th–11th Five-Year Plan	7th Five-Year Plan	8th Five-Year Plan	9th Five-Year Plan	10th Five-Year Plan	11th Five-Year Plan	2010
	(2001–2010)	(1986–1990)	(1991–1995)	(1996–2000)	(2001–2005)	(2006–2010)	
Beijing	0.46	0.49	0.54	0.53	0.46	0.45	0.46
Tianjin	0.52	0.52	0.57	0.53	0.54	0.48	0.47
Hebei	0.54	0.60	0.69	0.64	0.56	0.56	0.52
Shanxi	0.48	0.49	0.60	0.49	0.51	0.45	0.39
Inner Mongolia	0.49	0.46	0.61	0.67	0.52	0.50	0.54
Liaoning	0.54	0.48	0.49	0.45	0.51	0.53	0.58
Jilin	0.51	0.49	0.55	0.59	0.52	0.50	0.52
Heilongjiang	0.54	0.46	0.48	0.49	0.56	0.50	0.51
Shanghai	0.51	0.48	0.50	0.50	0.53	0.45	0.51
Jiangsu	0.51	0.45	0.50	0.56	0.49	0.55	0.51
Zhejiang	0.55	0.52	0.49	0.50	0.51	0.58	0.53
Anhui	0.52	0.40	0.53	0.48	0.50	0.49	0.50
Fujian	0.48	0.50	0.52	0.48	0.45	0.51	0.51
Jiangxi	0.49	0.39	0.48	0.51	0.47	0.52	0.50
Shandong	0.53	0.49	0.51	0.49	0.52	0.50	0.57
Henan	0.50	0.70	0.64	0.51	0.46	0.52	0.50
Hubei	0.53	0.52	0.54	0.51	0.48	0.50	0.50
Hunan	0.47	0.38	0.49	0.52	0.47	0.50	0.50
Guangdong	0.50	0.48	0.49	0.48	0.48	0.50	0.54
Guangxi	0.48	0.50	0.49	0.51	0.53	0.55	0.59
Hainan	0.50	na	0.49	0.48	0.49	0.51	0.52

Chongqing	0.49	0.52	na	na	0.48	0.50	0.50	0.52
Sichuan	0.45	0.51	0.35	0.47	0.43	0.46	0.50	0.51
Guizhou	0.50	0.49	0.38	0.50	0.47	0.49	0.50	0.43
Yunnan	0.49	0.48	0.63	0.54	0.55	0.48	0.50	0.50
Tibet	0.43	0.48	0.61	0.64	0.36	0.42	0.48	0.39
Shaanxi	0.50	0.50	0.54	0.56	0.60	0.49	0.51	0.54
Gansu	0.52	0.50	0.47	0.48	0.47	0.51	0.51	0.51
Qinghai	0.47	0.46	0.38	0.54	0.43	0.46	0.48	0.49
Ningxia	0.46	0.45	0.58	0.52	0.49	0.44	0.47	0.45
Xinjiang	0.49	0.50	0.48	0.48	0.54	0.48	0.50	0.50
Number of provinces with the indicator exceeding 0.5	12	19	12	20	16	14	22	24

Zhejiang are classified as the low energy and emission region but still have a good performance of low-carbon transformation. Of course, in the high energy and emission region, some provinces have a not bad performance of low-carbon development, such as Inner Mongolia and Gansu. But more provinces in the high energy and emission region have a bad performance of low-carbon transformation, such as Shanxi, Ningxia and Qinghai. Therefore, the conclusion that the high energy and emission region performs better than the low one in low-carbon transformation revealed by Figure 8.2 is too approximate, according to Table 8.2; we cannot conclude that one region performs consistently better than another and it is better to concretely analyse the low-carbon transformation for specific provinces. As analysed in Chapter 2, the reasonably good performance of low-carbon transformation in Shanghai is likely to be related to the government's investment driven growth model and the unreasonable specification of six pillar industries in the tenth Five-Year Plan.

8.5 Summary and comments

This chapter proposes a SBM-DDF-AAM based theoretical framework to analyse the low-carbon economic transformation and constructs an endogenously produced evaluation indicator to assess the process of low-carbon transformation for 31 provinces in China between 1986 and 2009, and forecasts the trend of low-carbon transformation in 2010. The estimated composite indicator reveals that the process of low-carbon transformation in China could be classified into four stages: the traditionally high-carbon growth in 1986–1990, excellent low-carbon transformation in the 1990s, the setback of low-carbon development in the early 2010s, and restarting the process of low-carbon transformation in recent years. The financial crisis has brought China a great opportunity to transform its extensive high-carbon model to a low-carbon one and it can be imagined that China will realize the substantial transformation of its economic development model during the period of the twelfth Five-Year Plan.

The GDP driven evaluation criterion is the root cause of traditional high-carbon growth in China. The evaluation indicator constructed in this chapter is the proportion of environmental TFP growth to GDP growth, which considers both the economic growth in China but also the growth quality, symbolized by environmental TFP and extremely important to low-carbon transformation. Therefore, the constructed evaluation indicator in this chapter could be used as the replacement of GDP to assess local governance performance. If the environmental productivity growth is not high, the high growth of GDP will further damage the process of low-carbon transformation, so in this case, the slowdown of GDP growth in the short run is very beneficial to sustainable development in long run by focusing on the improvement of growth quality; if the GDP growth is high and green TFP growth is higher, it is the ideal low-carbon transformation that we need and will lead to win-win development as proposed by Porter (1991). On average, low-carbon transformation in China has experienced regular change. But for different provinces there are different performances of low-carbon development. Therefore, it is necessary to analyse concretely different cases in different provinces to provide the specific environmental policies to push their respective low-carbon transformation forward. The changes shown by the evaluation indicator imply that low-carbon transformation is still not stable in China. In the future, the government should introduce more appropriate environmental regulations to support the process of low-carbon transformation, in order to realize final sustainable development in China in the long run.

9 Energy-saving and emission-abating regulations and win-win development simulations

9.1 Introduction

To achieve an agreement among countries to make a commitment to abate carbon dioxide after the expiry of the Kyoto Protocol in 2012, all countries launched a new round of negotiations. The negotiations were extremely difficult due to disputes about abatement obligations and worry about the slowdown of economic growth, especially during the period of the financial crisis. There was no substantial progress on how to extend the Kyoto Protocol in the Copenhagen and Cancun climate conferences in the last two years. In contrast with the avoidance of global responsibility, however, many countries believe that the low-carbon economy will lay the foundation for future growth and they are investing considerably in a green dimension of the stimulus packages drawn up to challenge the financial crisis, as stated previously. For example, an important component of the American Recovery and Reinvestment Act proposed by President Obama is to develop renewable energy. The House of Representatives also passed the landmark American Clean Energy and Security Act in 2009 in order to make renewable energy and low-carbon techniques a new economic driver. In 2009, UK also released legislative proposals outlining the national strategy, the 'Low Carbon Transition Plan', with goals to be achieved by 2020 to make progress to become a low-carbon country: cutting emissions, maintaining a secure energy supply and maximizing economic opportunities.

According to the HSBC bank's report (Robins *et al.*, 2009), with sizeable financial reserves and a tradition of long-term planning, in November 2008 China launched its RMB 4000 bn (US$ 584 bn) package. Almost 40 per cent of this was allocated to green themes, most notably rail, electricity grids and water infrastructure, along with dedicated spending on environmental improvement. Elsewhere in Asia, South Korea introduced a Green New Deal, with more than 80 per cent allocated to environmental themes. The new American Recovery and Reinvestment Plan commits US$ 787 bn to kick-start the economy, with US$ 94 bn for renewables, building efficiency, low-carbon vehicles, mass transit, grids and water. Although the green component is smaller than China's, it is more broadly based, and the only plan with a real boost to renewables. The existence of substantial automatic fiscal stabilizers in Europe has meant that the EU stimulus is, so far, smaller in size. However, the climate change dimension is greater than in the US, due to a focus on low-carbon investment in France, Germany and at the EU level.

So why do countries have totally distinct attitudes domestically and internationally towards the same issue? In fact, all countries are clear about the inevitability of energy saving and environment protection in the long run, since it is crucial for economic transformation and future competition in innovative technology. However in the short run, especially under the circumstance of the financial crisis, energy-saving and emission-abating will use up the limited resources which may be needed for other productions, this slowing the pace of economy resuscitation. That is the reason why all the countries hesitate on the promise for emission reduction in the international climate negotiations. As a matter of fact, there are also two opposite arguments on how energy saving and emission reduction may influence the economy in the academic field. On one side, the Porter hypothesis argues that energy saving and emission reduction can bring opportunities for win-win development, that is, simultaneous improvements in both environmental quality and productivity, meeting both social and economic goals. On the other side, some scholars raise doubts about the existence of this win-win development because, if it does exist, it will be unnecessary for governments to impose extra environmental protection costs on firms. Many researchers focus on empirical study of the existence of this win-win development possibility, which will be surveyed in Section 9.2 of this chapter.

China is the largest energy consumer in the world. Furthermore, the US and China are respectively the first and second largest coal consumers, which correspondingly makes the two countries the top two greenhouse gases emitters in the world. And in 2007, China's carbon dioxide emission exceeded that of the US, which brought for China a greater pressure about its abatement burden from the outside world. With China's proposal to take a scientific outlook on development, energy saving and emission abating has also become the propeller of China's economic structural adjustment and transformation of its development model (Cai *et al.*, 2008). Hence, an in-depth analysis is needed on both the positive and negative effects of energy saving and emission reduction on China's economy, especially the output growth and productivity of the real economy after the financial crisis. Searching for an optimal energy-saving and emission-abating path which can induce a win-win development for China in the following decades, a strategically critical period, is also a quite practical and pioneering issue, prompting the motivation for this chapter research. As is known to all, industry as a major part of China's real economy is the primary source of China's carbon dioxide emission. It counts for over 80 per cent of the total amount of emission, which makes it the primary target of energy saving and emission abating. As stated already, China's economy has been in the mid-industrialization stage which is characterized by a booming heavy industry with large energy consumption and pollutant emission. Energy and emission intensive industries (such as iron and steel, cement and chemistry industries) will continue to play pivotal roles in future economic growth. Thus, we can foresee that there will be more negative impact brought by energy-saving and emission-abating activities on China's industry, especially heavy industry. All in all, a correct understanding of the relationship between energy and environment, and industrial output and productivity, is tremendously meaningful for China's industrial economy and public decisions. This chapter still focuses on 38 sub-industries and newly proposes a dynamic version of the activity analysis model (AAM), modified from the directional distance function (DDF), to examine the existence of the Porter hypothesis in China. Based on this proposed model, we also attempt to search for an optimal energy-saving and emission-abating path which could lead to a win-win development possibility for China's industry from now on to the hundredth anniversary of the People's Republic of China.

The rest of this chapter is organized as below: Section 9.2 surveys the empirical studies to examine the existence of the Porter hypothesis; Section 9.3 designs different energy-saving and emission-abating paths, which will be added into the direction vector of DDF so as to extend the AAM into a dynamic version; Section 9.4 measures the magnitude of these win-win opportunities among a set of sub-industries corresponding to different paths designed in the former section to pin down an optimal path for China's industrial win-win development between 2009 and 2049; Section 9.5 concludes this chapter.

9.2 Literature review

In the recent 20 years, the relationship between energy, environment and the economy (3E) has always been a focal topic for scholars and policy makers. The traditional established notion of environmental protection is that the extra costs government imposes on firms can jeopardize their international competitiveness. Porter, however, first challenged this argument in his one-page paper published in 1991 (Porter, 1991). He regarded large energy consumption and pollutant emission as a form of economic waste and a sign of incompletion and inefficiency of resource usage. In his opinion, the amelioration of this inefficiency will provide firms with the win-win opportunity of improving both productivity and environment. And the efforts of environmental protection can help firms to identify and eliminate this production inefficiency and regulatory disincentives that prevent the simultaneous improvements in both productivity and environmental quality. Thus, whether these types of environmental policy initiatives are successful depends on the extent to which such inefficiencies are widespread in the sub-industries, particularly in the energy/pollution intensive industries. Deficient management systems may mean firms are not aware of certain opportunities, so environmental policy may open their eyes. Porter and van der Linde (1995) further emphasize that properly designed environmental protection policy, in the form of economic incentives, can trigger innovation that may partially or fully offset the costs of complying with them. Such innovation offsets occur mainly because pollution regulation is often coincident with improved efficiency of resource usage, and the inference is that stiffer environmental regulation results in greater productivity and competence. These arguments are titled as the Porter hypothesis

(Ambec and Barla, 2002). Admittedly, many scholars criticize the Porter hypothesis, arguing that it is a fundamental challenge to the efficient market hypothesis and neoclassical theory. They question why firms do not see these win-win opportunities by themselves, which at least implies that the argument does not have a general validity (Palmer *et al.*, 1995; Jaffe *et al.*, 1995; Faucheux and Nicolaï, 1998).

There are many empirical researches related to the Porter hypothesis. Combining the idea of ecological economics on capital substitution and the Porter hypothesis, Karvonen (2001) works on the development of Finland's capital intensive paper industry in the past 20 years, and reveals how the use of new technologies helped the industry achieve a win-win situation, and how man-made capital investments influenced the quality of natural capital. Mohr (2002) derives results consistent with Porter's hypothesis by employing a general equilibrium framework with a large number of agents, external economies of scale in production and discrete changes in technology. The model shows that endogenous technical change makes Porter's hypothesis feasible. However, a policy that produces results consistent with Porter's hypothesis is not necessarily optimal. Nugent and Sarma (2002) used an environmentally extended computable general equilibrium (CGE) model to analyse the case of India and found that a thorough integration of economic, distributional and environmental policies could collectively 'win' in achieving economic growth, distributional equity and environmental sustainability at the same time. Murty and Kumar (2003) estimated the output distance function of India's manufacturing industry using the stochastic parametric approach and came to the conclusion that the technical efficiency of firms increases with the intensity of environmental regulation and the water conservation efforts, which supports the Porter hypothesis about environmental regulation. Beaumont and Tinch (2004) find that abatement cost curve methodology proves to be a valuable management tool in identifying barriers to achieving a win-win state, or at least win-draw scenario for industry and the environment, and also in providing future direction for waste management strategy. Cerin (2006) supports the Porter hypothesis and finds private incentives to explore win-win development by applying the Coase theorem that emphasizes transaction costs and property rights. This paper argues that strong public support is needed to create private incentives for exploring economic and environmental win-win innovations. Greaker (2006) provided some support for the Porter hypothesis. The conclusion that policy should be more stringent when a well-developed market for new abatement equipment does not exist clearly has a general appeal. His simulations show that environmental policy has very little effect on export market share as long as the price of pollution abatement equipment is decreasing because of the stringency of environmental policy; thus, governments should a priori be less afraid of introducing a sufficiently stringent environmental policy. Managi (2006) tests the hypothesis that there are increasing returns to abating pollution. Empirical evidence on environmental risks in the US agricultural sector since 1970 support the existence of increasing returns. Kuosmanen *et al.* (2009) propose a new approach to environmental cost-benefit analysis (ECBA) which does not require prior valuation of the environmental impacts and is based on shadow prices; they conducted efficiency analysis of ten alternative GHG abatement timing strategies, taking into account the ancillary benefits. Groom *et al.* (2010) evaluated the impact of the Sloping Lands Conversion Programme (SLCP) on off-farm labour supply in China, and the results identify some support for the win-win hypothesis in the case of the SLCP, and show how the targeting of the programme can be improved. Reddy and Assenza (2009) emphasize that climate protecting policies based on market consideration can increase the opportunity for win-win development. In particular, the paper suggests the integration of climate policies with those of development priorities that are vitally important for developing countries and stresses the need for using sustainable development as a framework for climate change policies.

There are also a few papers whose conclusion is neutral or against the Porter hypothesis. Boyd and McClelland (1999) construct efficiency measures based on Shephard's distance function and view it as a test of the Porter hypothesis. The findings support aspects of both sides of the Porter debate; that is, there is evidence of a win-win potential to increase production and reduce pollution as well as evidence of losses to potential output due to environmental constraints. Thus, comparing the estimates with other studies must be approached with caution, since there can be substantial differences in methodologies. Xepapadeas and De Zeeuw (1999) isolated two effects resulting from the introduction of a stricter environmental policy in the form of a tax on emissions: a productivity effect and a profit-emission effect. The results indicated that, although a stricter environmental policy cannot be expected to provide a win-win situation in the sense of both reducing emissions and increasing profitability in an industry, you may expect increased productivity of the capital stock along with a relatively less severe impact on profits and more emission reductions, when the stricter policy induces modernization of the capital stock. By allowing for nonlinearities, Feichtinger *et al.* (2005) generalized Xepapadeas and De Zeeuw (1999) and determined scenarios in which their results do not carry over. The paper also focused more explicitly on learning and technological progress, and obtained results which showed that, in the presence of learning, implementing a stricter environmental policy with the aim of reaching a certain target of emissions reduction has a stronger negative effect on industry profits this implies quite the opposite to that described by the Porter hypothesis.

As stated previously, theoretical and empirical research have provided arguments for both positions and have not been conclusive so far, which may be due to different data sets used, the regulatory regime in different countries, the cultural settings, customer behaviour, the type of industries or size of companies analysed, the time span, and so on. However, the main reason for the conflicting results of the various empirical studies may be the lack of a reasonable theoretical framework within which to investigate the links between environmental regulation and economic performance (Schaltegger and Synnestvedt, 2002). For example, the commonly used CGE model fits static analysis well but its dynamic extension in empirical study is still rather scarce and too simple; the parametric macro-econometric model is restricted to its priori functional form and distribution assumption; environmental cost and benefit analysis needs the economic evaluation on environmental effects first, which is a technical challenge itself; the analysis based on the theories of property rights, externality and transaction cost cannot soundly quantify the economic influence of environmental regulation; the traditional Shepherd distance function cannot distinguish between the different characteristics of two outputs of both GDP and pollution, and so on. As discussed already, not until the presence of the directional distance function do we find a reasonable framework to capture the difference between GDP and environmental pollution. DDF allows for the type of inefficiency that is typified by the Porter hypothesis, which increases desirable output while decreasing undesirable output simultaneously, which means that DDF provides the most appropriate tool to examine the Porter hypothesis. By employing two kinds of DDF based on the strong and weak disposability of pollution, respectively, proposed by Boyd *et al.* (2002), this chapter attempts to measure the potential revenues and output loss, and corresponding change of production efficiency, technical progress and total factor productivity (TFP) resulting from energy-saving and emission-abating regulation. In order to forecast the win-win development possibility from now on to the year of 2049 and find the optimal environmental regulatory path, in particular, different energy-saving and emission-abating paths with the time lag operator are introduced into the direction vector of DDF to form a dynamic version of AAM. The methodology is described in Section 9.3.

9.3 Methodology

9.3.1 Designing the energy-saving and emission-abating paths

Different energy-saving and emission-abating paths will have obviously different impacts on the economy (Lee *et al.*, 2007; Kuosmanen *et al.*, 2009). This chapter designs five energy-saving scenarios and nine emission-reducing scenarios, totalling 45 policy paths combinations, and simulates their effect on potential output and productivity in the future so as to look for the best regulatory path leading to a win-win development possibility for Chinese industry.

The design of an energy saving scheme is based on the promissory targets to save energy stipulated in China's eleventh Five-Year plan in 2006, that is, to decrease energy consumption per unit of GDP (energy intensity) by 20 per cent between 2006 and 2010 (4 per cent per year). In view of the possibility of easier realization, in fact, this study just chooses a lower value of 3 per cent as the reduction rate annually for energy intensity. Based on this, if we assume that the average growth of China's gross industrial output value is likely to be one of five possibilities (4 per cent, 6 per cent, 8 per cent, 10 per cent, and 12 per cent) in the future, we can calculate that the corresponding average annual growth rate of energy consumption is 0.9 per cent, 2.8 per cent, 4.8 per cent, 6.7 per cent and 8.6 per cent, respectively.[1] Comparing this with an average annual 11.2 per cent and 6 per cent growth rate of industrial output and energy consumption, respectively,

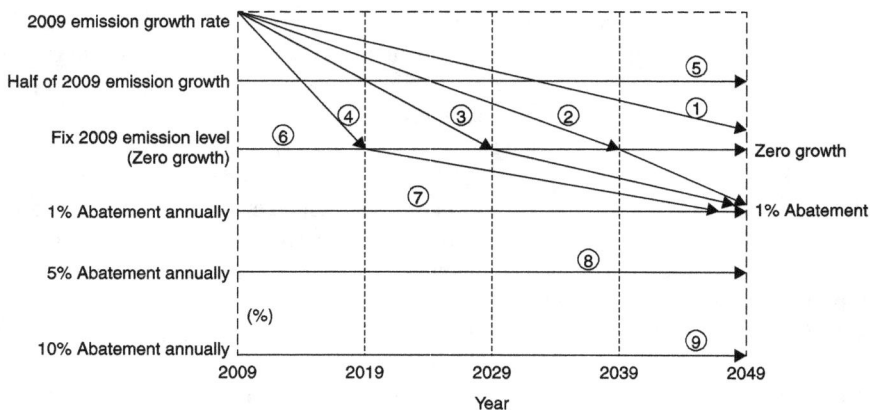

Figure 9.1 Design of carbon dioxide abatement paths (1–9) for Chinese industry (2009–2049).

achieved between 1981 and 2006, the five growth possibilities for output and energy consumption set previously look moderate and very likely to be realized.[2]

This chapter designs a scheme of emission reduction based on approaches from a gradual reduction to a sharp drop, the former of which caters for the national condition that China is a developing country whose major task is to develop. The design is also attributable to the generalized understanding of an emission abatement concept that emission reduction does not necessarily refer to an absolute decline in aggregate emission level; a declining emission growth rate, or declining relative to BaU, is also a type of emission abatement corresponding to the gradual or moderate principle. Therefore, as shown in Figure 9.1, the nine kinds of emission reduction paths from most moderate to the strongest abating intensity designed in this chapter are listed below: (1) The growth of carbon dioxide for different sub-industries evenly decreases from the respective growth rate of emission in the year of 2009 to zero growth in 2049, that is, the emission peak will appear in the middle of this century; (2) The emission growth of all sub-industries reduces from the 2009 growth level to zero growth in 2039, and after the emission peak, it decreases continuously and steadily to a −1 per cent growth rate in 2049 (i.e. the annual abating rate is 1 per cent in 2049); (3) The third and fourth paths are similar to the second path but the emission peak changes to the years of 2029 and 2019, respectively; (4) The emission growth of all sub-industries maintains half of their growth rate of emission in 2009 till 2049; (5) The carbon dioxide emission for each sub-industry remains the same as in 2009, that is, the emission growth is zero from 2009; (6) For paths 7, 8 and 9, the respective annual emission abating rates are 1 per cent, 5 per cent and 10 per cent, during the entire forecasting horizontal.

The 45 energy-saving and emission-abating policy paths designed above will be introduced into the dynamic activity analytical model through the direction vector, as we will discuss in the following subsections.

9.3.2 The Dynamic Activity Analysis Model (DAAM)

In this subsection, a novel dynamic activity analysis model (DAAM for short), not addressed so far, is proposed to simulate the effect of energy-saving and emission-abating regulation on the economy in the long run; this is extended from the standard DDF and AAM provided by Chambers *et al.* (1996a) and Chung *et al.* (1997) and applied by Färe *et al.* (2001), Jeon and Sickles (2004) and others. In this study, the DMUs are still 38 two-digit sub-industries (i = 1,2,. . .,38). The forecasting time span is from 2009 to 2049 (t = 2009, 2010,. . .,2049). For each sub-industry, there are three types of input (j = 1,2,3, corresponding to capital, labour and energy), one type of desirable output (gross industrial output value, GIOV), and one type of undesirable output (carbon dioxide emission, CO_2). The panel data for nearly 40 sub-industries, rather than aggregate data, significantly enhances the information that can be obtained to analyse microeconomic performance, particularly when examining the efficiency for each unit.

For ith sub-industry, the column vectors of $\mathbf{x}^i$, $\mathbf{y}^i$ and $\mathbf{b}^i$ represent the inputs, desirable output and undesirable output, respectively. Then the production technology for ith sub-industry at time point t can be described by its output set:

$$P\,(\mathbf{x}^i) = \{(\mathbf{y}^i, \mathbf{b}^i, -\mathbf{x}^i)\text{: } \mathbf{x}^i \text{ } can \text{ } produce \text{ } (\mathbf{y}^i, \mathbf{b}^i)\} \qquad (9.1)$$

As with the Shephard distance function, DDF is also the representative function to describe such a production technology. DDF is nonparametric frontier production function approach which assumes that some units are more efficient than others in production. The principle of DDF is illustrated in Figure 9.2. The general DDF starts at A and scales in the direction along ABC to capture the increase of desirable outputs (or goods) and decrease of undesirable outputs (or bads) simultaneously, which makes it possible to investigate the Porter hypothesis that allows for the possibility of crediting units for the reduction of pollutions. Formally, DDF is defined as:

$$\overrightarrow{D}_o\,(\mathbf{x}^i, \mathbf{y}^i, \mathbf{b}^i; \mathbf{g}^i) = \sup\{\beta : (\mathbf{y}^i, \mathbf{b}^i) + \beta \mathbf{g}^i \in P(\mathbf{x}^i)\} \qquad (9.2)$$

where $\mathbf{g}$ is the direction vector in which outputs are scaled.

9.3.2.1 Production inefficiency and loss due to environmental regulation

As shown in Figure 9.2, because the point A remains within the efficient production frontier, the inefficiencies resulting from such factors as high energy consumption and heavy emission give the producer the potential room to increase output, given the inputs and current output, by saving energy and abating emission.[3] But whether the observation vector projects from the point A to point B or C depends on the weak or free disposal assumption of undesirable output. If we assume that the undesirable output is strongly or freely disposable, that is, the disposability costs nothing, the producers will voluntarily get rid of the unwanted by-products,

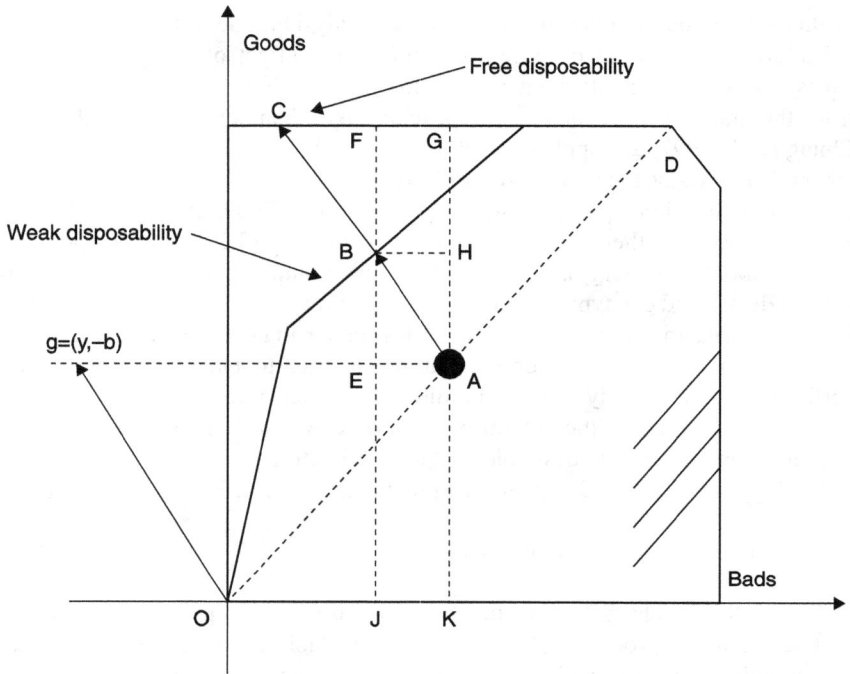

Figure 9.2 Principle of directional output distance function assuming free and weak
   disposability for undesirable outputs.

and then potential output growth based on current desirable output is maximized
which amounts to the distance function value β_s, i.e., the ratio of AC/Og. In this
case, energy and environment impose no restriction on output, then the production
in point C is the most efficient. However, it's impossible in reality to cost nothing
to reduce the undesirable output. The producers therefore are not willing to reduce
the bads because the cost makes use of the important inputs and then translates
into lost goods given inputs. The bads reduction only can be achieved by environmental regulation; corresponding with this, the more appropriate assumption is
weak disposability of undesirable output, the point A projecting into B on the
frontier, which is the standard DDF, the value being β_w (equal to AB/Og). In this
case, the potential goods growth is a tradeoff between more goods and less bads,
bound to less than the maximized β_s corresponding to highest level of inefficiency
under the strong disposability of bads.

The difference between β_w and β_s reflects the potential output loss caused by the
observable lack of free disposability (more vividly, due to enforced regulation), that
is, $l = \beta_w - \beta_s < 0$ (Boyd *et al.*, 2002). The value of l is analogous to the hyperbolic
output loss measure introduced by Färe *et al.* (1989) and used by Boyd and
McClelland (1999). The potential output loss l and potential output growth β_w

reveals the extent of the win-win potential for each sub-industry, given current output at some time point. If potential β_w exceeds or equals the absolute value of l, $|l|$, from the perspective of output, the win-win opportunity due to energy-saving and emission-abating regulation, described in the Porter hypothesis, happens, to some extent, suggesting that improved production efficiency can make up for the losses imposed by regulations; otherwise, environmental regulation does not lead to win-win development. This chapter will make use of this method to find the best energy-saving and emission-abating path that leads to the win-win development potentials.

9.3.2.2 Dynamic Activity Analysis Model (DAAM)

As stated previously, the direction vector in standard DDF is $\mathbf{g} = (\mathbf{y}, -\mathbf{b})$, and the value of DDF, β, captures the maximum feasible proportion that the goods $\mathbf{y}$ expand while the bads $\mathbf{b}$ contract, based on current output level $(\mathbf{y}, \mathbf{b})$ (the negative sign of $\mathbf{b}$ indicating the decline of bads). To simulate the dynamic process of energy-saving and emission-reducing activity, in this chapter, we introduce the time factor into the above standard direction vector and redefine the output direction vector as $\mathbf{g}^t = (\mathbf{y}^t, -\mathbf{b}^t) = [(1+u)\mathbf{y}^{t-1}, -(1+v)\mathbf{b}^{t-1}]$, where u and v respectively represent the varying rate of current goods and bads relative to previous time point (positive or negative, or increasing or decreasing) which do amount to the growth rate of gross industrial output value and the growth or abating rate of carbon dioxide emission from 2009 to 2049, designed in subsection 9.3.1 in this study. Similarly, the dynamic varying path for the jth input vector is defined as $\mathbf{x}_j^t = (1 + \sigma_j) \mathbf{x}_j^{t-1}$, where σ_j is respective varying rate – for energy, it equals the growth of energy consumption that matches the above GIOV growth so as to reduce the energy intensity by 3 per cent annually, as designed also in subsection 9.3.1; for the input of capital and labour, it is simply assumed that it keeps the historical average growth between 1981 and 2008.

In terms of the newly defined dynamic direction vector, the technology in t period and observation also in t period, the linear programming of two kinds of DDF, weak and strong disposability of undesirable output, is specified respectively for ith sub-industry, as below.

Directional distance function (weakly disposable bads):

$$\overline{D}_o^t (\mathbf{x}^{i,t}, \mathbf{y}^{i,t}, \mathbf{b}^{i,t}; \mathbf{y}^{i,t}, -\mathbf{b}^{i,t}) = \underset{\lambda,\beta}{Max} \ \beta_w$$

$$s.t. \qquad \sum_{i=1}^{38} \lambda^i \mathbf{y}^{i,t} \geq (1+\beta_w)(1+u)\, \mathbf{y}^{i,t-1}$$

$$\sum_{i=1}^{38} \lambda^i \mathbf{b}^{i,t} = (1-\beta_w)(1+v)\, \mathbf{b}^{i,t-1}$$

$$\sum_{i=1}^{38} \lambda^i \mathbf{x}_j^{i,t} \leq (1+\sigma_j)\, \mathbf{x}_j^{i,t-1} \quad (j=1,2,3)$$

$$\beta, \lambda^i \geq 0 \quad (i=1,2,\ldots,38) \tag{9.3}$$

In linear programming (9.3), $\beta = 0$ means that the sub-industry lies on the possibility frontier and its production is efficient; while $\beta > 0$ implies that the sub-industry is inefficient in production. The proportion of the sub-industries with $\beta > 0$ to all sub-industries shows us how widespread the inefficiencies are in the industry we study, which is related to the win-win opportunities offered by environmental regulation. The inequality for goods in (9.3) makes it freely disposable, which means that the goods can be disposed of without the use of any inputs and then without the decrease of bads. The bads is modelled with equality that makes it weakly disposable. The inequality specification of inputs illustrates also that the inputs are strongly disposable; that is, the increase of inputs will not cause the decrease of output. The intensity variable λ^i is the weight assigned to each sub-industry when constructing the production frontier. As shown in linear programming (9.3), the novel definition of the dynamic output and input direction vector not only introduces all kinds of possible energy-saving and emission-abating paths into DDF very well to capture the regulatory behaviour,[4] but also makes it possible to forecast the dynamic impact of energy-saving and emission-abating activity on the economy. Therefore, we abuse the terminology and refer to the extended DDF as the dynamic (environmental regulatory) activity analysis model (DAAM), which distinguishes it from the standard AAM in that it has introduced the lag operator into the direction vector and corresponding DDF. The DAAM constructed here has, to the best of our knowledge, not been addressed before our study.

Directional distance function (strongly disposable bads):

$$\overrightarrow{D}_o^t(\mathbf{x}^{i,t},\mathbf{y}^{i,t},\mathbf{b}^{i,t};\mathbf{y}^{i,t},-\mathbf{b}^{i,t}) = \underset{\lambda,\beta}{Max}\ \beta_s$$

$$s.t. \qquad \sum_{i=1}^{38}\lambda^i\mathbf{y}^{i,t} \geq (1+\beta_s)(1+u)\,\mathbf{y}^{i,t-1}$$

$$\sum_{i=1}^{38}\lambda^i\mathbf{b}^{i,t} \geq (1-\beta_s)(1+v)\,\mathbf{b}^{i,t-1}$$

$$\sum_{i=1}^{38}\lambda^i\mathbf{x}_j^{i,t} \leq (1+\sigma_j)\,\mathbf{x}_j^{i,t-1} \quad (j=1,2,3)$$

$$\beta,\lambda^i \geq 0 \quad (i=1,2,\ldots,38) \tag{9.4}$$

From the mathematical perspective, the equality constraint of undesirable output in linear programming (9.3) is changed into the same inequality constraint as on the desirable output, to reveal the strong disposability of undesirable output in linear programming (9.4). As mentioned above, the difference of the solutions between (9.3) and (9.4) measures the potential production loss due to energy-saving and emission-reducing activity.

The DAAM of DDF (9.3) with the weak disposal assumption of undesirable output models the energy-saving and emission-abating activity under environmental regulation; therefore, it can be used to measure the change of total factor productivity (TFP) and its decomposition, allowing for energy and environment restriction, by calculating the Malmquist–Luenberger Productivity Index (MLPI). The formulae used to calculate this and its decomposed two terms of the change of production efficiency (MLECH) and the change of technical progress (MLTCH); see the formulae (7.4)–(7.7) in Chapter 7.

9.4 Simulation analysis

9.4.1 Simulate the win-win prospect under different energy-saving and emission-abating paths

Table 9.1 reports the potential industrial output growth β_w, loss l and corresponding net value of loss averaged over the entire forecasting period under a total of 45 environmental regulatory paths combined by five energy-saving scenarios and nine emission-reducing scenarios.[5]

Seen from Table 9.1, the potential output growth, output loss and net value of loss caused by energy-saving and emission-reducing exhibit a quite regular pattern in distribution of their values. With the increase of GIOV growth, the magnitude of potential output growth decreases gradually from about 73 per cent in the group with 4 per cent of GIOV growth to a little more than 60 per cent with 10 per cent of GIOV growth. Therefore, the more rapid growth of industry will reduce the widespread extent of production inefficiencies, leading to the shrinking of improvement space for potential output growth. However, the potential output loss brought by energy saving and emission reduction keeps a roughly rising trend (though some values cross among the groups). Thus, the comparison between the potential output growth and loss enables us to see that the net value of output loss increases in fact from the range of [−18.12 per cent, −170.71 per cent] in the group with 4 per cent of GIOV growth to [−36.41 per cent, −185.23 per cent] in the group of 10 per cent GIOV growth. Such evidence implies that the optimal path of energy save and emission reduction must be in the group with a lower growth rate of gross industrial output value.

Table 9.1 classifies the former four moderate abating paths as the gradually abating group and the latter five strong abating paths as the sharp abating group. Obviously, as the abating strength enhances from the first path to the ninth path, the potential output growth varies not much but the output loss increases sharply from [−91.55 per cent, −97.09 per cent] for path 1 to [−244.64 per cent, 248.87 per cent] for path 9. The corresponding increase of net loss indicates that the optimal energy-saving and emission-abating path must be in the gradual abating group. The lowest value of net loss in each GIOV group is marked boldly in Table 9.1. It is easy to pin down that the lowest value among all groups is −17.93 per cent in the group with 4 per cent of GIOV growth, which corresponds to path 3. Considering that economic development is the prior task and so the growth rate of gross industrial output cannot be too slow in China, we finally choose the second path in the group with 6 per cent of GIOV growth, corresponding to the lowest net loss −22.95 per cent, as the best

Table 9.1 Win-win development forecasts corresponding to different energy-saving and emission-abating paths

Dioxide Carbon Abatement Paths			GIOV Growth, 4%			GIOV Growth, 6%		
			βw	l=β w-βf	Net Value	βw	l=β w-βf	Net Value
Gradual Abatement	Path 1	Emission Peak in 2049	73.43	−91.55	−18.12	69.46	−92.42	−22.96
	Path 2	Emission Peak in 2039	73.52	−91.46	−17.94	69.47	−92.43	**−22.95**
	Path 3	Emission Peak in 2029	73.52	−91.45	**−17.93**	69.46	−92.50	−23.04
	Path 4	Emission Peak in 2019	73.41	−91.59	−18.18	69.46	−92.48	−23.02
Sharp Abatement	Path 5	Half of 2009 Emission Growth	74.41	−106.58	−32.17	70.30	−101.82	−31.53
	Path 6	Keeping 2009 Emission Level	74.55	−155.68	−81.14	70.17	−152.12	−81.95
	Path 7	Abating 1% Annually	74.71	−174.94	−100.23	69.92	−176.01	−106.09
	Path 8	Abating 5% Annually	74.81	−201.17	−126.37	69.96	−203.87	−133.91
	Path 9	Abating 10% Annually	73.93	−244.64	−170.71	70.11	−248.87	−178.77

Dioxide Carbon Abatement Paths			GIOV Growth, 8%			GIOV Growth, 10%		
			βw	l=βw-βf	Net Value	βw	l=βw-βf	Net Value
Gradual Abatement	Path 1	Emission Peak in 2049	63.51	−92.95	−29.44	60.68	−97.09	**−36.41**
	Path 2	Emission Peak in 2039	63.62	−92.71	**−29.09**	60.67	−97.10	−36.43
	Path 3	Emission Peak in 2029	63.52	−92.81	−29.29	60.66	−97.35	−36.69
	Path 4	Emission Peak in 2019	63.48	−92.90	−29.42	60.68	−97.24	−36.56
Sharp Abatement	Path 5	Half of 2009 Emission Growth	67.46	−105.09	−37.63	64.38	−111.17	−46.80
	Path 6	Keeping 2009 Emission Level	65.20	−158.39	−93.19	62.49	−158.51	−96.02
	Path 7	Abating 1% Annually	65.59	−178.36	−112.77	62.72	−174.66	−111.94
	Path 8	Abating 5% Annually	65.72	−203.70	−137.98	63.19	−202.69	−139.50
	Path 9	Abating 10% Annually	65.53	−245.96	−180.43	63.17	−248.40	−185.23

energy-saving and emission-abating path for Chinese industry in the next forty years, in which, the matching growth of energy consumption consistent with the target of annual 3 per cent reduction in energy intensity is 2.8 per cent.[6] As stated previously, allowing for China's state condition fully, the chosen best path for sub-industries belongs to the gradual abating group that evenly abates the carbon dioxide emission from their respective emission growing rate in the year of 2009 to zero growth in 2039 (i.e. the peak of emission for all sub-industries) and then evenly to an abating rate of 1 per cent in 2049, the end of our forecasting period. Since all the potential net losses shown in Table 9.1 are negative, it seems that all paths cannot lead to the win-win development suggested by the Porter hypothesis, even though the best energy-saving and emission-abating path is chosen above.

The findings in Table 9.1 are consistent with most other researches. Schaltegger and Synnestvedt (2002) argue that not merely the level of environmental performance, but mainly the kind of environmental management approach with which a certain level is achieved, influences the economic outcome; thus, the economic success resulting from environmental protection finally depends on the chosen kind of regulatory approach rather than the level. It suggests that research and business practice should focus less on general correlations and more on the effect of different environmental management approaches on economic performance; this is consistent with the methodology used in our studies. Roughgarden and Schneider (1999) use the dynamic integrated climate-economy (DICE) model to calculate an optimal control rate or carbon tax and suggest that an efficient policy for slowing global warming would incorporate only a relatively modest amount of abatement of greenhouse gas emissions, via the mechanism of a small carbon tax. Chen *et al.* (2004) find that the earlier the emission reducing policy is implemented, the greater the GDP loss will be. If the start of the emission reductions is the year of 2030, 2020 or 2010 instead of 2040, then the undiscounted total GDP losses in the whole planning horizon would be 0.58–0.74, 1.00–1.32, or 1.10–1.83 times higher. Kuosmanen *et al.* (2009) suggest that if one is only interested in greenhouse gases (GHG) abatement at the lowest economic cost, then equal reduction of GHGs over time is preferred. These researches all support gradual or moderate emission abatement. Similar to the findings of our study, that there is a close relationship between emission reduction and development, Reddy and Assenza (2009) also suggest that the integration of climate policies with those of development priorities that are vitally important for developing countries and stress the need for using sustainable development as a framework for climate change policies.

Of course, the optimal path chosen here means that the combination of a relatively high growth of output and energy consumption may have a relationship with the traditional industry development model that it is hazardous to the win-win development of China's industry. An adjustment on the speed of output growth and a moderate reduction of energy consumption may be more beneficial to structural reconstruction, development model transformation and sustainable development for China's industry in the future.

9.4.2 *The impact of the best energy-saving and emission-abating path on future potential output*

Murty and Kumar (2003) point out that the win–win opportunities from environmental regulation could be found more in some industries and less in others, and studies of specific industries could help us to identify the industries with no such opportunities so that monitoring and enforcement could be directed to those industries in which incentives are absent. As a matter of fact, this is also the reason why we focus on the analysis of China's 38 two-digit sub-industries instead of merely the aggregated industry. Therefore, proceeding along the optimal path of energy saving and emission reduction chosen in the previous subsection, this subsection further simulates the potential output growth and loss for all sub-industries in the following 40 years. Figure 9.3 illustrates the forecasting prospects for each sub-industry.[7]

Table 9.1 shows that the average net losses brought by different regulatory paths are all negative, even for the best energy-saving and emission-abating path. However, if we analyse the individual sub-industry at different time points rather the aggregated industry, this will show totally a different story. The dashed line in Figure 9.3. represents the absolute value of potential output loss caused by energy saving and emission reduction and the real line represents potential output growth. What is found in this Figure is that the potential output loss exhibits a declining trend for all sub-industries and the potential output growth of most sub-industries does not change much. Except for six sub-industries (such as ferrous ores mining, apparel manufacturing, leather manufacturing, cultural articles manufacturing, plastic manufacturing and gas production and supply), the potential loss for all the other sub-industries decreases continuously and then appears to be smaller than potential output growth at some time point before 2049. This indicates that for most sub-industries, the energy-saving and emission-abating activity can bring the win-win development opportunity that the potential output growth exceeds the potential output loss. Even for the above exceptional six sub-industries, their potential output losses tend to decline, too, and are bound to be lower than the potential growth at certain time after the year of 2049, leading to the expectation of win-win development.

The reason why the average net values of potential loss for all paths, even the best one, are minus is that all the sub-industries (except medicine manufacturing, and communication equipment and computer manufacturing) have a large potential loss in the near future. It is thus clear that the aggregation analysis is unreliable and even leads you to the opposite conclusion. Specifically, the potential output loss of those energy and emission intensive sub-industries (such as petroleum extraction, ferrous ores mining, wood exploiting, and gas production and supply) are particularly large, which should be one of the causes of the negative weighted potential loss for aggregated industry. Moreover, what we feel matters about the energy saving and emission reduction is its final influential level instead of the accumulative effect; hence, the high potential loss in the nearer future is just meaningful for that period and useless for the analysis on the future opportunity

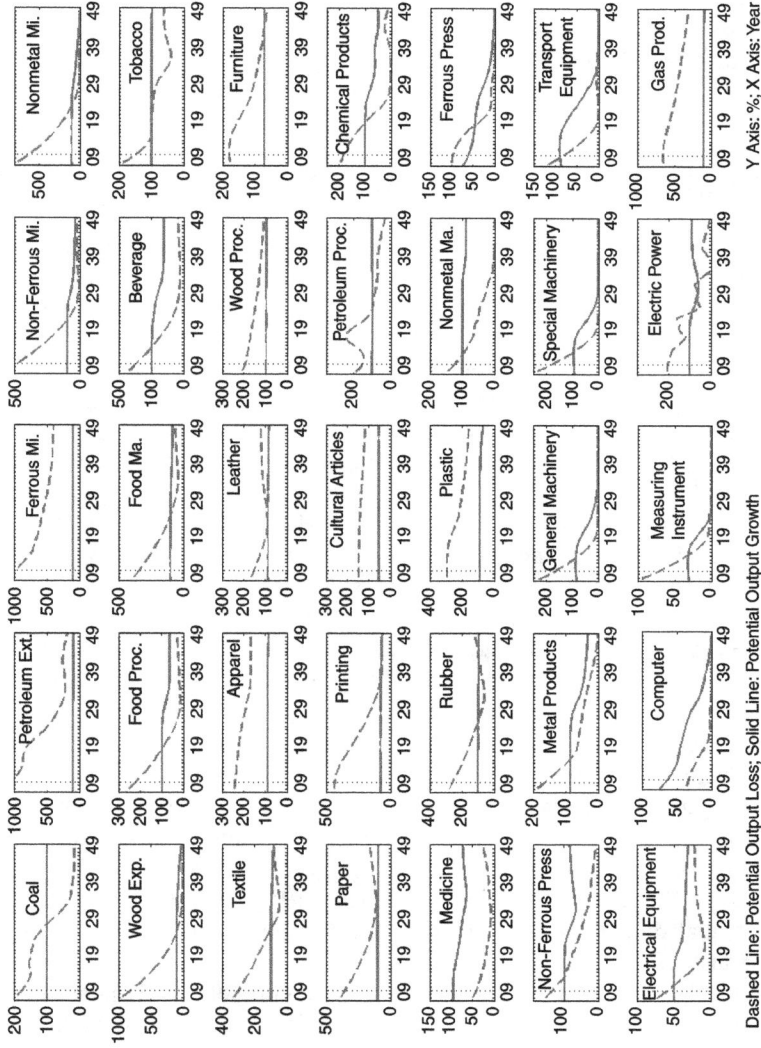

Figure 9.3 Sub-industrial win-win development forecasts under best energy-saving and emission-abating path (2009–2049).

Dashed Line: Potential Output Loss; Solid Line: Potential Output Growth

Y Axis: %; X Axis: Year

of win-win development. As mentioned above, the medicine manufacturing, and communication equipment and computer manufacturing are the only two sub-industries maintaining the win-win development over the whole predicting time span. As for general machinery manufacturing, special machinery manufacturing, transport equipment manufacturing and measuring instrument and machinery, they realize efficient production around the year of 2029, which means they are on the production frontier and have no space to improve the potential output growth, whether there is energy saving and emission reduction or not. All in all, the sub-industrial simulation results shown in Figure 9.3 manifests that, from the perspective of potential output, energy-saving and emission-abating can bring costs on output which means that the Porter hypothesis will not be satisfied in the very near future, but when time moves on, it will lead to the win-win development possibility for China's industry, finally supporting the Porter hypothesis.

According to the theory in Chenery *et al.* (1986) and current empirical work in Chen *et al.* (2011), the standard perception of industrialization is a general shift in relative importance from light to heavy industry. Light industry is of great importance normally at the early stage of industrialization and labour-intensive in nature with relatively low ratios of capital to labour; while heavy industry is at the middle or late stage and capital-intensive, with relatively high ratios of capital to labour. Therefore, we divide all sub-industries into light and heavy industrial groups according to the ranking of capital to labour ratio (K/L) in 2008. That is, the light industrial group corresponds to the top half of sub-industries with the lower K/L ratio, and the heavy industry to the last half of sub-industries with the larger K/L ratio. We refer to them as light industry and heavy industry in brief from now on in this chapter. Figure 9.4 depicts the weighted average potential output loss (bar with light colour) and output growth (deep colour) for light and heavy industry (panel a and b) and aggregated industry (panel c) corresponding to the best environmental regulatory path, in which the sub-industrial weight is its respective share of gross industrial output value.

Seen from Figure 9.4, in light industry, the averaged potential loss declines prominently from −173.95 per cent in 2009 to −46.85 per cent in 2049, while the potential output growth decreases less evidently from 80.51 per cent in 2009 to 46.86 per cent in 2049; in heavy industry, the corresponding varying range of averaged potential output loss, [−67.95 per cent, −16.32 per cent], and growth, [50.62 per cent, 27.75 per cent], is much less than that in light industry. Apparently, both the potential output loss and growth for light industry are high, whereas for heavy industries they are the opposite. The light industry does not reach a comparable level for potential output loss and growth until 2024 and maintains a similar situation to 2049, just meeting the win-win development condition. But for the heavy industry, the win-win situation is reached from the earlier year of 2014 and the potential output growth holds a large advantage over the loss. Therefore, heavy sub-industries are the beneficiaries of energy saving and emission reduction, but light sub-industries are also not the losers. For the aggregated industry, the potential output loss declines from −108.72 per cent in 2009 to −25.55 per cent in 2049, the potential output growth decreases from 62.11

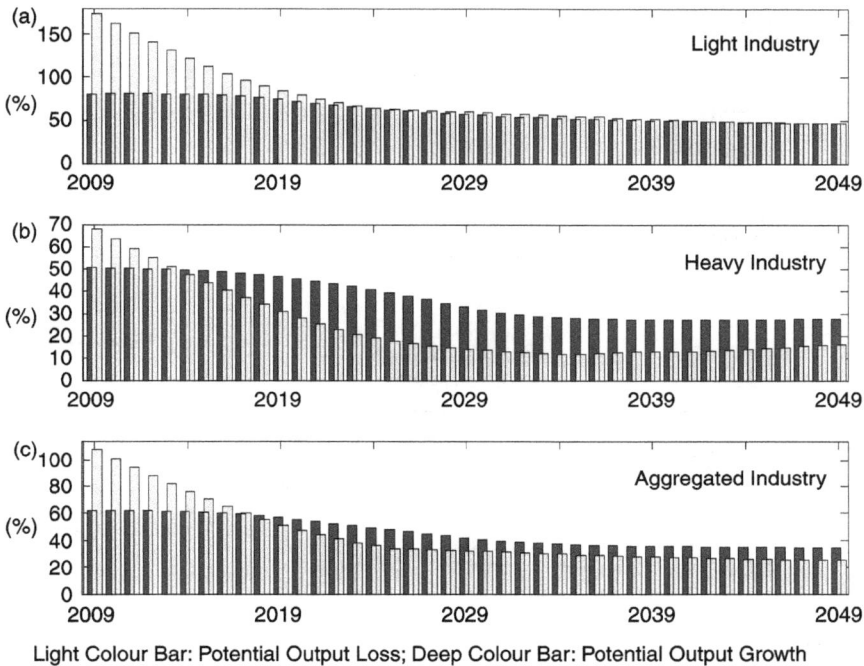

(a)

Light Industry

(b)

Heavy Industry

(c)

Aggregated Industry

Light Colour Bar: Potential Output Loss; Deep Colour Bar: Potential Output Growth

Figure 9.4 Averaged industrial win-win development forecasts under best energy-saving and emission-abating path (2009–2049).

per cent in 2009 to 34.97 per cent at the end of the forecasting period, being between that of light and heavy industry. Since heavy sub-industries have the higher weights, the varying pattern of the potential output for aggregated industry is dominated by and more similar to that of heavy industry – realizing win-win development in the year of 2018 with a clear advantage.

9.4.3 The effect of the best energy-saving and emission-abating path on future industrial productivity

Sickles and Streitwieser (1998) investigated the impact of the regulatory environment, such as the partial and gradual decontrol of natural gas prices, on both output change, technology and productivity in the interstate natural gas pipeline industry. Following this, this subsection also addresses the impact of optimal energy-saving and emission-abating policy on the foreseeable change of productivity, technical progress and efficiency, in addition to potential output, in Chinese industry. Adopting the same group classification and weights as in Figure 9.4, Figure 9.5 exhibits the averaged changing trends of total factor productivity (i.e. the Malmquist–Luenberger Productivity Index, MLPI) and its decompositions, MLECH and MLTCH, under

(a) Light Industry

(b) Heavy Industry

(c) Aggregated Industry

——— MLPI ——○—— MLECH ——□—— MLTCH

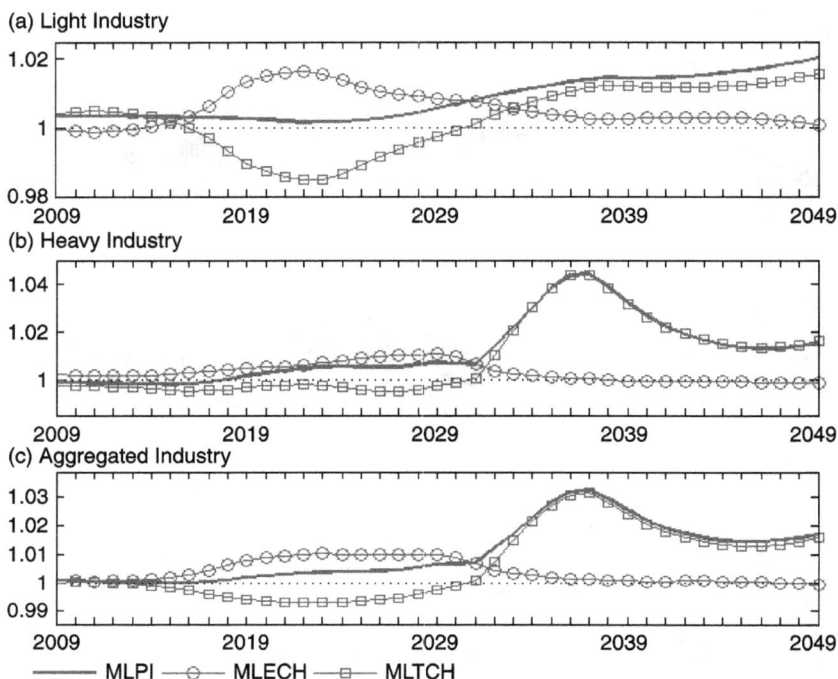

Figure 9.5 Averaged productivity forecasts and its decomposition under best energy-saving and emission-abating path (2009–2049).

the best path of energy saving and emission reduction for light and heavy industry (panel a and b) and aggregated industry, respectively. Three sub-figures show a similar pattern. That is, China's industrial TFP before the years of 2032 or 2033 is mainly influenced by production efficiency in which the catching-up effect of adoption of the frontier technologies due to the environmental regulation is very obvious. When production efficiency attaches its utmost limits and the catching-up energy is almost released, the technical progress begins to serve as the major propelling force through gradual accumulation and assimilation; that is, the change of TFP after the year of 2033 is mainly affected by technical progress. Therefore, as illustrated in Figure 9.5, the behaviour of the overall TFP index shows that industrial development has generally shifted in a win-win fashion.

More specifically, at the early stage, energy-saving and emission-abating policy mainly negatively affects industrial technical progress, more on light industry than on heavy industry. For instance, for light industry, the level of technical progress in 2023 is just 98.48 per cent of 2022, attaining the largest backward magnitude of the production frontier, −1.52 per cent, over the whole forecasting period; the largest backward extent of technical progress for heavy industry is −0.5 per cent in 2026 and the largest one for the aggregated industry is

−0.74 per cent in 2024. However, due to the obvious catching-up effect and improved production efficiency (at the peak of efficiency improvement, the index of MLECH of light industry in 2022 is 1.016 times of the previous year; that of heavy industry in 2029 is 1.011 times of that in 2028; that of the aggregated industry in 2023 is 1.01 times of the previous year), the TFP growth will keep an increasing trend from the earlier forecasting phase. After 2031, the negative effect of environmental regulation on technical progress fades gradually and begins to be positive; three or four years later, the catching-up effect disappears, especially in heavy industry; in around 2037, technical progress reaches its peak due to the long-term introduction, absorption, adoption and innovation of the advanced technologies – in particular in energy and emission intensive heavy industry, the technical progress in 2036 reaching 1.044 times of 2035 After that, technical progress and productivity for light industry keep stable with a slight increase, and those for heavy industry and the aggregated industry, dominated by heavy industry, will drop first and then rise more steadily due to the enhancement of abating strength after the peak of carbon dioxide emission in 2039. In a word, from the perspective of productivity, energy-saving and emission-abating activity has a negative impact on industrial technical progress at the earlier stage, but a positive effect on production efficiency and combined total factor productivity. During the entire forecasting period from 2009 to 2049, the TFP grows steadily with the annually averaged growth rate of 0.81 per cent for light industry, 1.11 per cent for heavy industry and 1 per cent for the aggregated industry. In the year of 2049, the growth rate for light, heavy and aggregated industry will reach 2.02 per cent, 1.54 per cent and 1.72 per cent, respectively, in terms of our forecasting. This is a win-win development prospect since productivity has grown and the targets of energy saving and emission reduction are also achieved.

9.5 Summary and comments

To challenge global warming and boost the development model transformation, energy-saving and emission-abating and developing the low carbon economy have become the necessary approach for all countries to achieve sustainable economic development. However, the energy saving and environment protection will seize important materials originally planned for normal production, causing the decline of desirable output and competitiveness, especially in the recovery period from the financial crisis. These conflicting views are also reflected in the academic area, that is, differences in whether researchers are in favour or against the Porter hypothesis. This chapter makes use of the directional distance function that precisely embodies the spirit of the Porter hypothesis – that the goods increase and bads decrease simultaneously – and proposes a novel dynamic activity analysis model (DAAM) to forecast win-win development possibilities for Chinese sub-industries between 2009 and 2049 and to investigate the existence of the Porter hypothesis in China.

From the perspective of potential output, the empirical results show that energy saving and emission reduction can cause bigger potential output loss in an early stage; but in long run, the loss will decline gradually and become lower than potential output growth eventually, achieving the win-win development prospect stated in the Porter hypothesis. Of course, compared with light industry, there exists a bigger win-win opportunity resulting from environmental regulation in heavy industry and aggregated industry. For example, the potential output loss and growth for aggregated industry in 2049 reaches −25.55 per cent and 34.97 per cent, respectively, finally leading to 9.42 per cent of net growth of potential output. From the viewpoint of productivity, the prediction analysis shows that energy-saving and emission-reducing policy will have a larger negative impact on industrial technical progress at an early stage, especially for light industry; however, due to the obvious catching-up effect and increasing production efficiency in the first half forecasting period and rising technical progress dominating in the second half period, the industrial TFP is not negatively influenced and always maintains a steadily but gradually increasing trend. During the whole forecasting period from 2009 to 2049, the annually averaged growth rate of productivity is 0.81 per cent for light industry, 1.11 per cent for heavy industry and 1 per cent for the aggregated industry. Overall, although energy-saving and emission-abating regulation will cause certain loss at an early stage, in the long run, it will not only reach the target of improving environment quality but also increase output and productivity, finally leading to win-win development in the next 40 years. Our forecasting analysis in this chapter favours the Porter hypothesis.

10 Double dividend forecasting and environmental taxation reform

Carbon tax case

10.1 Introduction

To transform the traditional development model and challenge global warming, in November 2009 the Chinese state council decided to abate the carbon dioxide (CO_2) emission per unit of GDP by 40 per cent–45 per cent until 2020 when compared with the benchmark level in 2005. Though it is only the relative abatement of CO_2 intensity, rather the absolute reduction of total discharge volume of CO_2 as used by most other countries, it is still challenging for China to realize this, due to its coal oriented energy consumption structure and extensive development model driven by high energy consumption, heavy pollution emission and other factors. Now, the natural question is how to reform the traditional environment regulatory policy in order to realize the promised carbon abating goal successfully?

Traditional environmental policy is normally implemented through administrative fiats in China. However, the economists have long argued that environmental policy must be based more firmly on the use of market-based mechanisms so as to introduce the cost of pollution clearly into economic analysis and impose ceaseless price pressure on the polluters to save energy and reduce pollutions (Bailey, 2002). Environmental tax and tradeable emission right permits are the main representatives of market-based environmental policy, based on Pigouvian Tax and Coase Theory, respectively. It's somewhat difficult to trade the emission rights, due to the big cost of protection and definition of property rights; therefore, environmental taxation has become the important instrument for many countries to protect the environment, including in China.

Environmental taxes levied in advanced countries include energy tax, carbon tax, sulphur tax, water pollution tax, solid waste tax and noise tax, and these have already played an important role in promoting sustainable development, which provides positive evidence for environmental tax reform for China. In fact, so far as we have been able to identify, environmental measures are implemented in China mainly by collecting pollution fees, rather than by taxation. Such few taxes are scattered around resources tax, consumption tax, value-added tax, vehicle and vessel tax and others, and these are not the precisely defined environmental taxation (Andrews-Speed, 2009). For example, pollution charges have been collected from 1982 in China and reach an annual amount of 20 billion RMB currently, but these are just the actual cost of dealing with the pollution without inclusion of external environmental cost. The resource tax already levied in China is only to adjust the resource differential income, not much correlated with environmental protection.

The situations described above show the urgency of environmental taxation reform in China. Of course, this is not to say that such environment regulatory policy as is carried out in China is anything but effective. The country has achieved a sustained decline of energy intensity in the period 1980–2001, in which the decline of energy intensity achieved the largest drop between 1997 and 2001, corresponding to ownership rights reform, and this caused the first reduction of total energy consumption; but this trend was reversed from 2002. Exemplified by the change of CO_2 emission reported in Table 7.5 of Chapter 7, relative to the positive growth for almost all the industrial sectors between 1981–1995, there are 32 sectors among all the 38 samples that decreased CO_2 emission by 5–73 per cent at the period of the ninth Five-Year Plan (1996–2000). Over the same period, the averaged output growth attained an annual level of 12.7 per cent, much greater than the 7.6 per cent averaged over the period 1981–1995. The substantial reduction of CO_2 makes it possible to estimate the MAC of CO_2 in China. Table 7.5 also reveals that the number of sectors with decreasing CO_2 emission fell to only nine during the period of the tenth Five-Year Plan (2001–2005) and five in 2006–2008. The rapid urbanization and update of the consumption structure, driven by the vigourous expansion of the housing and car industries, resulted in the reappearance of heavy industrialization in China from the start of the new millennium. In 2007, China became the largest emitter of CO_2 in absolute terms in the world, which meant that China faced continuous and increasing pressure from the rest of the world to abate carbon emission. There thus exists the possibility that the developed countries will impose carbon tariffs on imports from those countries without mandatory carbon abatement, such as China. In this case, as an example of environmental taxation, the levy of carbon tax is more urgent than other kinds of environmental taxation in China and could be appropriately regarded as the first step of environmental taxation reform. Though carbon tax has been levied in many counties, including Finland, Sweden, Norway, the Netherlands and Denmark, and its theoretical foundation is solid enough, it's still necessary to analyse its economic and ecological effect for the foreseeable future in China, and this is particularly useful for environmental policy makers. This background motivates the study in this chapter.

This chapter concentrates on industry in China because its output, energy use and carbon emission account for most of the state level. To analyse the potential impact of environmental taxation, the first problem is how to set the appropriate tax rate. As Bovenberg and Goulder (2002) and Zhang and Baranzini (2004) denoted, the optimal environmental tax rate should equal the marginal abatement cost (MAC) or shadow price of the environmental pollution. However, due to the lack of market pricing of pollution emission, the measurement of the abating cost and pollution price has been one of the greatest challenges in environmental economics. Following Boyd *et al.* (2002), this chapter will utilize two versions of the Directional Distance Function (DDF) to estimate the sectoral MAC of CO_2 emission over the entire reform period, and use the measured MAC as the pricing basis of industrial carbon tax rate to further evaluate the influence of carbon tax on the economy and ecology. The economic and ecological variables exemplified in this chapter are industrial value-added output and CO_2 emission intensity, respectively.

The structure of the rest of this chapter is arranged as below. Section 10.2 will survey the relevant literature. Section 9.3 describes the two stage analytical framework employed in this chapter, that is, two alternative DDFs used to measure the MAC of CO_2 at the first stage and Polynomial Dynamic Panel Model (PDPM) to evaluate the foreseeable effect of carbon tax on economy and ecology at the second stage. Section 9.4 and 9.5 discuss the varying pattern of measured MAC and the influence of carbon tax on target variables. For the concluding remarks, see Section 9.6.

10.2 Review of literature

Environmental tax, also referred to as an ecological or green tax, was first proposed by the British economist, Pigou, in his book *The Economics of Welfare* published in 1920. The tax on the negative externality is termed the Pigouvian tax, and that should equal the marginal environmental damage (MED). Without the environmental tax or government intervention, the marginal social cost (MSC) of the emitters is higher than their marginal private cost (MPC); the levy of environmental taxation may remove the difference between two types of cost so as to accurately reflect the social cost of production and internalize the cost into the market price. Therefore, the environmental tax rate should equal the difference between MSC and MPC, that is, the social cost due to per unit pollution abatement to improve environmental quality (also named the marginal abatement cost, MAC). At the equilibrium point, it equals the marginal social revenue (MSR) resulting from per unit abatement of environmental pollution. Figure 10.1 depicts the principle of the optimal environmental taxation. As shown in this figure, the difference between MSC and MPC is equal to MED due to the emission and MSR represents the marginal revenue (demand) curve. Therefore, the area of A equals the levied environmental tax, A+B represent the environmental improvement and B is the welfare gains; see Bovenberg and Goulder (2002) for the details.

Bovenberg and Goulder (1996) further point out that, in reality, the government is normally dependent on the distorted tax system to raise the financial revenue and budget. Under this second-best case, when setting the environmental tax, it is not necessary only to consider the mechanism of environmental taxation itself but also one should pay attention to the impact of other taxes, which can therefore lead to a smaller environmental tax rate than the Pigouvian Tax stated above.[1] They evaluate the optimal carbon tax rate in this case. In summary, under the second-best scenario of taxation distortion, the revenue-neutral environmental taxation reforms may cause both the negative tax-interaction effect (e.g. increasing

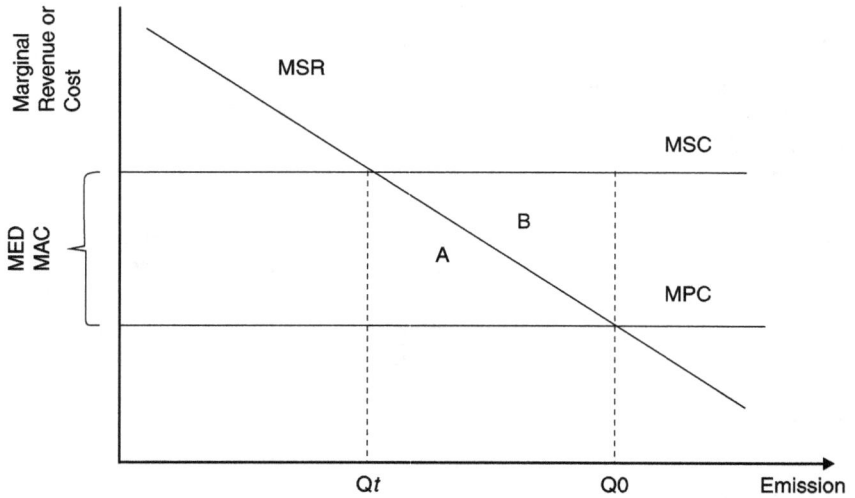

Figure 10.1 Principle of optimal environmental taxation.

the output price and decreasing the factor return) and the positive revenue-recycling effect (say, compensating the revenue loss); the absolute value of the former is often greater than the latter, leading to the relatively low efficiency of environmental tax when raising the tax revenue, leading to a reduced support for it to be levied, even though it is efficient in improving environmental quality. However, the pre-existing distorted tax system is of course inefficient, and that leaves the space for environmental taxation reform to adjust the tax distortion and remove the inefficiency. This naturally gives rise to heated debate on environmental taxation reform and its positive or negative effect in the field of international environmental policy (Bosquet, 2000; Patuelli *et al.*, 2005).

The discussion on environmental taxation reform was originally started by Tullock (1967), who suggested that environmental taxes must be levied in order to ensure optimal utilization of natural resources. The concept has been pushed further by Terkla (1984) and Lee and Misiolek (1986). Patuelli *et al.* (2005) analyse a large set of applied studies and offered a quantitative comparative study of the estimated performance of environmental tax policies based on meta-analytical principles. Though environmental tax may damage the economy, economic theory suggests that environmental taxation reform might in fact bring about a double dividend (DD), namely, the joint occurrence of a cleaner environment (the first dividend) and an economic improvement (the second dividend) under certain conditions (Pearce, 1991; Bovenberg and De Mooij, 1994). There is no standard definition of DD in the literature. For instance, the economic dividend is defined as the growth of employment or productivity by Carraro *et al.* (1996) and Jansen and Klaassen (2000); the effect of income distribution due to the change of salary or expenditure by Barker (1997) and Ekins and Speck (2000); the fiscal benefits by Morris *et al.* (1999); the economic growth represented by GDP and consumption by Garbaccio *et al.* (1999a); and the increase of output and economic welfare by Jorgensen and Wilcoxen (1993) and Peretto (2009).

Specifically, Bossier and Bréchet (1995) and Baranzini *et al.* (2000) describe the economic effect of environmental tax as the rise of currency inflation. Bovenberg and De Mooij (1997) explore how an environmental taxation reform impacts pollution, economic growth and welfare in an endogenous growth model with pre-existing tax distortions. They find that a shift in the tax mix away from output taxes towards pollution taxes may raise economic growth through two channels: an environmental production externality and a shift in the tax burden away from the net return on investment towards profits. Cremer *et al.* (2003) find that setting the environmental tax at its Pigouvian level benefits the high-income group at the expense of the low-income groups, and nonlinear taxation of polluting goods is a powerful redistributive mechanism. By using the computable general equilibrium (CGE) model, Van Heerden *et al.* (2006) find that ecological tax reform could lead to triple dividend of reduced emission, increased output and shrinking income inequality. Nakabayashi (2010) examines optimal tax rules and public sector efficiency, integrating them in a second-best world with pollution by using an overlapping generations model. He argues that environmental taxes going beyond Pigouvian ones may be welfare-improving if the dynamic efficiency of capital accumulation per unit labour (DECAL) is improved by the environmental tax. Even though based on the same definition of DD, however, different influential factors are likely to cause inconsistent conclusions in the applied literature.

Many papers have also discussed the relationships between carbon tax and the economy. Brännlund and Nordström (2004) argue that the overall welfare effects due to the carbon tax in Sweden are negative, where a carbon tax has existed since 1991. Zhang and Baranzini (2004) assess the main economic impacts of carbon taxes. Based on a review of empirical studies on existing carbon/energy taxes, it is concluded that their competitive losses and distributive impacts are generally not significant and definitely less than often perceived. Floros and Vlachou (2005) evaluate the impact of a carbon tax on energy-related CO_2 emissions in two-digit manufacturing sectors of Greece, based on the two-stage translog cost function. A carbon tax of US$50 per ton of carbon results in a considerable reduction in direct and indirect CO_2 emissions from their 1998 level, which implies that a carbon tax on Greek manufacturing is an environmentally effective policy for mitigating global warming, although a costly one. Scrimgeour *et al.* (2005) use the CGE model to assess the effect of environmental tax, in particular carbon and energy taxes, on the economy in New Zealand and conclude that carbon tax reduces carbon emissions more effectively than energy tax but adversely affects GDP. Wier *et al.* (2005) find that the carbon tax on income distribution in Denmark is negative, especially for the weak groups in the countryside. Based on CGE model, Fisher-Vanden and Ho (2007) find that imposing a carbon tax on the price of energy will result in two broad impacts on the economy: factor substitution (causing firms to substitute away from energy towards other factors of production) and output substitution (causing consumers to shift consumption towards less energy-intensive goods). Wissema and Dellink (2007) use the CGE model to quantify the impact of the implementation of carbon tax on the reduction of CO_2 in Ireland. They confirm that the reduction target for CO_2 emissions in Ireland of 25.8 per cent compared to 1998 levels can be achieved with a carbon tax of 10–15 euros per ton of CO_2. Callan *et al.* (2009) analyse the income distribution effect of carbon tax in Ireland and conclude that a carbon tax is regressive in the sense that, in absolute terms, a carbon tax of 20 euros per ton of CO_2 would cost the poorest households less than 3 euros each week and the richest households more than 4 euros each week. Kuosmanen *et al.* (2009) also employ the two-stage analytic method in which the shadow price of pollution emission is firstly estimated and then environmental cost-benefit analysis (ECBA) is utilized to investigate the influence of different emission-abating policies on the economy. For other research, see Nakata and Lamont (2001), Bruvoll and Larsen (2004), Kahn and Franceschi (2006), Voorspools and D'haeseleer (2006), Lee *et al.* (2007), Kerkhof *et al.* (2008), to name a few.

There also exists literature reporting studies of carbon taxes in China. For example, He *et al.* (2002) analyse the influence of carbon tax on the Chinese economy by using the input-output table in 1997 and the CGE model. Their results show that the levy of carbon tax influences GDP far less, and will reduce coal production and energy consumption and lead to the increase of coal and petroleum prices. By checking the impact of carbon tax under three scenarios also based on the CGE model, Wei and Glomsrod (2002) find that carbon tax will worsen economic growth but reduce CO_2 emission in China. Liang *et al.* (2007) simulate the impact of different carbon tax designs on macro-economy and energy-intensive sectors in China. Wang *et al.* (2009c) argue that a low rate carbon tax is a feasible option in China's near future. Lower carbon tax rates have a lower influence on economic development in China, but can lead to obvious CO_2 emission reduction.

Seen from these reviews, the most used methodology to study the influence of environmental tax on the economy and ecology is the CGE model. If you only have the input-output table at several discontinuous time points, the CGE model is undoubtedly the appropriate approach to deal with the limited data; if there exists the panel data, like the data in this chapter for the 38 sectors over continuous years from 1980 to 2008, the panel data model will be the more appropriate approach than the CGE model when forecasting the influence of carbon tax on the target variables. In addition, the CGE model only can analyse the influence of carbon tax under several fixed scenarios, which is also inferior to the DDF utilized in this chapter which makes it possible to forecast the time-varying and heterogeneous carbon tax rate. Therefore, the two-stage analytical framework, to be introduced in Section 10.3, will be adopted in this study. In brief, at the first stage, the sectoral MAC of CO_2 emission over the entire reform period is estimated by using two versions of DDF, which will be used to forecast the carbon tax rate likely to be levied in the foreseeable future; at the second stage, the Polynomial Dynamic Panel Model (PDPM) is employed to fit the historical relations between MAC and industrial value-added or

CO_2 emission intensity, based on the fit, to further evaluate the influence of carbon tax levy on the economic and ecological variables.

10.3 Methodology: two stage analytical framework

10.3.1 Directional distance function and marginal abatement cost

Not until the introduction of the directional distance function (DDF) do we find a reasonable framework to differentiate the desirable and undesirable outputs. Färe and Groffkopf (2000) proved that the DDF is dual to the profit function and provides the basis for computing the shadow prices, or MAC, of outputs. At the first stage, mainly following Boyd *et al.* (2002), two versions of DDF will be used to measure the MAC of undesirable CO_2 emission. See Figure 10.2 for the principle to measure MAC using two types of DDF.

As specified previously, also assume that there are n decision-making units (DMUs) at t time point, and there are k types of input, l types of desirable output, and m undesirable output for each DMU. For the ith DMU ($i = 1,2,...,n$), the column vectors x_i, y_i and b_i represent the inputs, desirable and undesirable outputs, respectively. And $X_{k \times n}$, $Y_{l \times n}$ and $B_{m \times n}$ are the input and output matrix for all the n DMUs. In this study, the DMU is also industrial sector; for each of them, $k = 3$, corresponding to capital, labour and energy, $l = 1$ being gross industrial output value (GIOV), and $m = 1$ carbon dioxide emission (CO_2). The sectoral input and output data sets between 1980 and 2008 come from Chen (2011). To make it possible to model the increase of desirable output and reduction of undesirable output simultaneously, Chambers *et al.* (1996a) and Chung *et al.* (1997) firstly put forward the weak disposability assumption of undesirable output based standard DDF (from the point A to B in Figure 10.2) to replace the traditional Shephard

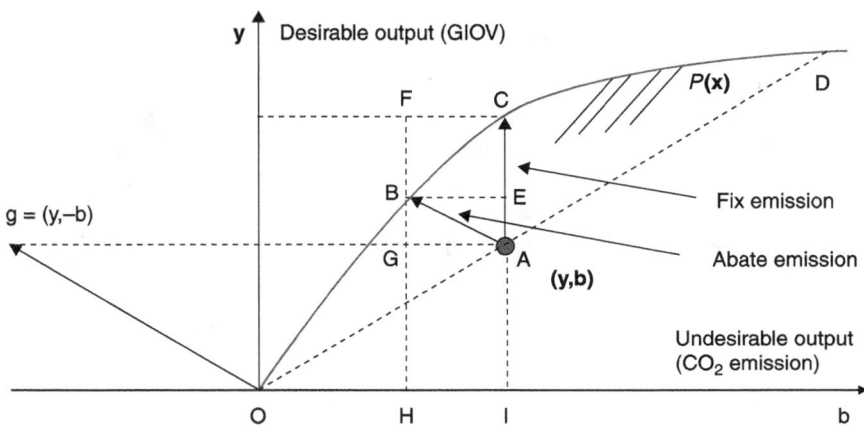

Figure 10.2 Principle to measure MAC using two types of directional distance function.

distance function (SDF, radially scales the original vector from A to D to describe the simultaneous increase of desirable and undesirable outputs) by Färe *et al.* (1994) and Boyd and McClelland (1999). The standard DDF can be estimated by resolving the the linear programming, specified already in formula of (7.3) in Chapter 7.

To measure the MAC of undesirable output like CO_2 emission, Boyd *et al.* (2002) and Jeon and Sickles (2004) define another DDF, or the variation of standard DDF, (from the point A vertically to C in Figure 10.2), which tell us the degree to which desirable output can be expanded, given inputs and undesirable output levels. It can also be referred to as fix emission based DDF, as shown in Figure 10.2, which is computed by resolving the following linear programming:

$$\overline{D}_o^t(\mathbf{x}_i^t, \mathbf{y}_i^t, \mathbf{b}_i^t; \mathbf{y}_i^t, 0) = Max_{\lambda, \beta} \; \beta$$

$$s.t. \;\; \mathbf{Y}\lambda \geq (1+\beta)\mathbf{y}_i; \mathbf{B}\lambda = \mathbf{b}_i; \mathbf{X}\lambda \leq \mathbf{x}_i; \lambda \geq 0 \tag{10.1}$$

The DDF modelled by LP (10.1) is consistent with what is proposed in the Kyoto Protocol which controls the CO_2 emission at the 1990 level, and also for the case of carbon emission quota. The comparison of two alternative DDFs can estimate the MAC of CO_2 emission. Let the value of LP (7.3) and (10.1) be β_1 and β_2, respectively. Then $GB = \beta_1 \mathbf{y}$ and $GA = \beta_1 \mathbf{b}$ are the solution to LP (7.3), while $AC = GF = \beta_2 \mathbf{y}$ corresponds to the fix emission based DDF. The difference labelled $BF = (\beta_2 - \beta_1)\mathbf{y}$ is the additional output that is forgone to reduce the emission by $\beta_1 \mathbf{b}$. Therefore, the ratio of $(\beta_2 - \beta_1)\mathbf{y}$ to $\beta_1 \mathbf{b}$ will tell us how much GIOV is given up for a unit reduction in CO_2 emission, which can be used to measure the MAC of CO_2 emission and approximate the shadow price of CO_2 (unit in this chapter: 10 thousand RMB/ton of CO_2). Since there is no market price for CO_2 emission, measuring MAC of CO_2 emission is important for policy makers, particularly when setting the appropriate carbon tax rate.

10.3.2 Polynomial Dynamic Panel Model and forecasting scheme

At the second stage, the MAC of CO_2 emission measured at the first stage will be used as the proxy of policy variable (PV), i.e., carbon tax. To analyse the influence of environmental tax reform on economy and ecology, we firstly fit the historical relationships between PV and the target variable (TV) and then, based on the estimated model, forecast the future trend of TV. The estimating and forecasting model used in this chapter is the Polynomial Dynamic Panel Model (PDPM) specified as below:[2]

$$\ln TV_{it} = \beta_0 + \beta_t t + \beta_{lag} \ln TV_{i,t-1} + \beta_1 \ln PV_{it} + \cdots + \beta_n \ln^n PV_{it} + u_i + \varepsilon_{it} \quad (n \leq 5) \tag{10.2}$$

in which, TV and PV take the form of natural logarithm. The inclusion of dynamic term, i.e., the lagged dependent variable, $\ln TV\_lag1$, as the independent variable extracts the entire history of the right-hand side variables on which all the results are conditionally based; in this case, PV of carbon tax and its higher-degree terms, especially the issue in this chapter, represent the current shock of new policy information. The notation of u_i controls for the heterogenous characteristics and t captures the time trend of most sectors influenced by some general economy-wide factors such as the common macroeconomic policy and environmental regulations, which represent the mushroom effect and yeast effect of the target variables vividly described in Harberger (1998). The stochastic disturbance ε_{it} is assumed to follow the normal distribution of white noise.

In both the fixed and random effects settings, the difficulty is that the lagged dependent variable is correlate with the disturbance, even if it is assumed that the disturbance is not itself autocorrelated. There also exists possible endogeneity for PV and its higher degree terms because PV is also constructed by the desirable and undesirable outputs using the DDF approach. To control for the possible endogeneity, the method of Hausman and Taylor Instrumental Variable (HT/IV) is chosen to estimate PDPM (10.2). It is possible to estimate the time in varying variable u_i while still maintaining the assumption that the sectoral effect is correlated with the explanatory variables. Another method of System GMM by Arellano and Bover is also employed to estimate PDPM (10.2) for robustness check. According to the reviews in Section 10.2, the target economic and ecological variables chosen in this chapter are industrial value-added and CO_2 emission intensity, the former of which is the preferably concerned growth indicator in the developing countries and the latter is the relative carbon abating indicator formally employed by the Chinese government.

The period in which the target variables are forecasted under the scenarios of carbon tax levy is from 2009 to 2020, which is the span for the coming twelfth and thirteenth Five-Year Plan, and that for the realization of the officially proposed carbon abating goal. To forecast the dependent variable, the explanatory variables of carbon tax rates in the following years are firstly forecasted based on the historical values of MAC estimated over 1980–2008 by adopting the time trend quadratic polynomial model. When forecasting the dependent variable, the one-year-ahead short-term recursive forecasting scheme is employed with an updating sample window. Specifically, the first estimating period is from 1980 to 2008 and the value of TV of 2009 could be forecasted based on the estimated Equation (10.2); the estimating and forecasting process is carried out recursively by updating the sample with one observation each time, the last estimating period is from 1980 to 2019 and the value of TV of 2020 will be forecasted last. Thus, 12 one-period-ahead forecasting values of TV over 2009–2020 are obtained. As a polynomial model, of course, the appropriate degrees of Equation (10.2) must be chosen in advance for different dependent variables to avoid the possible overfitting problem. By following Chen *et al.* (2010), we divide the sample into two parts: training sample and validating sample. Based on the training sample and the short-term recursive forecasting scheme described previously, we could forecast the value of TV corresponding to the validating sample for polynomial models from first to fifth degree, and then calculate five mean absolute forecasting errors (MAE). Here, the first training sample is spanned from 1980 to 2003 and the last is 1980–2007. The validating sample corresponds to 2004–2008 and 38 sectors each year, including a total of 190 observations.

To test for the forecasting accuracy of polynomial models for different degrees, we use the two-sided DM test statistic proposed by Diebold and Mariano (1995) for the difference of MAE loss function. The null hypothesis is H0: $MAE_1-MAE_0=0$, where the subscript 0 denotes the benchmark model and 1 the target model. The DM tests in this study are investigated in a robust form, by simply scaling the numerator by a heteroscedasticity and autocorrelation consistent (HAC) (co)variance matrix calculated according to Newey-West procedures (Newey and West, 1987). We use Andrews' (1991) approximation rule to automatically select the number of lags for HAC matrix. In the case of large sample, the DM statistic converges in distribution to a standard normal. Based on MAE and DM test, the appropriate degrees of polynomials with the best generalization ability and out-of-sample performance can be determined to meet different dependent variables when using Equation (10.2).

10.4 Measurement of marginal abatement cost for industrial carbon dioxide

Figure 10.3 depicts the measured sectoral MAC of CO_2 emission over 1980–2008 and the forecasted carbon tax rate in 2009–2020, in which the dots present the measured MAC, the solid line is the fitted line of MAC, and three dashed lines are the forecasted carbon tax rate based on MAC and its forecasting intervals at the confidence coefficient of 95 per cent. The sub-figure in the lower right corner of Figure 10.3 displays the averaged MAC and carbon tax rate for aggregated industry, light and heavy industry in which the weight is sectoral GIOV share. Similar to Chapter 9, the light and heavy industry is classified according to the sectoral ranking of capital to labour ratio from the lowest to highest in 2004, in which the former 19 sectors with a lower capital to labour ratio are classified into the light industrial group and the second half of sectors with a higher ratio belong to heavy industry.[3]

Seen from Figure 10.3, on average, the appropriate levying range of carbon tax rate forecasted over 2009–2020 for aggregated industry is [0.1868, 0.4583] with the unit of ten thousand RMB per ton of CO_2, which the probability that the forecasting intervals contain is 95 per cent. MAC represents the internal valuation of pollution emission by societies, in which the pollution with more negative externality should be valued more. Thus, CO_2 is expected to have the lowest (by absolute value) MAC, nitrogen oxides (NO_2) to have the highest one and sulphur dioxide (SO_2) to be somewhere inbetween. The averaged shadow price of industrial SO_2 estimated by Tu (2009) over 1999–2005 in China is 8.26 ten thousand RMB per ton of SO_2, which is reasonably larger than that of CO_2 estimated in this chapter. The forecasted MAC of CO_2 in this chapter is similar to that estimated by Gao *et al.* (2004) using MARKAL-MACRO model. They also found that the MAC of CO_2 in China is in fact considerably higher; when abating rate equals 45 per cent, the MAC of CO_2 attains 250 US\$/ton of CO_2. Of course, many

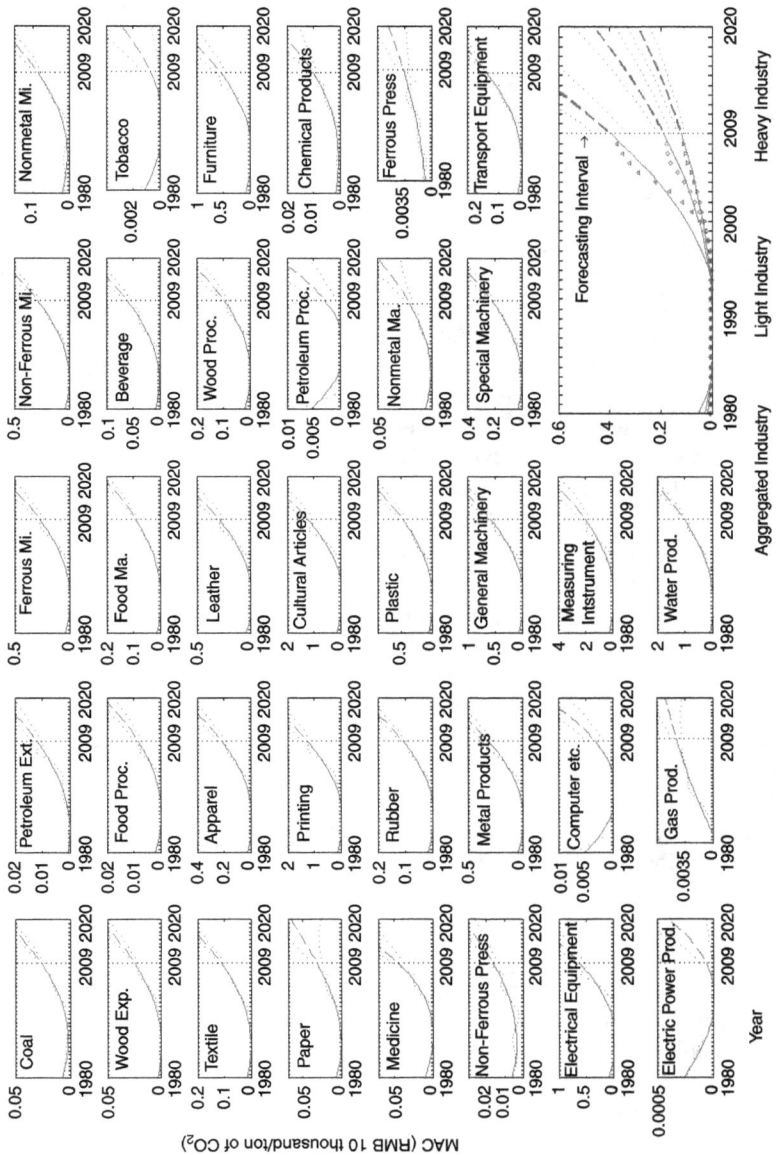

Figure 10.3 Estimates and forecasts of marginal abatement cost for sub-industrial CO_2 emission.

measures in the literature are lower than the measures in this chapter. For example, the estimated carbon tax rates are 39.2–399.3 RMB/ton of CO_2 by He *et al.* (2002), 41.5–83.0 by Wei and Glomsrod (2002), 99–727 by Wang *et al.* (2005b), 20–200 by Wang *et al.* (2009b), and so on. There exists reasons to underestimate the carbon tax. Reddy and Assenza (2009) argue that the integration of climate policies with those of development priorities that are vitally important for developing countries and the need for using sustainable development as a framework for climate change policies should be stressed. Based on this, lower carbon tax obviously favours economic development. The carbon tax levied in developed countries is also not high. For example, it's about 5.5–11.1 euros/ton of CO_2 in Finland and Denmark and 37.9 euros in Sweden. That's also not surprising. Barrett (1994) and Rauscher (1994) denote that all the governments have an incentive to distort the environmental tax downward from the Pigouvian level in order to lower the costs for domestic firms and to shift profits to them, which is sometimes referred to as ecological dumping.

Though low carbon tax has its rationality, the analysis of the rest of this chapter is still based on the real but higher carbon tax forecast in Figure 10.3. This can be attributed to the following points. The forecast carbon tax in this chapter approximates more to the actual cost of abating CO_2 in reality; thus, levying a carbon tax based on this will lead to a better policy effect. Zhang and Baranzini (2004) denote that the carbon tax currently levied in the developed countries is too low to stabilize the CO_2 concentration in the atmosphere; if the carbon tax is the only policy to abate CO_2, its rate should become higher. As opposed to other environmental policies, carbon tax provides the pollution emitters with the economic incentives to change their behaviour through the market mechanism. Thus, to fully reflect the institution value of environmental tax or carbon tax, the carbon tax should be high enough to influence the emitters' behaviour and the additional social cost must suffice to stimulate the emitters' awareness of environmental protection.

Figure 10.3 also provides us with two meaningful conclusions. The first is that the MAC or carbon tax rate increases over time, implying that the levy of carbon tax should increase the tax rates in the future. Zhang and Baranzini (2004) obtain a similar conclusion based on the review of literature. That is, the carbon tax rate should rise over time if it has to reflect the rising costs of damage from the accumulation of CO_2 concentration in the atmosphere; if it has to give the markets the signal that CO_2 emissions will eventually be heavily taxed; and if there are few economically feasible substitutes available. This signal strengthens the incentive for the technical innovation needed to make more stringent future emissions targets affordable. Another conclusion is that the MAC or carbon tax varies considerably across industrial sectors. On average, the MAC or carbon tax of light industry is larger than that of heavy industry. For example, the sectors with the carbon tax rate above 10 thousand RMB/ton of CO_2 in the 2015 mostly belong to the light industry, including Manufacture of Furniture, Printing and Reproduction of Recording Media, Cultural Articles Manufacturing, General Machinery Manufacturing, Electrical Equipment Manufacturing, Measuring Instrument and Machinery; only Production and Supply of Water is in heavy industry. This is consistent with the findings in many studies, too. Hoel (1996) denotes that, though indifferentiated carbon tax should be levied according to standard welfare theory, it has becomes not preferred due to the incomplete international climate agreement and the incentive for some countries to be a free rider. Therefore, all users of carbon should not face the same carbon tax; for instance, carbon intensive tradeable sectors should face a lower carbon tax than other sectors of the economy. Zhang and Baranzini (2004) also point out that there would be significant variation in the size of the carbon taxes among countries and regions, given that the marginal cost of abating CO_2 emissions substantially differs across countries and regions. Lee *et al.* (2007) argue that, since the implementation of a carbon tax is an complex problem that will most certainly not result in blanket reductions of CO_2 emission for all countries, it might well be that it should be implemented on a case-by-case basis, involving at most a few countries from specific regions rather than a one-for-all policy. Wang *et al.* (2005b) also find that the MAC in heavy industry (like electric power, coal, and petrochemical sectors) is lower, indicating that there is much scope to abate CO_2 in heavy industry. Though the carbon tax rate is relatively low in heavy industry, the total amount of carbon tax gathered is still larger than that in light industry due to its huge CO_2 emission.

The estimated MAC serves as a reference value not only for environmental taxes but also for the pricing of the pollution emission permit trade. For example, one party is willing to purchase the emission permit that is lower than its own MAC to emit additional pollution, while the other party shrinks its pollution by selling the permit at the price higher than its own MAC; the trade of permits continues until MAC across countries, regions or sectors equalizes. In a word, as opposed to administrative fiats and emission permit trade, environmental taxes, including carbon tax, are more flexible and endow firms with more choice. The firm can choose to emit and pay taxes or reduce emission and avoid the payment of carbon tax according to its own MAC of carbon emission. Therefore, environmental taxes make it possible for the firms to react to the market signals in an economic way and make a choice between the payment of taxes and the reduction of emission. Technological progress or innovation necessary for the emission reduction will make additional profits; thus, the firms always have the motivation to further increase their ability to reduce emission.

10.5 Forecasting the influence of environmental tax on economy and ecology

Table 10.1 reports the out-of-sample one-period-ahead forecasting accuracy, the value of MAE, of two target variables by polynomials of Equation (10.2) from the first to fifth degree and the corresponding p-values of Diebold-Mariano (DM) test for the MAE difference, which are defined as the significance levels at which the null hypothesis under investigation can be rejected. In calculating the DM statistic, the null hypothesis of equal forecasting ability is related to five benchmark models: linear, quadratic, cubic, quartic and quintic polynomials, referred to as DM1, DM2, DM3, DM4 and DM5, respectively. For instance, DM1 presents the

Table 10.1 Comparison of forecasting error and choice of forecasting model

Target Variables	Model Specification	MAE	DM1	DM2	DM3	DM4	DM5
Industrial Value-Added	Linear Polynomial	171.64		0.9686	0.5766	0.0581	0.0024
	Quadratic Polynomial	137.52	0.0314		0.0497	0.0079	0.0000
	Cubic Polynomial	160.09	0.4234	0.9503		0.0350	0.0125
	Quartic Polynomial	200.01	0.9419	0.9921	0.9650		0.0276
	Quintic Polynomial	360.75	0.9976	1.0000	0.9875	0.9724	
CO_2 Emission Intensity	Linear Polynomial	0.8747		0.0309	0.0499	0.0075	0.0123
	Quadratic Polynomial	1.0174	0.9691		0.8538	0.0114	0.0195
	Cubic Polynomial	0.9546	0.9501	0.1462		0.0104	0.0250
	Quartic Polynomial	2.4126	0.9925	0.9886	0.9896		0.5874
	Quintic Polynomial	1.7514	0.9877	0.9805	0.9750	0.4126	

test results for linear polynomial, where a p-value no greater than 0.05 indicates that linear polynomial yields a higher forecasting error (in terms of absolute error) relative to the competing model at 5 per cent significance level, a p-value no smaller than 0.95 means that linear polynomial produces a lower forecasting error at 5 per cent level, while a p-value between 0.05 and 0.95 implies that the benchmark and competing model have the equivalent forecasting accuracy from the viewpoint of statistics. The same interpretation applies to the p-values reported for DM2-DM5.

According to Table 10.1, for the target variable of industrial value-added, quadratic PDPM produces the lowest value of MAE but quintic one yields the largest MAE. Can we obtain the statistical evidence to support the quadratic PDPM to be the best model when forecasting industrial value-added? DM2 statistic (quadratic polynomial is the benchmark model) shows that the forecasting ability of quadratic polynomial outperforms that of linear and cubic polynomials at 5 per cent significance level, and quartic and quintic ones at 1 per cent level. Other DM tests not only conclude the same thing but also tell you the result compared pairwise between any two models except the quadratic one. For instance, the linear polynomial is better than the quartic polynomial at 10 per cent significance level; linear and cubic polynomials have equal forecasting accuracy. Therefore, quadratic PDPM is chosen as the best model in this study to forecast the influence of carbon tax variable on industrial value-added in the future. In a similar way, linear PDPM is selected as the best model when forecasting the ecological variable of CO_2 emission intensity, which not only has the lowest forecasting error but is also supported statistically.

Table 10.2 presents the regression results of two target variables on MAC of CO_2 based on the best forecasting model selected above. HT/IV estimation, the main method used in this study, shows that all the coefficients are statistically significant at least at the 5 per cent level. As the secondary method, dynamic GMM estimation tells us that its coefficients estimators have the similar absolute

Table 10.2 Effect of MAC on economic and ecological variables: Polynomial Dynamic Panel Model (1980–2008)

lnTV	Industrial Value-Added (100 million RMB)				CO₂ Emission Intensity (tons/10 thousand RMB)			
	HT/IV		Dynamic GMM		HT/IV		Dynamic GMM	
	Coef.	s.e.	Coef.	s.e.	Coef.	s.e.	Coef.	s.e.
Constant	0.1992	0.0548	0.2199	0.0625	0.3297	0.0684	0.2760	0.1574
lnTV_lag1	0.9564	0.0108	0.9440	0.0165	0.9540	0.0103	0.9157	0.0215
t	0.0081	0.0012	0.0087	0.0019	−0.0085	0.0011	−0.0114	0.0021
lnPV	−0.0038	0.0014	−0.0022	0.0012	−0.0087	0.0042	−0.0099	0.0032
lnPV square	0.0011	0.0005	0.0011	0.0005				
Sectoral effect: General F Test	694.16		572.36		367.19		404.68	
Overall significance: *Wald* Test	91027		79000		97171		61124	

Note: The null hypothesis of least square restricted general F test for sectoral effect is $\beta_i = 0$ for all the sectors.

value and sign to those of HT/IV method, and are also significant at least at 10 per cent level. Thus, the results estimated by the specified HT/IV method are robust enough and can be used to forecast the target variables in the future. Wald statistics reveal that four models specified in this study are overall significant. General F tests show that there exists the heterogenous effect across sectors and then the two-digit sectoral panel data analysis in this study, rather aggregation analysis, is very necessary. Seen from Table 10.2, industrial value-added and CO₂ emission intensity have very high and significant values of positive autocorrelation coefficients (0.92–0.96), indicating that the lagged target variables contain the extremely plenty of historical information which will play the most important role in forecasting the target variables. Obviously, the introduction of dynamic explanatory variable is indispensable to the estimation and prediction of the economic and ecological variables in this study; such a conditional forecasting based on all the historical information will have relatively high forecasting accuracy. The coefficients of time trend variables are small but extremely significant – industrial value-added of all the sectors increases while CO₂ intensity decreases over time. Thus, the yeast effect exists in Chinese industry in the sense that some common economy-wide factors tend to affect most sectors at the same time, rather than a limited number of sectors. Of course, the most important issue in this chapter is the influence of PV on TV. As the new information, the coefficients of PV are much less than the slopes of historical information variable (lnTV_lag1) and intercepts but significant too. Due to the application of double-log models here, the elasticity analysis becomes the most convenient. For the model of CO₂ emission intensity, the slope of PV is just the elasticity value because the linear polynomial is utilized. 1 per cent increase of the MAC of CO_2 or carbon tax of PV significantly decrease CO_2 emission intensity by 0.0087 per cent–0.0099 per cent, implying that the levy of carbon tax will play an effective role in realizing the abating goal of carbon intensity. Because the quadratic polynomial is chosen for the target variable of industrial value-added, the elasticity of PV is not straightforward and must be calculated in terms of $\partial(\ln TV_{it})/\partial(\ln PV_{it}) = \beta_1 + 2\beta_2 \ln PV_{it}$. According to the forecasted output elasticity of aggregated industrial carbon tax, at the first stage a 1 per cent rise of carbon tax will decrease the industrial value-added by 0.0034 per cent; then the value of elasticity increases gradually, turns to positive from negative in 2016, and attains 0.00027 per cent in 2020. The signs of estimated coefficients in quadratic polynomial of industrial value-added also tell us that, with the increase of carbon tax rate, industrial value-added will fall firstly and then rise. It is thus evident that, in the short run, the levy of carbon tax will increase energy price, influence the input of energy factors, further increase energy-intensive products' price and decrease their international competitiveness – leading to the negative impact on output; in the long run, the negative influence will disappear and carbon tax levy will promote output growth. However, even if the negative influence of carbon tax on output exists in the short run, its absolute

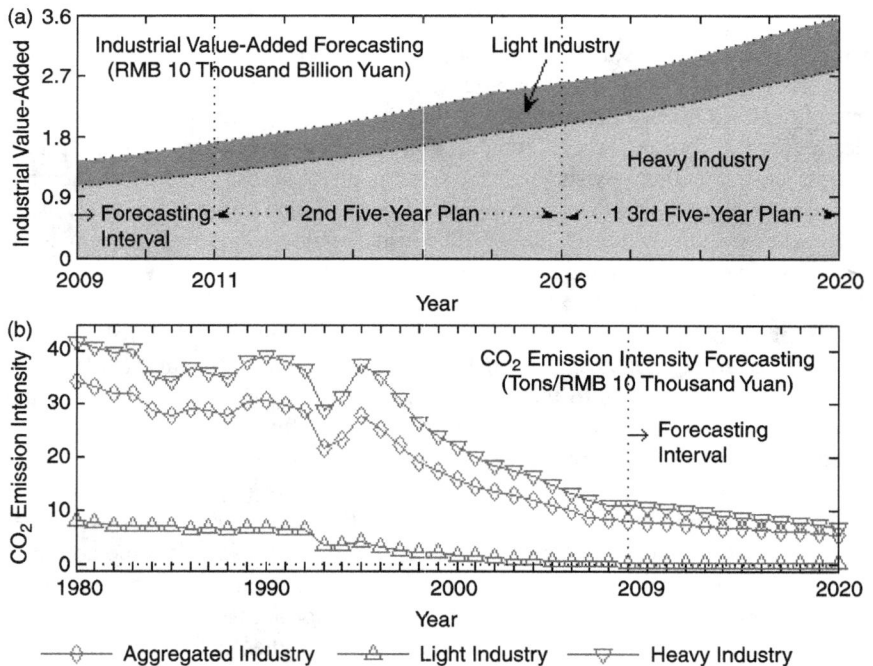

Figure 10.4 One-year-ahead forecasting of value-added and CO_2 intensity under scenario of carbon tax levy.

value is of no consequence. The factor that plays the absolute role in forecasting industrial value-added is its historical information; the very high positive value of autocorrelation coefficient says that, as a whole, industrial value-added will still keep strong growing trend regardless of the levy of carbon tax.

Beginning with the estimated coefficients reported in Table 10.2, the recursive one-year-ahead forecasting values of industrial value-added and CO_2 emission intensity for 38 sectors and over the forecasting interval of 2009–2020 are obtained and depicted in Figure 10.4, in which sectoral values are condensed into light and heavy industrial ones. Specifically, industrial value-added is the absolute value, thus, its area graph for light and heavy industry are drawn in panel (a); while CO_2 emission intensity is the intensity indicator and its weighted averages for light, heavy and aggregated industry are depicted respectively in panel (b) in which the weights are just the actual and forecasted sectoral value-added behind panel (a). Figure 10.5 further illustrates the sectoral CO_2 emission intensity over the estimating interval of 1980–2008 and forecasting interval of 2009–2020, respectively.

Seen from panel (a) in Figure 10.4, even under the scenarios of carbon tax levy, aggregated industrial value-added still increases from 14 thousand billion RMB in 2009 to 35.6 in 2020; its averaged growth rate is 8.6 per cent annually, just a little less than the historical averaged rate of 11.9 per cent over 1980–2008. Though the output elasticity of carbon tax is negative in the beginning and turns to positive only at the end of forecasting period, its influence on output growth is very small; dominated more by its historical data, the growth rate of aggregated industrial value-added attains 9.2 per cent per year in the period of the twelfth Five-Year Plan, greater than the 7.9 per cent forecast in the thirteenth Five-Year Plan. Industrial value-added in heavy industry is higher than that in light industry; in 2009, the forecast heavy industrial value-added is 10.5 thousand billion RMB while light industrial one is just 3.5; in both light and heavy industry, the output growth in the twelfth Five-Year Plan is also higher than that in the thirteenth Five-Year Plan; until 2020, the forecast heavy industrial value-added attains 28 thousand billion RMB while light industrial value-added attains 7.6. As for the sectoral forecasting values of value-added,[4] the light sectors with higher output growth in the sample period still have higher growth during the forecasting period. For example, the averaged growth of value-added for Manufacture of Communication Equipment, Computers and Other Electronic Equipment is 25 per cent, 17.3 per cent, 18.4 per cent and 15.4 per cent corresponding to different period of 1980–2008, 2009–2020, the twelfth and thirteenth Five-Year Plans, respectively, ranked the first among all the sectors in each period. The value-added growth for Manufacture of Transport Equipment is all ranked second in each period, attaining 16.8 per cent, 10 per cent, 11.3 per cent and 7.9 per cent, respectively. Manufacture of Medicines and Manufacture of Electrical Machinery and Equipment are ranked the third and fourth according to their value-added growth. The sectors with lower historical output growth are normally in heavy industry, which still have low even negative growth in the forecasting period. Extraction of Petroleum and Natural Gas, for example, has the slowest value-added growth (1 per cent) in 1980–2008; it still has the largest fall in growth (i.e., −11.4 per cent) in 2009–2020, in which it decreases by 14.2 per cent in the period of the thirteenth Five-Year Plan, bigger than the reduction of 9.8 per cent in the twelfth Five-Year Plan. The output growth of Petroleum Processing is 2.1 per cent in the estimating period, ranked second last among all the sectors, and decreasing by 8.3 per cent annually during the forecasting period, still ranked second last. Therefore, though heavy industrial value-added keeps growing as a whole in the forecasting period, some heavy sectors decrease in output, in addition to above two, including Mining and Processing of Nonmetal Ores, Production and Supply of Electric Power, Heat Power, Gas and Water, Mining and Washing of Coal, Manufacture of Non-metallic Mineral Products, Smelting and Pressing of Ferrous Metals. Though the output of these heavy sectors is influenced more by the levy of carbon tax, according to the output elasticity of carbon tax, their output growth will turn to positive in not long future. As shown in Figure 10.3, the carbon tax rate levied in light industry is higher than that in heavy industry but light industrial output is influenced less by carbon tax due to its low total taxes; heavy industry emits more than 95 per cent of total industrial CO_2, leading to larger total carbon taxes though its tax rate is lower than light industry.

Examine panel (b) of Figure 10.4 again. Corresponding to negative CO_2 intensity elasticity of carbon tax in Table 10.2, the aggregated industrial CO_2 emission intensity decreases with a big fluctuation from 34 tons of CO_2/10 thousand RMB of industrial value-added in 1980 to 27.9 in 1995; it decreases more sharply and more smoothly in 1996–2008, and, in 2008, it falls to only 8.4 tons of CO_2 per 10 thousand RMB; it decreases by 4.9 per cent annually over the entire sample period. During the forecasting period, aggregated industrial CO_2 intensity still keeps the steadily decreasing trend from 8.15 ton of CO_2 per 10 thousand RMB in 2009 to only 5.6 in 2020; it falls by an annual 3.3 per cent, in which it is 3.2 per cent in the twelfth Five-Year Plan and 3.8 per cent in the thirteenth Five-Year Plan. Relative to aggregated industrial CO_2 intensity in 2005 (11.2 tons/10 thousand RMB), under the scenarios of carbon tax levy in this study, it has decreased by 50 per cent in 2020, somewhat greater than 40 per cent–45 per cent, the abating goal of CO_2 intensity officially announced by Chinese central government, indicating that the levy of carbon tax is beneficial to the successful realization of binding abating goal. As opposed to heavy industry, light industry has the much lower value of CO_2 intensity, leading to a relatively small influence of light industry on weighted aggregated industrial CO_2 intensity though its output weights are big; as shown in Figure 10.4, aggregated industrial CO_2 emission intensity is dominated by heavy industry. In 1980–2008, the annual decreasing rate of CO_2 intensity in light industry is 8.76 per cent, faster than the 4.58 per cent in heavy industry; in 2009–2020, due to the specified carbon tax levy, the annual decreasing rate of CO_2 intensity in heavy industry is 3.85 per cent, faster than the 2.5 per cent in light industry. Specifically, CO_2 intensity in heavy industry decreases more rapidly in the period of the thirteenth Five-Year Plan than in the twelfth Five-Year Plan, while CO_2 intensity in light industry decreases more slowly in 2016–2020 than in 2011–2015. The forecasted values of CO_2 intensity for light and heavy industry in

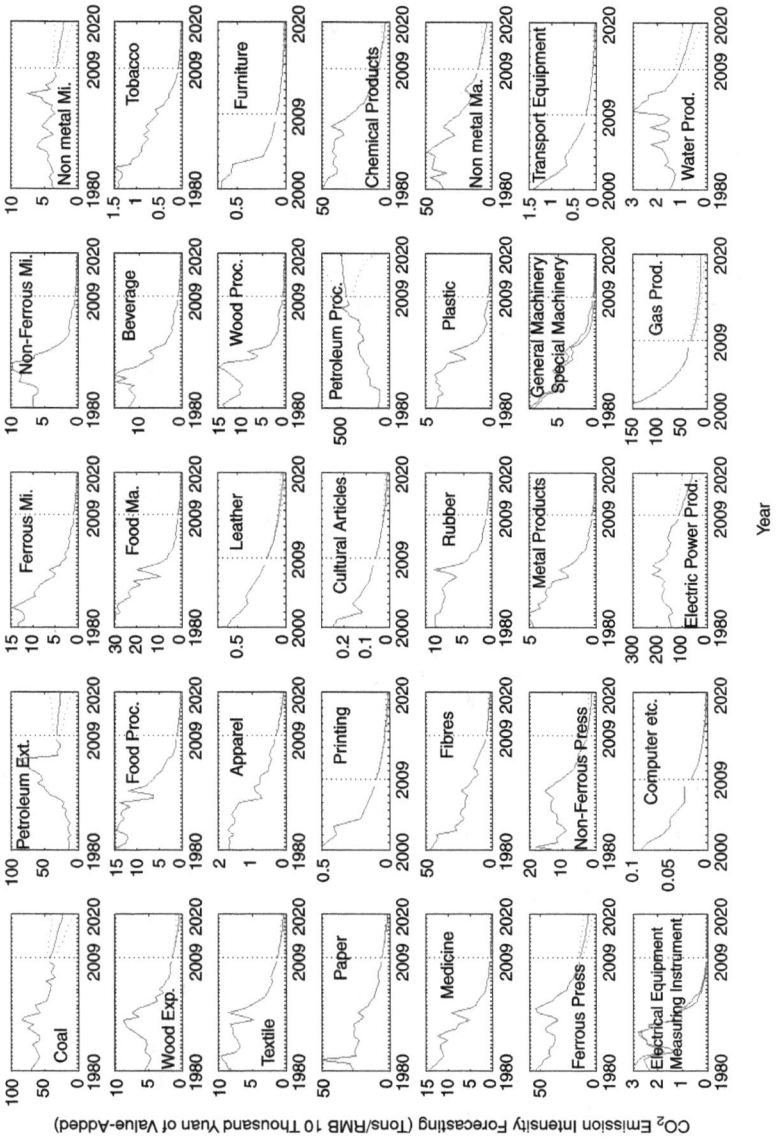

Figure 10.5 Forecasting sectoral CO$_2$ emission intensity in Chinese industry (2009–2020).

2020 are 0.47 and 6.99 tons of CO_2 per 10 thousand RMB, which decreases by 48 per cent and 53.5 per cent, respectively, relative to the absolute value of 0.9 and 15 in the year of 2005. Though the panel (b) of Figure 10.4 shows that light, heavy and aggregated industry maintain a similar decreasing trend as a whole, it is not the story if we look at sectoral forecasting in Figure 10.5 further. Extraction of Petroleum and Processing of Petroleum are the only two sectors whose CO_2 intensity increases in the sample period (2.8 per cent and 4.6 per cent); they are also the only two sectors whose CO_2 intensity remains almost unchanged or even rises in the forecasting period (-1.9 per cent and 1.1 per cent), which maybe relates to the largest reduction of industrial value-added for the two sectors in this period, and also implies that only levying the carbon tax is not enough to reduce the CO_2 intensity of the two sectors and other effective abatement policies should be taken into account. Several high-tech sectors (such as Manufacture of Communication Equipment and Computers, Manufacture of Transport Equipment, Manufacture of Measuring Instruments, Manufacture of Electrical Machinery and Equipment) not only have the largest reduction of CO_2 intensity in 1980–2008 (this varies from -18.8 per cent to -13.3 per cent) but also reduce by the largest amount, 13.35 per cent–6.6 per cent in the forecasting period, different from the findings that averaged CO_2 intensity of light industry decreases more slowly than that of heavy industry in the forecasting period. In a word, the levy of carbon tax plays a obvious role in abating industrial CO_2 intensity for either emission-intensive heavy sector or light industry with a low emission. The scenarios of carbon tax levy specified in this chapter lead to the similar reduction of CO_2 intensity to the official goal and are consistent with the requirement of development priorities in developing countries like China. Many studies have also found that gradual abating way is the preferable choice in the developing countries (Roughgarden and Schneider, 1999; Kuosmanen *et al.*, 2009; Reddy and Assenza, 2009).

10.6 Summary and comments

This chapter estimates the marginal abatement cost of CO_2 emission for industrial sectors over the reform period in China by using the directional distance function, and utilizes them to forecast the carbon taxes likely to be levied in the foreseeable future. Polynomial dynamic panel data model based forecasting tells us that the levy of carbon tax negatively influences the industrial value-added in the short run, but such a negative influence will disappear in the long run. However, the influence of carbon tax on output from negative to positive is extremely small; driven more by the historical information, aggregated industrial value-added remains the increasing trend with an annual growth of 8.6 per cent in the forecasting period, in which its annual growth is 9.2 per cent and 7.9 per cent in the period of the twelfth and thirteenth Five-Year Plans, respectively.

Though the carbon tax rate of heavy industry is low, heavy industrial value-added is influenced more by the levy of carbon tax than light industry, due to its huge total CO_2 emission. Carbon tax levy is obviously beneficial to the abatement of CO_2 emission intensity. The forecasting reveals that CO_2 intensity of aggregated, light and heavy industry in 2020 will reduce by 50 per cent, 48 per cent and 53.5 per cent relative to that in 2005, exceeding the binding abating goal of 40 per cent–45 per cent announced by China at the end of 2009. For some specific heavy sectors such as Extraction of Petroleum and Processing of Petroleum, however, the levy of carbon tax is not enough on its own to reduce their carbon intensity, and more effective environmental regulations should be implemented.

In sum, the ecological effect of carbon tax levy is obvious because the carbon tax can promote CO_2 reduction from two approaches: carbon tax can directly promote carbon intensity abatement by increasing energy price and improving energy efficiency, and indirectly by redistributing the carbon tax income, reinvesting in low carbon technology, adjusting the distortion of traditional tax system, and so on. Therefore, based on the influential analysis of carbon tax on economy and ecology in this study, the gains from a carbon tax levy outweigh its disadvantages, and environmental taxation reform in China should be implemented, beginning with the levy of carbon tax as early as possible.

The design of carbon tax in China may take the following points into account. The levy of carbon tax is urgency needed in China to challenge climate change and achieve an agreement among countries to commit to abate carbon after the expiry of the Kyoto Protocol in 2012, as already denoted. Low carbon investment, almost 40 per cent of the RMB 4000 bn recovery package in 2008, will reduce the shock at the beginning of the carbon tax levy in China. The carbon tax should be levied from the production units that emit CO_2 directly to the atmosphere and be based on their total CO_2 emission. It's very convenient to levy a carbon tax in such a manner because the calculation of taxes is very simple, due to the fixed carbon emission coefficients of fossil fuels and electricity. It's also easier to urge the firms to develop low carbon technology and make big efforts to abate carbon when the carbon tax levy is according to the carbon content of fossil fuels.

The rate of carbon tax could be based on the MAC of CO_2 emission estimated in this study that should increase over time and vary across industrial sectors.[5] The fixed carbon tax rate as discussed in many studies is thus not preferable. As tested in this study, the levy of such a carbon tax rate will lead to a reduction of CO_2 intensity similar to the officially announced abating goal. Though the forecast carbon tax rate is somewhat large, it does reflect the actual abating cost of CO_2 emission in reality. As stated previously, to fully reflect the institutional value of environmental taxation, the environmental tax rate should be high enough to influence the emitters' behaviour and stimulate their awareness of environmental protection.

Of course, the levy of carbon tax is only the first step of environmental taxation reform in China. Environmental taxation reform is surely a systematic scheme and should be understood as the process of moving from the second-best tax system to the best one, by continuously adjusting or removing the tax distortion. Further studies on environmental taxation reform along these lines will be considered next.

11 Conclusions

The issues of energy, environment and economics (3E) have become more and more important because the energy crises and environmental disasters have happened with increasing frequency, which has seriously affected stable economic growth in the world since the 1970s. To challenge this, the concepts of sustainable development and the low carbon economy have been proposed during the past two decades. In recent years, the increasingly powerful economic crisis has already destroyed the traditional growth model and now provides great opportunities for many countries, either developed or developing, to transform their economic growth model from high-carbon and extensive in nature to be low-carbon based, intensive and sustainable. For example, both the American Recovery and Reinvestment Act and the American Clean Energy and Security Act aim to develop renewable energy and low-carbon techniques and plan to make them a new economic driver. In 2009, the UK also released its white papers outlining national strategy, the Low Carbon Transition Plan, with targets to 2020. The topics of energy, environment and economic low-carbon transformation are more important and urgent in China. The basic fact is that China is the second largest economy in the world, just behind the US, but it has become the largest energy consumer and carbon emitter at the present time.

This book focuses on the energy and environmental constraints and their impact on economic transformation in China. In the first chapter, the Introduction, the panel data for industrial sectors and provinces in China, especially including low-carbon factors such as energy consumption and environmental pollutants emissions, were constructed to provide readers with the basic statistical information on economic growth in China. The database is also the research sample for the analysis of the following chapters. In Chapter 2, we further described the energy consumption and its induced CO_2 emissions for six industries such as agriculture, industry, construction, transport, commerce and the household sector in Chinese provinces between 1995 and 2007. The factors which influence the CO_2 emissions and the measure of the CO_2 shadow price are addressed in Chapters 3 and 4. Traditional economic growth in China is often regarded as high-carbon driven and an unsustainable model by consuming high resources of energy and capital and emitting heavy pollutants, and so on. This is discussed in Chapter 5, together with the corresponding energy and environmental policies which result in this situation. Economic transformation implies the update of the economic structure. In Chapter 6, we investigated the impact of structural reform, such as energy structure, light and heavy industrial structure and so on on the growth of productivity and output in China. The key in this book is how to include low-carbon factors, such as energy and environment, into the analysis of economic development, and how to evaluate the process of economic transformation. This is analysed in detail in Chapters 7 and 8. The more appropriate directional distance function (DDF) approach and its extension into the SBM-DDF-AAM approach are chosen and adopted as the theoretical framework of low-carbon transformation in this book. Accordingly, the produced environmental total factor productivity (TFP) is used to measure the quality of economic growth within the restrictions of energy and environment, and its contribution to output growth is selected as the evaluation indicator to dynamically assess low-carbon economic transformation; this differs from the traditional statistical indicator to describe green economic development. Chapter 7 empirically assesses the economic transformation of industrial sectors and Chapter 8 concentrates on the 31 provinces in China. In Chapter 9, we designed several energy-saving and emission-abating paths and simulated their impact on the win-win development possibilities in the long run. Chapter 10 discusses the market mechanism based environmental taxation reform, which is very crucial to low-carbon transformation in China. The basic conclusions and policy implications are summarized as follows:

1 *The reform of resources prices* As stated previously, the over-dependence of economic growth on energy consumption is due to the long-term distortion of the resources price system. Therefore, during the period of the twelfth Five-Year Plan, China should accelerate the price reform of energy products and make the energy factor markets play the real role in allocating resources well. The energy price is the important variable to determine energy demand. Energy prices which are too low are the cause of the over-use of fossil energy in China for a long time, and the appropriate increase of the energy price is bound to decrease energy demand and drive the energy- and emission-intensive enterprises to develop energy-saving and emission-abating techniques to transform their development model.

   In fact, at the end of the eleventh Five-Year Plan, central government has started energy price reform. With the decrease of both product prices and international energy prices, the state council carried out the reform of taxes and fees for refined oil on 1 January 2009. That is, six types of charge for the refined oil price were cancelled and the consumption tax rate levied within the refined oil price increased in order to reflect the extent of scarcity of petroleum resources and production cost and complete the price formation system of refined oil. The rise of the proportion of imported gas lead to the birth of gas price reform. In June 2010, China started to adjust the producer reference price of gas domestically-produced. China has implemented low electricity prices for the household sector for a long time. Though tight energy supply has caused a rise of electricity price at the current time, the scale and frequency of changes to electricity prices in the household sector are lower than for other sectors. In other countries, the electricity price in the household sector is about 1.5–2 times that in industry, but in China the electricity price for households is lower than that for industry and is normally subsidized by the latter. In October 2010, China began to implement the ladder-type electricity price in the household sector, indicating the end of the period of low electricity prices. The pricing reform for refined oil, implemented since the beginning of 2009, is still at an early stage. The monopoly of the China National Petroleum Corporation and Sinopec Group result in the diesel oil shortage of 2010. Thus, it is not the time to give pricing rights to the enterprises and the government should continue to nurture the refined oil market in order to finally realize the marketization of the oil market.

2   *The reform of environmental taxation*   Traditional environmental policy is normally implemented through administrative fiats in China. It is not efficient for the economy because it does not consider the concrete case for different sectors and increases the implementation cost. It is also not efficient for the environment because it does not provide enterprises with further motivation to abate emission. In fact, so far as is known, environmental measures are implemented mainly by collecting pollution fees, less by taxation, in China. Such few taxes are scattered around in such examples as resources tax, consumption tax, value-added tax, vehicle and vessel tax, and so on; these are not precisely defined environmental taxation. For example, pollution charges have been collected from 1982 in China and have reached an annual total of 20 billion RMB currently; this is simply the actual cost of dealing with pollution without the inclusion of an external environmental cost. The resource tax already levied in China is only to adjust resource differential income, and is not greatly correlated with environmental protection.

Economists argue that environmental policy must be based more firmly on the use of market-based mechanisms so as to introduce the cost of pollution clearly into economic analysis and impose ceaseless price pressure on the polluters to save energy and reduce pollution. Environmental taxes and tradeable emission right permits are the main examples of market-based environmental policy, based on Pigouvian Tax and Coase Theory, respectively. It is somewhat difficult to trade emission rights due to the big cost of the protection and definition of property rights; therefore, environmental taxation becomes the important instrument for many countries to protect the environment, and this is so in China. Environmental taxes levied in advanced countries include energy tax, carbon tax, sulphur tax, water pollution tax, solid waste tax and noise tax. These have already played an important role in promoting sustainable development, which has provided a positive experience of environmental tax reform for China. In recent years, China has become the largest emitter of CO_2 in absolute terms in the world, and that means China faces continuously increasing pressure from the rest of the world to abate carbon emission. There thus exists the possibility that the developed countries will impose carbon tariffs on imports from those countries without mandatory carbon abatement, such as China. In this case, as an example of environmental taxation, the levy of a carbon tax is more urgent than other kinds of environmental taxation in China, and could be appropriately regarded as the first step of environmental taxation reform. Though a carbon tax has been levied in many counties including Finland, Sweden, Norway, the Netherlands and Denmark, and its theoretical foundation is solid enough, it's still necessary to analyse its economic and ecological effect for the foreseeable future in China, as studied in Chapter 10 in this book. The basic conclusion is that the environmental taxation reform in China should be implemented with the levy of carbon tax as early as possible, ideally during the period of the twelfth Five-Year Plan. Of course, the levy of a carbon tax is only the first step of environmental taxation reform. Environmental taxation reform is surely a systematic scheme and should be understood as the process from the second-best tax system to the best one, by continuously adjusting or removing the tax distortion.

3   *The reform of economic structure*   Economic transformation is dependent on the update of the economic structure. According to the studies in this book, the structural reform in China in the future could be carried out in the following several ways. First, it is necessary to reform the heavy industrial sectors by eliminating and clearing the extensive production capabilities, controlling the production of energy- and emission-intensive products (such as steel and iron, rubber and non-ferrous metals etc.), developing energy-saving and emission-abating techniques, and so on. Second, the type composition of energy consumption should be adjusted substantially. Though, in the short run, it is not easy to change the energy structure due to the different energy resource endowment for different countries, in the long run, it is possible to adjust the energy structure from being coal dominated to a higher proportion of clean energy and renewable energy, such as clean coal, nuclear, water, wind and solar energy. Third, continue to develop light and heavy industry evenly. China is still at the mid-stage of industrialization and urbanization, and the heavy industrial sectors will continue to play a fundamental role in pushing the process forward. But the industrialization will evolve from heavy industrialization to high-processing industrialization and technique intensification, reflecting the objective process from labour-intensive to capital- and technique-intensive industry. The light and advanced technique industry will play a part in this process. The highest growing sector of Manufacture of Communication Equipment, Computers and Other Electronic Equipment has experienced very high productivity, strong labour absorbability, low capital demand, and the lowest energy and emission per value-added, which seems to be very instructive for China's new-style industrialization in the future. Finally, according to the experience of industrialized economies with a hump-shaped pattern, it is necessary to develop the tertiary industry, especially the productive tertiary industry, to optimize

the economic structure. Different regions should undertake the appropriate structural adjustment policy to develop the industries with comparative advantages and avoid the duplication of similar projects.

4   *The application of new evaluation criterion, rather than GDP only*   The traditional high-carbon extensive growth model in China is mainly attributable to the GDP driven evaluation system, which only concerns quantitative expansion without consideration of the growth quality, energy utilization efficiency, environmental pollution and so on. According to modern growth theory, the input factors driven economic growth is not sustainable because it is bound to be restricted by the law of diminishing marginal returns. Only technical progress and the corresponding increase of the total factor productivity (TFP) pave the way in the direction of sustainable development. Thus, the quality effect reflected by TFP growth should be adopted to evaluate the transformation of the development model. This raises challenging questions for the researchers. One is how to include the low-carbon factors, such as energy and environment, into the measure of real TFP growth in China, and another is the choice of new evaluation criterion for local governments instead of GDP. As analysed in this book, the (SBM-)DDF-AAM approach resolves this problems and the resulting environmental TFP growth's contribution to output growth becomes a more appropriate evaluation indicator than GDP. The advantage of the new evaluation indicator is based on its format. It is the ratio of environmental TFP growth to output growth. Thus, it still meets the basic requirement that output growth is the main task for the developing countries, like China. More importantly, the introduction of environmental TFP growth as the numerator also places the growth quality under the constraints of energy and environment into account, which is crucial to sustainable development. The new evaluation indicator vividly reveals the relationships between economic growth and quality considerations. Now it is easy to answer the question whether the low-carbon transformation will slow down economic growth. The new evaluation indicator is also different from the statistical indicator because it is endogenously produced from economic theory and could be employed to dynamically assess whether the economic transformation has happened or not. The empirical quantification of low carbon transformation in China reveals that the process of low carbon transformation in China is still not stable and that central and local government should implement more appropriate and effective environmental regulations to support the low carbon transformation in China in the future.

Appendix

The creation of the sub-industrial and provincial panel database

This book constructs the input and output panel database of the two-digital industrial sectors where the industrial sectors are classified according to the new version of the National Standard of Industrial Classification (GB/T4754), revised in 2002 in China. Data available for the period between 1978 and 2010 allow an analysis to be undertaken for 38 different industrial sectors, which belong to three bigger categories: mining; manufacturing; electric power, gas and water production and supply. This sub-industrial database is the sample for the analysis in most of the chapters in this book, such as Chapters 1, 3–7, 9 and 10. To this end, the sub-industries must be reclassified and recombined to match one another, the scope of all the industrial variables are adjusted to the same statistical content, and some missing data is added rationally. These constructed variables include gross output value of industry, industrial value-added, two sets of capital stock, two sets of labour force, energy consumption, intermediate input and carbon dioxide emission, in which the capital stock and CO_2 emission cannot be obtained directly and need estimating, as explained later. All value-type variables are calculated based on 1990 price levels. The codes and names of the 38 industrial sectors are reported in Table A.1. The sub-industrial gross output value, CO_2 emission, one set of capital stock and labour, and energy consumption are listed in Tables A.2–A.6, respectively. The details can be seen in the author's paper 'Reconstruction of Sub-industrial Statistical Data in China (1980–2008)' published in *China Economic Quarterly* (volume 10, issue 3, pp. 735–776), which was awarded the best paper prize (2010–2011) by the journal in 2012.

Table A.1 Codes and names of 38 industrial sectors analysed in this book

Number	Two-digital Codes	Full Name of Sectors	Abbreviations
1	6	Mining and Washing of Coal	Coal Mi.
2	7	Extraction of Petroleum and Natural Gas	Petroleum Ext.
3	8	Mining and Processing of Ferrous Metal Ores	Ferrous Mi.
4	9	Mining and Processing of Non-Ferrous Metal Ores	Non-Ferrous Mi.
5	10	Mining and Processing of Nonmetal Ores	Nonmetal Mi.
6		Exploiting of Wood and Bamboo	Wood Exp.
7	13	Processing of Food from Agricultural Products	Food Proc.
8	14	Manufacture of Foods	Food Ma.
9	15	Manufacture of Beverages	Beverage
10	16	Manufacture of Tobacco	Tobacco
11	17	Manufacture of Textile	Textile
12	18	Manufacture of Textile Wearing Apparel, Footware and Caps	Apparel Leather
13	19	Manufacture of Leather, Fur, Feather and Related Products	
14	20	Manufacture of Wood, Bamboo and Straw Products	Wood Proc.
15	21	Manufacture of Furniture	Furniture
16	22	Manufacture of Paper and Paper Products	Paper
17	23	Printing, Reproduction of Recording Media	Printing
18	24	Manufacture of Articles For Culture, Education and Sport Activities	Cultural Articles
19	25	Processing of Petroleum, Coking, Processing of Nuclear Fuel	Petroleum Pro.
20	26	Manufacture of Raw Chemical Materials and Chemical Products	Chemical
21	27	Manufacture of Medicines	Medicine
22	28	Manufacture of Chemical Fibres	Fibres
23	29	Manufacture of Rubber	Rubber
24	30	Manufacture of Plastics	Plastic
25	31	Manufacture of Non-metallic Mineral Products	Nonmetal Ma.
26	32	Smelting and Pressing of Ferrous Metals	Ferrous Press
27	33	Smelting and Pressing of Non-ferrous Metals	Non-Ferrous Pr.
28	34	Manufacture of Metal Products	Metal Products
29	35	Manufacture of General Purpose Machinery	General Mac.
30	36	Manufacture of Special Purpose Machinery	Special Mac.
31	37	Manufacture of Transport Equipment	Transport Eq.

32	39	Manufacture of Electrical Machinery and Equipment	Electrical Eq.
33	40	Manufacture of Communication Equipment, Computers and Other Electronic Equipment	Computer etc.
34	41	Manufacture of Measuring Instruments and Machinery for Cultural Activity and Office Work	Measuring Inst.
35	44	Production and Supply of Electric Power and Heat Power	Electric Power
36	45	Production and Supply of Gas	Gas Prod.
37	46	Production and Supply of Water	Water Prod.
38	11, 42, 43	Manufacture of Artwork and Other Manufacturing	Others

Table A.2 Sub-industrial gross output value (1990=100, unit: 100 million RMB)

	1980	1981	1982	1983	1984	1985	1986	1987	1988	1989	1990	1991	1992	1993	1994
Coal Mi.	223	237	251	267	285	301	327	349	405	489	516	517	489	483	502
Petroleum Ext.	241	255	259	259	345	312	320	395	392	417	427	422	393	338	289
Ferrous Mi.	18	17	18	22	24	26	30	36	42	43	45	45	49	61	86
Non-Ferrous Mi.	60	64	67	69	78	89	92	81	102	113	111	120	122	183	235
Nonmetal Mi.	107	110	118	119	128	141	157	179	226	240	210	233	269	361	429
Wood Exp.	84	85	87	90	115	102	118	126	118	103	96	88	86	88	131
Food Proc.	522	588	609	636	731	753	825	903	1088	1083	1155	1323	1269	1408	1692
Food Ma.	206	248	268	268	276	282	309	340	405	401	436	499	494	649	541
Beverage	135	195	233	231	231	250	300	317	394	383	415	493	535	668	786
Tobacco	124	187	202	208	234	230	262	307	379	454	512	536	533	579	742
Textile	1248	1411	1430	1575	1518	1958	2132	2367	2489	2488	2520	2685	2796	2890	2833
Apparel	171	226	221	233	215	287	315	364	412	434	506	598	633	816	1018
Leather	96	110	107	105	114	144	170	200	227	224	241	284	289	457	680
Wood Proc.	81	157	175	177	117	123	141	137	137	131	139	137	153	201	333
Furniture	53	67	69	66	79	104	95	106	123	114	108	125	139	172	242
Paper	184	146	159	170	242	304	343	405	461	454	462	506	512	613	736
Printing	93	102	118	117	131	160	177	195	203	189	201	250	245	302	354
Cultural Articles	32	35	38	40	43	55	73	86	93	96	107	135	143	195	271
Petroleum Pro.	510	395	407	442	683	635	372	418	473	529	579	690	602	550	541
Chemical	748	760	818	920	1080	1100	1132	1315	1464	1537	1611	1754	1790	2057	2359
Medicine	106	106	136	152	170	211	234	281	337	321	364	485	524	625	677
Fibres	44	70	75	85	78	123	138	180	219	236	276	333	342	414	534
Rubber	135	137	137	153	176	215	241	264	300	309	306	324	332	400	455
Plastic	123	127	158	169	184	254	287	338	404	396	445	527	592	747	878
Nonmetal Ma.	562	510	580	580	805	883	967	1107	1260	1220	1260	1424	1539	1776	2163
Ferrous Press	646	624	673	721	844	878	1068	1178	1285	1322	1350	1363	1359	1510	1596
Non-Ferrous Pr.	271	261	271	291	343	402	432	459	505	528	544	603	609	615	763
Metal Products	289	279	322	350	377	488	510	602	664	662	695	782	838	1045	1277
General Mac.	748	755	835	888	985	1203	1197	1463	1697	1610	1495	1765	2089	1685	1753
Special Mac.	417	409	492	490	537	632	621	901	1052	1000	921	1059	1154	994	1141
Transport Eq.	234	230	284	297	364	462	481	545	698	713	741	929	1171	1642	1973
Electrical Eq.	314	326	425	442	488	714	758	871	990	913	868	1074	1211	1670	1859
Computer etc.	76	70	145	144	161	269	278	394	552	563	608	821	906	1243	1835
Measuring Inst.	78	78	97	97	111	142	139	129	141	125	118	145	172	318	383
Electric Power	266	268	285	306	347	370	389	420	493	601	680	711	724	762	767
Gas Prod.	50	58	59	87	86	78	63	65	62	62	60	57	59	33	45
Water Prod.	18	18	22	249	27	29	30	35	38	42	46	57	56	59	71
Others	123	131	139	117	164	214	270	319	399	408	467	527	718	1132	1034

1995	1996	1997	1998	1999	2000	2001	2002	2003	2004	2005	2006	2007	2008	2009	2010
559	616	637	737	696	694	735	797	905	1220	1360	1593	1913	2310	2487	2982
245	255	274	311	296	270	244	249	257	277	291	293	308	322	345	330
81	112	137	170	145	155	160	171	222	292	339	469	626	807	1055	1364
215	264	315	400	414	410	397	407	441	495	581	680	816	916	1050	1175
444	524	619	730	653	601	556	552	576	598	656	817	990	1153	1375	1650
104	129	145	173	153	143	140	150	155	156	151	145	144	142	138	136
1800	2047	2330	2666	2616	2773	2860	3193	3778	4274	5308	6400	7523	8879	10679	12489
565	658	773	913	932	1066	1127	1304	1442	1685	2147	2626	3269	3816	4473	5288
830	1033	1165	1359	1402	1451	1473	1575	1722	1817	2270	2832	3624	4269	5034	5959
776	846	892	941	910	936	1071	1244	1356	1558	1674	1884	2205	2608	2848	3364
3083	3277	3370	3611	3794	4047	4339	5038	5827	7279	8800	10345	12467	13930	15116	17168
1108	1289	1376	1564	1536	1657	1810	1972	2229	2482	3084	3770	4602	5574	6152	7117
652	677	731	812	800	854	944	1040	1262	1460	1771	2083	2510	2777	3065	3682
373	520	681	949	961	1020	1053	1109	1233	1558	1947	2443	3306	4195	4932	6048
246	306	369	446	436	448	470	517	649	938	1110	1436	1792	2163	2372	2963
787	805	899	1007	1080	1245	1363	1549	1838	2340	2805	3321	4071	4721	5173	6226
300	333	391	462	509	537	628	722	911	1073	1269	1472	1780	2156	2342	2735
319	348	377	429	459	510	563	667	811	987	1170	1360	1587	1841	1919	2218
605	654	702	919	968	1120	1082	1126	1225	1500	1698	1812	2027	2130	2214	2549
2491	2792	3125	3631	3853	4245	4635	5222	6294	7795	8998	11108	13909	15718	19241	22944
714	863	1018	1165	1318	1663	1936	2282	2791	3182	4095	4888	6050	7192	8584	10306
562	687	835	1055	1217	1309	1149	1316	1585	1928	2457	2982	3709	3592	3827	4338
489	561	602	680	714	745	814	967	1162	1554	1781	2097	2551	2945	3306	3914
975	1190	1382	1647	1762	1930	2100	2440	2891	3571	4033	4955	6093	7129	8108	9888
2453	2740	3004	3403	3400	3440	3528	3826	4484	5408	6458	7909	10113	12317	14291	17656
1705	1806	1970	2208	2389	2630	3052	3407	4615	6368	7682	9440	11572	12792	14488	16352
794	936	1049	1253	1351	1439	1634	1838	2347	3247	3832	5070	6170	7365	8622	9999
1451	1742	1942	2245	2285	2493	2687	3014	3334	3947	4750	6025	7763	9411	10253	12446
1921	2174	2389	2646	2661	2885	3179	3711	4757	6568	7906	10056	13107	16516	18281	23110
1260	1359	1413	1482	1542	1710	1829	2194	2951	3780	4420	5656	7355	9676	11081	13943
2270	2505	2749	3025	3307	3791	4617	6055	8160	10057	11548	15007	19906	24053	29998	39610
2425	2698	2952	3243	3697	4512	5189	5947	7730	10445	12462	15086	19138	23862	27716	34329
2244	2857	3984	5679	6997	9380	11691	15548	22964	33476	42609	54064	65772	74913	79453	99694
386	562	727	925	864	992	1010	1120	1622	2078	2654	3386	4143	4752	4863	6140
941	959	998	1122	1211	1345	1431	1621	1845	3816	4426	5217	6268	6988	7595	9034
35	33	34	40	52	65	67	77	89	132	155	207	267	384	460	578
71	73	65	72	71	67	65	66	71	80	89	106	115	131	144	157
1073	1162	1252	1440	1618	1662	1779	1826	1830	2037	2647	3114	3931	4576	4938	6188

Table A.3 Sub-industrial carbon dioxide emission (unit: 10 thousand tons)

	1980	1981	1982	1983	1984	1985	1986	1987	1988	1989	1990	1991	1992	1993	1994
Coal Mi.	8754	9032	9311	9589	9867	10174	11386	12466	12360	13826	15519	17154	16660	20889	18625
Petroleum Ext.	2313	2209	2236	2285	2461	3234	3030	3453	3859	4131	4318	5020	5512	6895	7519
Ferrous Mi.	118	133	148	163	178	215	207	197	187	178	168	158	154	154	171
Non-Ferrous Mi.	197	207	217	227	237	275	297	335	375	414	454	493	513	535	544
Nonmetal Mi.	276	286	296	306	316	347	385	454	515	577	599	660	702	703	788
Wood Exp.	296	308	321	333	345	359	390	399	407	414	422	428	434	449	434
Food Proc.	1213	1254	1302	1348	1397	1779	1977	2176	2374	2471	2530	2727	2925	2930	3007
Food Ma.	1017	1083	1155	1227	1300	1583	1682	1724	1780	1822	1861	1881	1977	2027	2213
Beverage	654	736	815	895	978	1129	1148	1640	1778	2035	1915	1875	1816	1919	1930
Tobacco	110	131	153	174	194	204	188	211	224	240	268	266	245	306	342
Textile	2910	3058	3215	3371	3533	3712	3862	4147	4500	4688	4667	4699	4883	5000	5009
Apparel	69	75	81	87	93	151	153	158	172	187	197	217	227	231	232
Leather	174	178	181	185	188	234	249	257	260	264	270	276	286	294	284
Wood Proc.	356	358	360	363	367	423	423	424	444	464	493	523	543	543	630
Furniture	75	81	88	94	101	111	118	114	109	103	107	111	114	113	119
Paper	1585	1651	1727	3203	1887	2483	2676	2866	3128	3253	3244	3353	3533	3747	3982
Printing	138	138	138	138	138	101	102	103	114	118	128	138	148	154	163
Cultural Articles	26	37	47	58	71	47	52	51	55	59	65	67	71	72	69
Petroleum Pro.	24181	24186	24534	24933	25306	26622	29581	31125	32833	34957	34488	34544	38754	40748	43459
Chemical	12719	13078	13439	13800	14149	14159	15156	17323	18311	19430	19779	20928	21721	22176	23358
Medicine	515	534	554	574	595	742	817	982	1003	1097	1090	1209	1304	1237	1525
Fibres	854	934	1014	1094	1205	1298	1473	1603	1619	1761	2017	2288	2460	2938	3061
Rubber	515	534	554	575	595	688	733	713	733	753	792	812	833	828	977
Plastic	158	168	178	188	198	303	321	395	415	434	454	474	494	486	551
Nonmetal Ma.	9114	9785	10468	11151	11836	17087	18188	19597	20856	21180	19776	20483	21363	21820	24196
Ferrous Press	12022	12289	12564	12859	13151	13097	14423	14802	15486	15590	16223	17506	18483	20159	22860
Non-Ferrous Pr.	835	1304	889	915	943	1023	1101	1376	1470	1505	1615	1741	2130	2720	2993
Metal Products	496	536	575	615	658	696	758	741	749	744	741	739	737	741	833
General Mac.	1677	1709	1713	1733	1754	1737	1794	1889	1965	1873	1738	1848	1919	1881	1884
Special Mac.	1101	1115	1130	1145	1160	1136	1148	1264	1325	1268	1274	1282	1286	1240	1285
Transport Eq.	1409	1413	1422	1431	1440	1422	1470	1608	1649	1613	1620	1634	1639	1596	1593
Electrical Eq.	373	387	401	416	431	414	507	643	688	679	665	706	746	741	749
Computer etc.	299	325	351	382	417	337	328	519	537	515	456	343	457	435	441
Measuring Inst.	84	91	100	109	118	86	100	130	143	140	132	134	148	130	135
Electric Power	25349	26294	27634	29096	30506	33350	36741	40860	45993	49795	53902	59156	65931	69158	77913
Gas Prod.	987	1085	1184	1283	1381	1441	1480	1520	1539	1559	1579	1618	1579	1618	1868
Water Prod.	16	17	18	19	20	21	28	28	30	32	36	39	47	46	50
Others	1198	1272	1346	1420	1497	1596	1636	1675	1715	1755	1795	1894	1994	2085	2455

1995	1996	1997	1998	1999	2000	2001	2002	2003	2004	2005	2006	2007	2008	2009	2010
16361	18476	18917	15899	13990	13285	13574	14365	20478	20479	25943	28446	32710	36260	40055	43851
7530	7870	10467	10577	11551	12871	13129	13872	16003	6385	6740	6054	6355	6839	6972	7104
187	174	178	149	146	130	128	121	196	184	222	221	240	360	332	304
358	314	285	251	195	165	175	163	168	178	180	180	180	190	187	184
873	898	888	825	884	796	953	997	1441	922	1134	1052	1155	1195	1321	1446
419	420	365	322	288	203	236	194	178	156	148	144	142	148	138	128
3485	3476	3591	3293	3180	2668	2781	2643	2893	2105	2282	2305	2521	3252	3317	3381
2400	2173	1807	1527	1450	1200	1205	1133	1071	1555	1639	1672	1753	2161	2117	2072
1943	1857	1470	1584	1382	1140	1214	1129	1234	1391	1360	1341	1389	1713	1640	1567
378	372	356	313	349	234	257	249	266	255	219	217	200	196	187	179
5097	4419	3989	3424	2874	2618	2656	2518	2828	3941	4239	4556	4739	5024	4832	4640
232	301	267	287	284	220	241	215	243	341	381	406	426	458	439	420
472	214	169	190	169	126	131	124	128	167	167	169	168	170	171	171
716	633	588	572	466	415	413	403	495	698	691	701	715	874	880	886
124	116	100	96	102	76	84	77	91	53	52	52	53	75	68	62
4210	4166	3819	3652	3221	3394	3345	3456	3632	5365	5989	6592	6686	7640	7930	8221
172	125	129	127	108	92	98	94	122	76	76	77	78	92	89	86
66	75	103	42	40	31	35	21	34	33	33	33	33	36	37	38
50921	51953	51387	53356	57410	62421	64150	69776	80295	109683	117520	133295	144111	148388	158631	168874
26011	27874	25041	24336	22574	22596	21789	22843	25953	28020	33196	33755	36194	42641	42141	41642
1813	1380	1217	1171	1143	995	999	975	1099	1080	1155	1205	1180	1456	1428	1399
3184	3100	3242	3367	3208	3457	3436	3406	3676	3713	3685	3704	3830	1538	1484	1431
1122	958	768	749	678	490	522	502	536	710	727	724	745	919	912	905
617	611	531	407	352	268	272	209	276	454	455	462	462	641	755	869
26714	27039	25481	23259	21914	19834	18181	17731	22111	32639	33695	33821	34484	46496	47878	49260
25588	26024	25070	22306	22615	22038	21295	23468	29088	32156	38097	42077	44723	47984	52812	57640
2673	2728	2737	2602	2697	2363	2514	2597	2844	4193	4510	4876	5325	6649	6240	5831
922	873	782	692	606	436	457	447	422	550	556	541	545	721	715	710
1624	1972	1570	1240	947	667	691	657	649	701	716	756	748	982	972	962
1337	1283	1098	877	811	640	631	576	782	932	966	991	1008	1193	1240	1288
1714	1780	1632	1504	1551	1319	1376	1380	1361	1665	1559	1577	1620	1903	1936	1968
697	670	588	521	495	364	355	338	336	316	302	307	314	406	823	1240
302	273	256	238	226	205	206	220	255	375	376	375	394	504	472	440
141	132	106	89	85	58	51	50	63	42	41	41	42	58	58	57
86664	98177	100345	100918	102916	108825	113811	128980	154594	186702	208519	237977	261913	271463	288712	305967
1519	1651	1703	1642	2342	2133	1906	2151	2224	2445	2727	3017	3106	2461	2592	2722
74	54	41	59	102	85	81	71	74	61	62	62	62	72	59	46
1844	1107	1244	929	657	577	475	458	622	1218	1000	968	907	993	930	866

Table A.4 Sub-industrial capital stock (1990=100, unit: 100 million RMB)

	1980	1981	1982	1983	1984	1985	1986	1987	1988	1989	1990	1991	1992	1993	1994
Coal Mi.	690	803	801	849	901	1029	1099	1167	1229	1326	1435	1484	1567	1597	1617
Petroleum Ext.	222	281	319	380	449	493	608	695	812	967	1080	1227	1351	1594	1604
Ferrous Mi.	54	53	53	51	52	59	56	59	62	63	70	73	76	75	104
Non-Ferrous Mi.	86	96	112	128	140	140	144	158	167	176	181	198	210	224	248
Nonmetal Mi.	115	130	138	143	156	157	165	171	198	210	227	244	266	317	308
Wood Exp.	138	153	157	166	169	176	183	196	204	208	213	217	221	217	231
Food Proc.	154	153	173	213	283	339	376	426	474	523	567	621	661	714	844
Food Ma.	71	87	103	116	126	151	167	189	209	230	249	274	291	369	424
Beverage	53	61	84	103	121	161	194	237	282	314	337	377	401	454	537
Tobacco	13	19	27	33	39	43	55	68	80	96	116	143	166	187	218
Textile	311	333	357	411	660	741	832	981	1116	1258	1371	1689	1616	1794	2023
Apparel	31	42	53	62	75	82	96	112	128	140	156	186	214	268	338
Leather	29	35	42	47	52	58	65	74	83	93	106	117	130	166	210
Wood Proc.	40	43	53	56	60	79	84	93	104	109	117	135	143	168	202
Furniture	24	28	33	39	41	45	49	52	55	59	60	65	69	77	95
Paper	114	114	114	112	112	182	202	230	258	282	310	333	362	424	457
Printing	49	47	45	51	68	90	105	120	127	132	141	154	164	200	234
Cultural Articles	13	15	18	21	23	25	28	34	38	41	47	53	59	73	96
Petroleum Pro.	122	132	149	175	194	202	198	228	333	378	418	508	517	595	758
Chemical	696	775	805	863	928	956	972	1028	1106	1235	1346	1447	1547	1670	1837
Medicine	49	56	66	74	83	92	103	119	137	154	173	209	235	271	316
Fibres	69	82	91	165	160	170	205	250	266	276	288	322	357	422	474
Rubber	50	55	64	69	76	82	93	102	116	127	132	143	157	171	197
Plastic	59	81	104	123	141	155	180	211	232	255	282	317	347	401	476
Nonmetal Ma.	534	651	765	848	887	951	1036	1147	1259	1335	1401	1495	1580	1803	2228
Ferrous Press	766	776	811	853	870	942	1116	1181	1282	1369	1441	1575	1741	1939	2228
Non-Ferrous Pr.	216	228	241	249	258	269	295	319	360	397	419	444	472	569	542
Metal Products	151	172	191	205	217	228	245	279	307	328	349	381	414	499	644
General Mac.	680	708	752	782	846	839	868	911	948	971	994	1044	1092	1109	1206
Special Mac.	462	491	524	568	584	580	601	634	662	678	693	725	751	759	822
Transport Eq.	390	402	419	429	446	484	509	547	573	596	619	669	742	864	974
Electrical Eq.	153	161	180	207	229	233	267	313	366	401	417	468	510	594	711
Computer etc.	136	153	162	168	177	219	246	266	281	294	323	401	428	480	609
Measuring Inst.	72	72	77	80	83	90	93	101	106	110	112	120	123	163	177
Electric Power	988	996	997	1056	1126	1536	1640	1760	1913	2097	2289	2460	2735	3225	3856
Gas Prod.	28	30	33	38	43	48	51	65	87	107	123	151	174	170	197
Water Prod.	68	74	96	105	113	132	146	166	178	193	222	246	265	285	342
Others	43	48	65	76	89	100	134	132	155	177	189	209	232	297	322

1995	1996	1997	1998	1999	2000	2001	2002	2003	2004	2005	2006	2007	2008	2009	2010
1811	1905	2013	1891	2055	2059	2205	2377	2454	2632	2809	3229	3571	4175	4715	5208
1714	1666	1795	1820	2271	2529	2730	2730	2929	3371	3405	3795	4542	5022	5412	5681
110	124	122	123	115	126	129	147	172	238	270	315	405	523	582	848
257	256	254	252	262	266	268	256	252	248	255	294	336	434	471	582
371	379	380	365	373	483	485	492	484	471	452	418	419	493	526	570
237	228	220	216	212	210	207	204	200	204	209	213	218	222	226	230
1015	1114	1188	1129	1134	1115	1073	1085	1144	1279	1359	1510	1704	2122	2516	2965
528	564	608	585	598	612	619	671	656	745	784	847	935	1094	1221	1393
608	674	746	778	846	841	834	849	863	829	860	908	1005	1158	1278	1424
267	315	343	369	409	416	419	417	420	429	434	425	438	460	495	534
2184	2273	2324	2299	2306	2246	2244	2268	2388	2569	2626	2781	2940	3256	3351	3648
387	422	450	447	443	461	479	495	518	553	594	661	732	896	919	1028
248	252	261	256	253	253	262	265	284	318	343	365	392	450	478	543
244	263	317	291	290	322	350	357	372	430	449	491	551	697	744	875
109	118	135	123	121	137	143	156	181	217	239	288	335	369	389	452
586	635	728	792	821	973	1047	1076	1148	1310	1475	1551	1648	1844	1947	2157
282	291	316	322	357	364	424	410	465	525	556	589	619	710	750	810
114	116	130	137	128	135	137	153	163	194	203	213	231	274	285	305
874	967	1040	1233	1520	1625	1716	1683	1604	1745	1943	2156	2235	2453	3026	3537
2160	2343	2734	2812	3030	3159	3265	3366	3522	3695	4047	4666	5046	5831	6700	7809
361	398	441	470	498	550	600	675	759	864	986	1057	1155	1305	1432	1589
573	591	594	673	660	631	628	615	605	584	653	681	716	721	733	769
230	263	296	316	330	322	340	349	382	451	473	516	622	701	762	837
585	614	675	668	721	763	812	850	907	1103	1113	1187	1234	1432	1536	1738
2658	2944	3021	3000	3061	3111	3019	3072	3214	3582	3709	3925	4174	4911	5515	6385
2641	2742	2860	3110	3395	3521	3645	3582	3869	4262	4717	5432	6064	6939	7879	8955
732	811	835	965	954	979	1012	1059	1133	1349	1515	1697	1993	2519	2758	3354
725	807	893	867	874	905	941	938	892	985	1052	1167	1327	1675	1878	2164
1409	1552	1597	1540	1557	1598	1580	1604	1684	1871	1964	2128	2381	3026	3336	4100
933	1006	1007	989	986	977	952	940	1144	1186	1218	1344	1491	1901	2105	2469
1210	1435	1631	1703	1773	1864	1930	1987	2132	2350	2658	2942	3405	4175	4696	5450
834	950	1037	1085	1106	1136	1174	1178	1226	1372	1480	1623	1823	2329	2710	3308
703	772	878	1015	1074	1187	1363	1514	1732	2217	2499	2795	3279	3799	4040	5315
209	233	241	250	241	238	248	256	299	328	352	390	416	482	523	630
4350	4418	5313	6158	6549	7773	8606	9357	10832	12505	13481	15619	17824	20313	22328	24592
212	216	235	253	262	263	253	272	302	375	369	429	445	491	572	652
420	491	515	554	661	645	703	789	870	1043	1035	1227	1257	1423	1571	1750
340	356	347	337	337	339	327	317	310	419	418	464	491	587	639	676

Table A.5 Sub-industrial labour forces (unit: 10 thousand workers)

	1980	1981	1982	1983	1984	1985	1986	1987	1988	1989	1990	1991	1992	1993
Coal Mi.	496	492	492	502	543	587	541	552	626	648	673	692	698	720
Petroleum Ext.	40	41	42	44	48	52	53	59	71	76	73	72	75	119
Ferrous Mi.	29	28	26	26	29	34	32	33	39	38	39	41	44	46
Non-Ferrous Mi.	53	56	59	59	59	61	64	68	72	76	77	81	83	86
Nonmetal Mi.	150	129	110	106	137	184	196	205	214	214	212	216	224	230
Wood Exp.	101	103	105	105	106	108	109	116	123	121	118	116	116	111
Food Proc.	192	179	172	170	212	262	256	261	275	276	277	278	286	274
Food Ma.	101	112	127	121	129	162	176	190	208	209	209	214	223	227
Beverage	68	80	94	93	100	116	125	136	148	150	148	152	159	176
Tobacco	14	16	20	20	21	23	25	26	28	29	30	31	32	33
Textile	560	576	609	633	687	767	673	751	984	1017	1029	1051	1091	1054
Apparel	186	173	164	164	203	258	272	278	285	287	296	315	337	361
Leather	77	78	80	77	85	105	116	123	127	126	128	137	148	170
Wood Proc.	73	62	51	51	72	103	111	112	115	116	118	119	121	136
Furniture	69	53	38	38	57	81	86	88	89	85	81	80	81	76
Paper	113	102	94	93	111	146	128	145	198	207	210	217	228	235
Printing	80	97	115	117	107	102	109	115	120	121	121	125	129	137
Cultural Articles	40	45	50	49	50	49	48	55	60	60	64	71	75	89
Petroleum Pro.	36	40	45	48	46	44	39	46	61	66	70	76	84	75
Chemical	300	270	241	236	291	359	332	355	430	453	468	487	509	529
Medicine	45	52	61	63	61	60	55	62	79	84	88	95	102	108
Fibres	18	23	29	32	29	26	27	32	39	41	43	46	50	54
Rubber	61	71	83	85	79	77	64	70	95	99	100	104	110	112
Plastic	127	129	136	134	150	176	128	136	202	203	201	210	225	233
Nonmetal Ma.	738	656	586	569	726	975	862	953	1225	1218	1174	1151	1166	1217
Ferrous Press	227	231	237	238	246	264	260	273	310	328	336	340	356	427
Non-Ferrous Pr.	66	63	59	59	68	79	84	91	99	104	109	113	117	132
Metal Products	287	256	227	228	273	326	234	258	388	387	381	387	399	417
General Mac.	476	452	432	434	494	577	602	637	690	701	688	701	737	651
Special Mac.	409	332	257	259	334	410	412	413	414	415	415	415	417	397
Transport Eq.	250	266	283	285	282	287	298	305	311	317	322	333	349	426
Electrical Eq.	193	199	206	207	217	239	259	275	288	293	297	310	326	356
Computer etc.	116	122	129	130	134	142	148	155	163	166	172	185	196	206
Measuring Inst.	67	69	71	72	74	78	80	81	82	82	82	83	85	114
Electric Power	103	101	102	108	118	126	128	133	144	153	163	175	184	199
Gas Prod.	10	10	11	11	11	13	14	15	17	18	20	21	22	20
Water Prod.	13	18	22	23	21	19	20	21	23	24	26	28	30	35
Others	229	239	248	220	270	313	291	336	373	371	360	363	381	405

1994	1995	1996年	1997年	1998	1999	2000	2001	2002	2003	2004	2005	2006	2007	2008	2009	2010
690	684	677	667	658	614	570	533	535	527	540	575	581	553	573	551	551
117	120	121	120	119	130	69	73	69	92	98	104	106	98	115	103	107
63	60	50	48	47	48	49	50	51	57	62	79	81	82	96	84	92
86	87	87	84	81	75	69	63	60	57	54	58	63	77	76	71	80
236	230	226	217	208	183	161	144	131	118	113	106	109	115	133	135	138
131	130	114	213	397	233	150	107	79	67	56	54	54	53	52	52	53
297	334	357	371	386	344	306	292	291	293	296	340	360	393	461	487	525
211	187	179	182	185	170	156	148	156	156	160	178	186	194	219	228	243
187	190	177	174	171	161	152	138	130	124	122	124	131	145	165	177	197
39	42	35	33	31	28	26	25	23	21	20	20	19	19	20	20	21
1009	963	921	834	755	721	687	685	693	728	763	835	837	821	825	754	766
381	380	375	359	343	355	365	389	424	448	482	511	547	590	642	618	605
230	221	204	190	177	179	181	201	220	254	276	330	337	336	341	308	317
178	177	143	142	141	139	137	133	128	150	158	178	186	205	242	231	241
101	104	86	83	79	76	74	75	80	95	108	135	148	151	162	144	154
244	254	255	231	210	201	191	193	195	194	201	213	212	210	222	216	216
139	137	133	123	114	109	103	104	108	119	127	135	136	140	156	154	156
113	122	110	101	92	95	98	102	117	136	149	165	164	164	174	154	155
88	90	87	89	90	87	78	72	69	74	78	89	89	91	94	90	95
514	537	546	531	516	490	464	432	426	433	445	471	487	509	565	570	604
123	124	126	122	119	117	116	119	120	130	132	138	146	155	170	182	196
71	74	63	65	68	61	55	49	45	39	43	48	48	50	50	45	48
116	116	113	111	108	104	100	95	99	102	109	126	122	123	129	123	124
231	237	247	246	246	240	235	241	260	276	292	332	346	366	398	387	404
1261	1279	1277	1194	1116	1027	944	878	844	839	840	844	843	868	946	946	993
450	438	402	389	377	352	328	308	291	307	309	331	333	334	335	338	353
175	182	144	141	137	135	132	137	129	135	147	162	165	183	212	198	208
439	424	403	390	377	356	337	331	338	323	350	392	420	446	515	486	507
622	603	591	567	544	515	486	464	452	485	528	587	604	649	737	704	758
418	423	385	362	340	323	306	274	264	304	310	322	340	367	436	433	463
462	490	473	445	418	407	395	384	387	409	433	459	480	516	589	612	694
388	382	371	350	331	333	336	332	353	394	445	521	547	582	655	637	691
226	230	221	221	222	236	250	263	296	355	435	550	607	679	753	711	800
107	108	113	118	124	112	102	95	93	110	114	126	136	144	153	144	156
212	221	213	225	238	251	264	262	268	276	279	294	302	300	303	324	322
27	27	16	20	26	22	22	19	19	18	18	18	18	20	22	22	24
41	45	45	48	51	52	54	52	56	57	58	61	64	60	68	74	80
419	410	410	400	390	341	300	265	236	212	209	234	245	240	255	235	233

Table A.6 Sub-industrial energy consumption (unit: 10 thousand tce)

	1980	1981	1982	1983	1984	1985	1986	1987	1988	1989	1990	1991	1992	1993	1994
Coal Mi.	2596	2683	2769	2856	2942	2965	3180	3433	3534	3815	4159	4599	4411	4660	4780
Petroleum Ext.	1200	1151	1102	1053	1004	1496	1555	1643	1828	1825	1930	2064	2220	3088	3150
Ferrous Mi.	250	256	263	269	275	288	316	330	320	310	290	300	305	297	283
Non-Ferrous Mi.	270	283	295	308	320	326	347	355	365	375	390	405	410	417	487
Nonmetal Mi.	200	208	215	223	230	226	287	350	400	450	490	480	500	554	554
Wood Exp.	160	165	170	175	180	185	183	190	200	210	220	240	260	266	243
Food Proc.	778	820	862	903	945	1012	1122	1200	1235	1300	1250	1300	1481	1268	1620
Food Ma.	400	425	450	475	500	650	600	800	850	900	1000	1000	1000	1102	1155
Beverage	220	265	310	355	400	550	680	750	880	900	850	900	920	999	1000
Tobacco	80	85	90	95	100	100	120	150	170	181	179	181	180	186	205
Textile	1992	2114	2236	2358	2480	2410	2467	2656	2875	3015	3034	3113	3325	3347	3439
Apparel	51	56	60	65	69	97	96	100	120	140	160	180	200	223	276
Leather	88	89	89	90	90	119	119	150	170	190	180	200	220	232	261
Wood Proc.	390	399	408	416	425	222	210	220	240	260	280	300	320	322	351
Furniture	60	63	65	68	70	74	74	75	85	80	90	95	100	106	106
Paper	894	928	962	995	1029	1293	1376	1484	1583	1697	1694	1735	1895	1824	1981
Printing	40	45	50	55	60	68	98	105	110	120	140	160	180	187	195
Cultural Articles	40	40	40	40	40	39	37	40	45	42	49	47	55	60	61
Petroleum Pro.	1220	1253	1285	1318	1350	1409	1703	1878	1990	2200	2208	2428	2834	3215	3591
Chemical	8394	8594	8794	8993	9193	8104	8569	9636	10194	10946	10986	11531	12019	16571	16196
Medicine	400	413	425	438	450	452	490	568	597	646	653	738	801	920	1061
Fibres	300	325	350	375	400	512	575	689	701	729	749	856	932	958	993
Rubber	280	298	315	333	350	378	401	420	460	480	520	550	600	617	630
Plastic	150	165	180	195	210	233	238	250	290	350	400	450	500	540	541
Nonmetal Ma.	4265	4587	4908	5230	5551	7960	8564	9248	9925	10206	9722	10198	10904	12054	12556
Ferrous Press	7784	7904	8024	8144	8264	7304	8280	8890	9445	9980	10555	11154	11922	12144	15339
Non-Ferrous Pr.	700	800	900	1000	1100	1163	1277	1552	1687	1797	1891	2047	2297	2269	2555
Metal Products	500	550	600	650	700	703	760	762	765	770	790	790	850	860	927
General Mac.	1350	1368	1386	1403	1421	1270	1300	1450	1475	1500	1480	1500	1620	1886	1768
Special Mac.	700	706	713	719	725	665	680	800	900	850	830	850	900	1117	1103
Transport Eq.	600	606	613	619	625	565	580	700	730	750	730	750	800	1053	1215
Electrical Eq.	250	256	263	269	275	235	250	350	390	360	340	360	400	610	619
Computer etc.	130	131	133	134	135	113	120	190	200	225	230	250	300	421	371
Measuring Inst.	120	125	130	135	140	111	114	120	125	120	123	131	150	154	149
Electric Power	1875	1960	2044	2129	2213	2528	2757	2968	3366	3657	3867	4237	4941	5027	5790
Gas Prod.	200	208	215	223	230	240	250	260	270	280	300	320	350	340	371
Water Prod.	150	158	165	173	180	190	228	235	250	270	290	320	350	387	438
Others	511	575	639	702	766	817	867	918	948	969	999	1020	1120	1223	1493

1995	1996	1997	1998	1999	2000	2001	2002	2003	2004	2005	2006	2007	2008	2009	2010
5500	5367	5791	5545	4324	4081	4053	4242	5396	6343	6917	6787	7171	9356	10207	11057
2813	2731	3565	3353	3535	3749	4006	4518	4618	3626	3761	3626	3677	4210	3946	3682
268	322	350	364	328	334	351	400	554	692	946	1112	1314	1408	1251	1094
557	615	468	375	359	376	408	427	571	623	664	719	820	863	833	803
553	578	536	531	543	576	622	655	778	857	863	897	947	1028	1096	1163
220	234	213	188	168	148	141	126	120	115	110	105	105	110	100	90
1973	1907	1833	1889	1611	1404	1479	1605	1548	1821	2034	2156	2336	2731	2795	2859
1208	1181	980	975	983	858	879	947	862	1026	1169	1256	1322	1545	1563	1582
1000	988	772	835	696	618	634	663	703	849	880	949	980	1162	1191	1221
224	330	258	245	297	252	265	259	265	238	238	232	230	233	234	235
3531	3332	3080	2842	2503	2497	2679	2984	3469	4550	4978	5756	6208	6396	6251	6106
329	306	278	335	312	298	333	355	399	473	547	623	676	725	713	701
290	188	160	202	188	174	189	210	243	279	310	347	375	389	384	380
380	360	339	310	306	286	321	324	421	552	691	767	829	982	1049	1116
106	165	162	80	92	84	95	88	108	111	129	143	148	182	184	186
2138	2200	1943	1916	1741	1827	1937	2181	2371	3081	3274	3444	3343	3999	4101	4203
203	194	176	172	176	179	202	197	365	338	274	295	324	350	357	365
62	86	79	139	104	110	131	155	147	184	196	199	208	220	215	209
5567	3665	7389	6868	7085	7411	7837	8479	8991	12174	11882	12360	13177	13747	15328	16910
15822	20118	15702	14045	12885	12701	12886	14508	17108	20347	22494	24779	27245	28961	28946	28931
1201	1008	835	827	797	760	841	845	1026	1041	1122	1161	1183	1360	1355	1349
1278	1080	1435	1600	1538	1678	1705	1943	2200	1303	1342	1424	1554	1449	1437	1425
644	702	624	661	586	578	646	644	738	884	1079	1180	1259	1336	1345	1354
542	644	684	627	565	614	659	703	819	1129	1447	1558	1626	1852	1895	1938
13058	13747	12318	11635	10962	10101	9981	10625	12656	18088	18850	19948	20355	25461	26882	28304
18533	18214	18156	17012	16961	16792	17136	19327	24070	29702	35988	42812	47774	51863	56404	60946
2842	3040	3292	3385	3540	3605	3893	4373	5409	6404	7189	8633	10686	11288	11401	11515
994	1105	1045	1037	1037	1064	1229	1482	1699	1967	2220	2576	2832	3024	3038	3052
1651	1762	1559	1369	1156	1088	1162	1325	1523	1706	1982	2336	2587	2758	2985	3212
1089	1011	889	826	788	744	754	782	924	1147	1243	1344	1441	1630	1672	1713
1376	1431	1519	1442	1299	1278	1441	1556	1654	2080	1950	2134	2377	2733	3032	3331
629	635	647	618	573	558	594	725	890	1119	1191	1323	1543	1791	1854	1918
321	344	493	514	591	628	683	799	1044	1272	1474	1743	2007	2197	2216	2235
143	150	83	133	140	137	146	169	199	172	194	228	259	285	292	299
7053	8431	10076	9346	9537	9689	9727	11151	13277	14578	15803	17417	18475	18676	19575	20473
341	370	427	511	672	560	469	548	513	536	629	646	616	635	566	498
489	472	558	542	581	563	573	544	550	653	691	749	802	834	875	916
1264	1034	1368	1113	1240	1238	1259	1318	1517	1299	1416	1476	1461	1643	1728	1813

1 The principles for creating the sub-industrial database.

In the course of creating the industrial sectoral databases, several problems face us. The first problem is the inconsistency of data in the1990s compared with earlier data, especially 1980–1984, due to different industrial classification. We corrected these data by re-classifying and re-combining them, according to the corresponding relationship between the old and new industrial classification criterion provided by the fourth appendix of the 1988 China Industry Economy Statistical Yearbook (p. 373). The second issue is the inconsistency of the statistical scope. Before 1997, the sectoral data came from the industrial enterprises using independent accounting systems, including both urban industry and rural industry at the township level (*xiangjixiangyishang*), while the sectoral data since 1998 included state-owned and non-state-owned industrial enterprises only above a designated size, those with annual revenue from the principal business of more than 5 million RMB. Fortunately, the China Statistical Yearbook also provides the data of industrial sectors at the village level of rural industry (*cunbangongye*) before 1997 and the China Economic Census Yearbook reports the sub-industrial data of non-state-owned industrial enterprises with annual revenue from the principal business below the designated size of 5 million RMB (*guimoyixia*) in 2004 and 2009. We added the former into the database information before 1997 and used the latter, calculating the expanding proportion, to adjust the data after 1998, in order to form the input and output databases with the same statistical scope at the level of all industrial enterprises for the comparability of the data over the entire reform period.

2 Estimate of Stock Capital.

One of the two sets of capital stock is estimated according to the underlying relationships behind the original and net value of fixed assets provided by the China Statistical Yearbook, following these steps.

Step 1: Calculate the rate of depreciation δ for each sector over time using the expressions:

$$cd_t = ovfa_t - nvfa_t; \quad CD_t = cd_t - cd_{t-1}; \quad \delta_t = CD_t / ovfa_{t-1}$$

where cd is the value of cumulative depreciation, $ovfa$ is the original value of fixed assets, $nvfa$ is the net value of fixed assets and CD represents the current depreciation.

Step 2: Calculate the gross investment at 1990 price using:

$$inv_t = ovfa_t - ovfa_{t-1}; \quad I_t = inv_t / P_{K,t}$$

where inv is gross investment at current price for each year, I is gross investment at constant 1990 price depreciated according to the price indices of investment in fixed assets, P_K.

Step 3: Determine the original capital stock in the first year of 1980.

Set the net value of fixed assets in 1980 at the 1990 price level to be the original capital stock for that same year of 1980.

Step 4: Estimate the capital stock according to the perpetual inventory approach:

$$K_t = I_t + (1 - \delta_t) \times K_{t-1}$$

3   Estimates of energy-induced carbon dioxide (CO_2) emissions.
According to the World Bank definition, CO_2 emissions are those stemming from the burning of fossil fuels and the manufacture of cement, the former of which accounts for at least 70 per cent of total CO_2 emission. Therefore, the definition of CO_2 emission used in this book is only related to fossil energy combustion; that is to say, CO_2 emission is computed from the consumption of primary solid coal, liquid oil, and gas fuels by using the following expression:

$$CO_2 = \sum_{i=1}^{3} CO_{2,i} = \sum_{i=1}^{3} E_i \times NCV_i \times CEF_i \times COF_i \times (44/12)$$

where CO_2 represents the flow of carbon dioxide with the unit of 10 thousand tons, $i = 1,2,3$ corresponding to three types of primary energy (coal, oil and gas), and E is their respective consumption. NCV is net calorific value provided by the China Energy Statistical Yearbook in 2007, CEF is carbon emission factor provided by the IPCC (2006) and COF is the carbon oxidization factor set to be 1 for both oil and gas and 0.99 for coal in this study. Therefore, the calculated CO_2 emission coefficients for coal, oil and gas are 2.763, 2.145 and 1.642 tons of CO_2 per ton coal equivalent, respectively, in the case of China.

This book also constructs the input and output database with three dimensions, that is, for six industries such as agriculture, industry, construction, transport, commerce and the household sector for 30 Chinese provinces between 1995 and 2007, which is used in Chapter 2. We will not report the sample of the three dimensions here. To evaluate the regional low-carbon transformation process in China in Chapter 8, the provincial input and output panel data for 31 provinces over the reform period was also constructed. The capital stock and CO_2 emission could not be observed directly and needed estimating, following the same approach as that for creating the industrial sectoral database. Specifically, the constructed variables include two desirable outputs (regional GDP and regional gross output value), five undesirable outputs (carbon dioxide emission; wasteful water; wasteful gas; sulphur dioxide, SO_2; chemical oxygen demand, COD), and seven inputs (energy consumption, electricity consumption, investment in the treatment of industrial pollution, investment in the treatment of environmental pollution, area of afforestation, labour force, capital stock). The variables of regional GDP, three environmental pollutants of CO_2, COD and SO_2, capital stock, labour and energy consumption are listed in Tables A.7–A.13, respectively. All the value-type variables are depreciated at the 2005 price level.

Table A.7 Provincial GDP (2005=100, unit: 100 million RMB)

Provinces	1980	1981	1982	1983	1984	1985	1986	1987	1988	1989	1990	1991	1992	1993	1994
Beijing	594	591	635	739	868	943	1019	1116	1259	1315	1383	1520	1692	1900	2160
Tianjin	338	354	369	400	477	528	558	601	636	646	681	722	806	904	1033
Hebei	728	735	822	916	1048	1179	1239	1383	1570	1666	1762	1956	2261	2662	3058
Shanxi	374	377	435	496	603	646	688	724	780	821	862	898	1010	1142	1260
Inner Mongolia	234	259	307	337	391	459	486	530	581	597	642	690	766	856	951
Liaoning	863	849	894	1013	1183	1340	1451	1656	1850	1907	1924	2042	2289	2630	2924
Jilin	342	362	390	475	534	570	612	727	842	821	849	899	1009	1137	1247
Heilongjiang	778	808	861	935	1039	1101	1140	1238	1344	1429	1512	1612	1717	1844	1998
Shanghai	810	855	917	989	1103	1251	1306	1404	1546	1592	1648	1765	2026	2332	2670
Jiangsu	911	1009	1109	1245	1441	1690	1865	2116	2531	2593	2723	2949	3704	4437	5168
Zhejiang	630	702	782	845	1028	1251	1403	1568	1744	1733	1801	2121	2525	3080	3696
Anhui	436	514	562	611	734	847	941	987	1040	1094	1126	1116	1303	1545	1770
Fujian	333	385	421	447	527	620	655	744	850	917	985	1125	1354	1660	1997
Jiangxi	391	413	451	482	556	638	681	738	822	872	911	986	1132	1287	1400
Shandong	1063	1125	1252	1426	1674	1865	1982	2256	2538	2639	2779	3185	3723	4482	5209
Henan	807	870	907	1123	1236	1403	1468	1688	1853	1983	2072	2215	2519	2917	3319
Hubei	601	640	717	759	918	1066	1125	1220	1315	1374	1443	1539	1756	1985	2256
Hunan	725	765	836	913	999	1119	1210	1322	1431	1482	1542	1663	1848	2077	2297
Guangdong	881	960	1075	1153	1332	1572	1772	2120	2455	2631	2935	3454	4218	5188	6210
Guangxi	392	423	476	492	526	584	621	679	709	735	786	886	1048	1240	1429
Hainan												171	196	278	335	373
Chongqing															
Sichuan	733	763	846	939	1054	1180	1245	1354	1455	1501	1638	1787	2012	2275	2533
Guizhou	213	227	263	296	355	383	404	448	486	508	530	579	625	691	749
Yunnan	330	356	411	445	510	576	601	675	783	828	900	960	1064	1182	1326
Tibet	25	30	30	29	36	42	38	38	39	43	47	47	50	58	67
Shaanxi	333	348	379	407	479	559	607	668	808	835	863	925	1002	1123	1220
Gansu	195	178	194	223	254	287	319	348	395	430	454	484	531	593	657
Qinghai	71	70	78	87	98	109	118	124	134	136	141	147	158	173	188
Ningxia	63	64	69	80	91	107	116	125	141	151	157	165	179	197	212
Xinjiang	222	240	264	300	342	400	447	492	539	572	639	730	826	910	1021

1995	1996	1997	1998	1999	2000	2001	2002	2003	2004	2005	2006	2007	2008	2009
2419	2637	2903	3179	3525	3941	4403	4909	5449	6217	6970	7876	9018	9838	10842
1187	1357	1521	1662	1828	2026	2269	2557	2935	3399	3906	4480	5174	6028	7022
3483	3953	4448	4924	5372	5882	6394	7007	7820	8829	10012	11354	12807	14101	15511
1411	1578	1756	1930	2071	2265	2494	2816	3236	3727	4231	4772	5531	6001	6325
1047	1198	1328	1470	1599	1772	1961	2220	2618	3154	3905	4651	5544	6531	7634
3132	3401	3704	4011	4340	4726	5152	5677	6330	7140	8047	9190	10568	11985	13555
1368	1553	1693	1847	1998	2182	2385	2612	2878	3230	3620	4163	4834	5607	6370
2182	2405	2645	2865	3080	3332	3642	4014	4423	4941	5514	6181	6923	7739	8622
3052	3452	3894	4295	4742	5263	5816	6473	7269	8301	9248	10422	12006	13171	14251
5964	6693	7493	8319	9158	10126	11154	12455	14149	16243	18599	21370	24554	27672	31104
4317	4865	5405	5956	6552	7273	8044	9057	10389	11895	13418	15283	17529	19300	21017
2023	2279	2546	2757	3008	3258	3548	3889	4254	4820	5350	6019	6874	7747	8746
2288	2592	2955	3274	3599	3933	4276	4712	5253	5873	6555	7525	8669	9795	11000
1495	1670	1876	2009	2165	2339	2544	2812	3177	3596	4057	4556	5157	5838	6603
5938	6656	7395	8194	9013	9941	10936	12215	13852	15971	18367	21067	24058	26945	30233
3811	4340	4792	5214	5636	6171	6727	7366	8154	9271	10587	12112	13880	15560	17256
2554	2849	3188	3464	3732	4053	4412	4818	5287	5879	6590	7460	8549	9695	11004
2534	2840	3142	3409	3695	4027	4390	4785	5244	5879	6596	7440	8556	9746	11081
7176	7984	8878	9837	10833	12075	13342	14993	17218	19765	22557	25896	29754	32849	36035
1591	1723	1862	2048	2212	2387	2585	2859	3151	3523	3984	4526	5209	5876	6693
387	405	433	469	509	555	606	664	734	813	898	1017	1177	1298	1450
			1768	1902	2064	2250	2482	2767	3104	3468	3898	4517	5172	5943
2805	3102	3429	3760	4009	4349	4739	5225	5818	6559	7385	8382	9597	10653	12198
805	876	955	1036	1128	1222	1330	1451	1597	1779	2005	2262	2597	2890	3220
1482	1646	1807	1954	2096	2254	2407	2624	2855	3179	3462	3863	4335	4794	5374
79	89	100	112	126	139	157	177	198	222	249	282	321	354	398
1347	1494	1653	1845	2035	2247	2467	2741	3064	3460	3934	4481	5188	6039	6861
725	811	885	971	1059	1162	1275	1401	1551	1730	1934	2156	2422	2666	2941
203	220	240	261	283	308	344	386	431	484	543	616	699	793	873
232	257	278	302	330	363	400	441	497	552	613	690	778	876	980
1113	1186	1285	1382	1484	1613	1752	1896	2108	2348	2604	2891	3243	3600	3892

Table A.8 Provincial carbon dioxide emission (10 thousand tons)

Provinces	1980	1981	1982	1983	1984	1985	1986	1987	1988	1989	1990	1991	1992	1993	1994
Beijing	4814	4821	4882	5062	5482	5668	6144	6354	6723	6809	6981	7400	7694	8445	8827
Tianjin	3981	4110	4238	4367	4495	4624	4752	4880	5009	5137	5235	5111	5499	5889	6024
Hebei	8289	9764	10507	11228	11952	12233	13626	14839	16043	16606	16457	17411	18480	21122	22009
Shanxi	8296	7998	8570	8427	8783	9543	9979	10230	11167	9630	9213	9322	9602	10202	10581
Inner Mongolia	2620	2699	2778	2858	2937	2970	2948	3033	3053	3414	3535	3691	3655	7159	7646
Liaoning	13183	12772	10428	13608	13933	16199	16278	16686	17797	18373	18642	18754	18741	22459	23892
Jilin	4856	4531	5314	5601	6235	6567	6649	7279	8013	8581	7474	7566	7604	8083	8039
Heilongjiang	8960	9216	9827	10452	11404	11443	11722	12454	13001	13599	14042	14313	13975	12758	14560
Shanghai	5810	5936	6062	6188	6315	6431	7208	7547	7701	7903	8247	8949	9438	10070	10532
Jiangsu	9848	9975	10101	10227	10354	10412	11143	12690	14372	14669	14638	15165	16580	17343	19219
Zhejiang	3620	3833	4046	4259	4472	4685	4898	5111	5324	5537	5820	6651	7420	8613	9576
Anhui	4757	5022	5286	5550	5814	6079	6343	6607	6872	7136	7318	7692	8370	8856	9802
Fujian	1469	1405	1521	1683	1873	2096	2335	2587	2835	3017	3079	3437	3350	3926	4019
Jiangxi	2724	2848	2972	3096	3220	3328	3700	3999	4349	4382	4291	4442	4635	4810	5002
Shandong	9847	10624	11402	12179	12956	13734	14511	15289	16066	16843	17698	18657	19435	20730	21767
Henan	9096	9681	9509	10769	11960	12300	12506	13309	14038	13533	13788	14255	14852	15583	16509
Hubei	4595	4973	5101	5473	5888	6648	7126	7936	8445	8616	8251	9096	9542	10523	11758
Hunan	7588	7703	7818	7933	8048	8163	8278	8393	8508	8623	8786	8841	8899	8956	9014
Guangdong	4164	3934	4200	4507	4815	5111	5465	6168	7141	7978	7679	8798	9693	10769	12057
Guangxi	1548	1453	1546	1621	1710	1736	1769	2079	2274	2418	2593	2830	3196	3337	3820
Hainan											250	316	386	478	542
Chongqing															
Sichuan	5338	5424	5685	6005	6285	6721	6876	7539	8237	8514	8217	8448	8664	8902	9150
Guizhou	1335	1254	1342	1459	1578	1976	2170	2824	3132	3215	3109	3373	3744	4006	4570
Yunnan	2022	1994	2229	2466	2715	2827	3040	3423	3716	3796	4263	4080	4302	4493	4633
Tibet	183	190	191	210	220	251	256	258	291	310	339	401	424	452	428
Shaanxi	3141	3179	3343	3813	4156	4674	4929	5368	5453	5566	5710	6006	6206	6514	6835
Gansu	3040	3192	3344	3446	3547	3628	3932	4337	4559	4517	4660	5119	5351	5523	5922
Qinghai	566	550	597	631	663	695	649	700	752	803	862	710	685	710	820
Ningxia	2090	2039	1986	2075	2249	2534	2513	2640	2727	2738	1266	1185	1158	1127	1710
Xinjiang	2564	2666	2671	2938	3078	3509	3582	3605	4067	4340	4740	5101	5545	6050	6317

1995	1996	1997	1998	1999	2000	2001	2002	2003	2004	2005	2006	2007	2008	2009
9168	9611	9574	9808	10005	10612	10827	11223	11820	12943	13830	14660	15542	15687	15833
6598	6422	6299	6427	6508	7086	7393	7694	8178	9465	10363	10920	11726	12512	13490
23958	24088	24350	24637	25270	30245	32805	36242	41508	46875	53440	58800	63659	65630	67623
13588	14102	13732	13603	11404	11959	13759	15996	17917	18282	18179	19722	21665	22194	22793
7707	8462	9900	9423	9924	10519	11895	13793	17901	23200	27992	32569	37644	41791	45950
24344	24592	24169	23384	23283	25749	26757	26974	29864	32634	35221	39047	43222	46240	49296
8077	8305	8460	7159	7150	6795	7142	8248	9492	10531	15281	17114	18413	20470	22353
16332	15839	16748	16772	15901	13938	14308	15347	15656	18920	19101	19237	20066	21192	22182
11312	11647	11991	12263	13039	13891	14841	15487	16630	17507	18873	19725	20909	22289	23470
20667	20649	20407	20462	20937	21668	21843	23596	26836	32171	40836	44724	48457	49103	51744
10332	11001	11600	12047	12693	13971	15447	17633	20280	23053	25623	29203	32829	33237	34520
11229	12101	11803	12241	12529	13119	13766	14259	14642	16148	17443	18988	20719	22158	23724
4406	4879	4643	4931	5483	5899	5991	7406	9029	10429	13503	14825	17142	18130	19592
5790	5225	5024	4820	5132	5813	6154	6943	8689	9039	10161	11129	12303	12720	13482
22751	23062	23322	23353	23485	32894	36242	42454	47870	56089	66180	74320	80615	84444	88713
17190	17670	17848	19245	19598	21017	22077	23637	27845	34289	38479	42789	47131	49641	52133
12600	13356	13962	13792	14060	13949	14082	15138	17001	18861	20783	23446	26030	25448	26947
9071	9129	9186	9244	9301	9360	10613	11228	12767	14950	21229	22523	24518	25837	27109
13039	13261	13087	13749	14035	15462	16433	18160	21196	23393	28413	31427	34206	36295	38159
3973	4125	3815	4051	4292	4506	5129	5157	6402	7831	9592	9506	12251	12728	13707
643	701	746	838	880	965	1032	1237	1441	1565	1653	1805	2137	2245	2394
			5277	5674	5801	6127	6387	6836	7856	9703	10531	11918	12662	13431
9508	10019	10872	11336	12186	11746	12021	13822	17686	19504	19740	21945	24087	25362	26883
4883	5261	5695	6096	10480	10676	10297	10777	13954	16255	18254	20740	22353	23303	24086
5308	5557	7318	7196	6806	6657	7400	8057	8704	10470	11988	14197	15111	15640	16475
416	435	434	447	430	435	436	439	465	509	519	515	559	556	575
7605	7984	8138	7648	6699	6599	7641	8678	9982	11964	13761	15377	16621	17917	19356
6200	6500	5911	6097	6581	6827	6945	7258	8071	8974	9940	10904	11682	12164	12761
928	938	1222	1149	1374	1184	1207	1310	1493	1804	2174	2733	2934	3523	3776
1778	1842	1847	1827	1815	1907	2054	2377	3949	6043	6564	7350	8029	8412	8833
6645	7397	7694	7830	7666	7952	8212	8573	9537	11306	12868	14097	15395	16650	17812

Table A.9 Provincial COD emission (unit: 10 thousand tons)

Provinces	2000	2001	2002	2003	2004	2005	2006	2007	2008	2009
Beijing	18	17	15	13	13	12	11	11	10	10
Tianjin	19	11	10	13	14	15	14	14	13	13
Hebei	71	65	64	64	66	66	69	67	61	57
Shanxi	32	31	31	36	38	39	39	37	36	34
Inner Mongolia	26	28	24	27	28	30	30	29	28	28
Liaoning	70	68	59	55	50	64	64	63	58	56
Jilin	48	41	36	37	37	41	42	40	37	36
Heilongjiang	52	53	51	51	51	50	50	49	48	46
Shanghai	32	31	33	34	29	30	30	29	27	24
Jiangsu	65	83	78	77	85	97	93	89	85	82
Zhejiang	63	58	58	56	56	60	59	56	54	51
Anhui	44	42	41	41	43	44	46	45	43	42
Fujian	32	31	28	35	36	39	40	38	38	38
Jiangxi	39	42	39	42	45	46	47	47	45	44
Shandong	100	92	86	83	78	77	76	72	68	65
Henan	82	76	74	71	70	72	72	69	65	63
Hubei	70	67	66	63	61	62	63	60	59	58
Hunan	67	71	74	81	85	90	92	90	89	85
Guangdong	95	111	95	98	93	106	105	102	96	91
Guangxi	103	83	85	93	99	107	112	106	101	98
Hainan	9	7	7	7	9	10	10	10	10	10
Chongqing	26	25	25	26	27	27	26	25	24	24
Sichuan	98	99	94	94	88	78	81	77	75	75
Guizhou	23	21	21	22	22	23	23	23	22	22
Yunnan	30	31	30	29	29	29	29	29	28	27
Tibet	4	1	1	1	1	1	2	2	2	2
Shaanxi	33	33	32	32	34	35	36	35	33	32
Gansu	14	12	13	16	16	18	18	17	17	17
Qinghai	3	3	3	3	4	7	8	8	8	8
Ningxia	18	19	11	10	7	14	14	14	13	13
Xinjiang	20	20	21	23	26	27	29	29	29	29

Table A.10 Provincial SO$_2$ emission (unit: 10 thousand tons)

Provinces	2000	2001	2002	2003	2004	2005	2006	2007	2008	2009
Beijing	22	20	19	18	19	19	18	15	12	12
Tianjin	33	27	24	26	23	27	26	25	24	24
Hebei	132	129	128	142	143	150	155	149	135	125
Shanxi	120	120	120	136	142	152	148	139	131	127
Inner Mongolia	66	65	73	129	118	146	156	146	143	140
Liaoning	93	84	79	82	83	120	126	123	113	105
Jilin	29	27	27	27	29	38	41	40	38	36
Heilongjiang	30	29	29	36	37	51	52	52	51	49
Shanghai	47	47	45	45	47	51	51	50	45	38
Jiangsu	120	115	112	124	124	137	130	122	113	107
Zhejiang	59	59	62	73	81	86	86	80	74	70
Anhui	40	40	40	46	49	57	58	57	56	54
Fujian	23	20	19	30	33	46	47	45	43	42
Jiangxi	32	31	29	44	52	61	63	62	58	56
Shandong	180	172	169	184	182	200	196	182	169	159
Henan	88	90	94	104	126	163	162	156	145	136
Hubei	56	54	54	61	69	72	76	71	67	64
Hunan	77	76	74	85	87	92	93	90	84	81
Guangdong	91	97	97	108	115	129	127	120	114	107
Guangxi	83	70	68	87	94	102	99	97	93	89
Hainan	2	2	2	2	2	2	2	3	2	2
Chongqing	84	72	70	77	80	84	86	83	78	75
Sichuan	122	114	112	121	126	130	128	118	115	114
Guizhou	145	138	133	132	132	136	147	138	124	118
Yunnan	39	36	36	45	48	52	55	53	50	50
Tibet	0	0	0	0	0	0	0	0	0	0
Shaanxi	62	62	64	77	82	92	98	93	89	80
Gansu	37	37	43	49	48	56	55	52	50	50
Qinghai	3	4	3	6	7	12	13	13	14	14
Ningxia	21	20	22	29	29	34	38	37	35	31
Xinjiang	31	30	30	33	48	52	55	58	59	59

Table A.11 Provincial capital stock (2005=100, unit: 100 million RMB)

Provinces	1980	1981	1982	1983	1984	1985	1986	1987	1988	1989	1990	1991	1992	1993
Beijing	316	348	381	449	551	707	913	1186	1491	1771	2083	2364	2730	3191
Tianjin	591	635	704	789	905	1051	1194	1325	1445	1527	1598	1730	1871	2023
Hebei	1278	1318	1421	1536	1638	1803	1979	2177	2420	2623	2823	3106	3466	3850
Shanxi	819	840	890	972	1118	1328	1520	1693	1808	1882	1961	2069	2179	2311
Inner Mongolia	340	358	392	451	530	625	693	765	866	933	999	1082	1206	1397
Liaoning	2118	2105	2160	2240	2393	2641	2921	3277	3665	3972	4270	4635	4993	5575
Jilin	514	544	594	644	718	843	948	1076	1214	1278	1360	1482	1630	1892
Heilongjiang	870	995	1174	1313	1522	1773	2000	2224	2432	2556	2676	2812	2977	3153
Shanghai	779	895	1053	1211	1398	1620	1873	2174	2515	2759	2993	3224	3564	4031
Jiangsu	816	908	1114	1339	1603	1967	2449	2980	3587	4048	4544	5154	6168	7323
Zhejiang	188	204	224	244	279	342	412	491	569	627	687	764	882	975
Anhui	590	600	678	783	952	1171	1387	1600	1776	1925	2092	2226	2384	2645
Fujian	2094	2002	1937	1903	1890	1906	1960	2032	2076	2098	2115	2178	2290	2490
Jiangxi	528	550	593	653	719	798	915	993	1031	1125	1204	1276	1421	1621
Shandong	1982	2166	2418	2695	3083	3568	4080	4714	5303	5759	6224	6846	7583	8395
Henan	1180	1264	1379	1556	1731	1954	2200	2404	2739	2989	3216	3489	3756	4062
Hubei	1014	1026	1088	1171	1303	1475	1625	1813	1987	2100	2223	2369	2553	2824
Hunan	1296	1356	1452	1596	1708	1877	2109	2375	2661	2758	2814	2931	3123	3353
Guangdong	935	1054	1230	1396	1608	1847	2105	2365	2690	3016	3403	3857	4583	5690
Guangxi	941	979	971	990	1047	1120	1226	1332	1380	1417	1432	1494	1606	1813
Hainan											574	646	761	874
Chongqing														
Sichuan	1986	2076	2197	2358	2561	2818	3056	3367	3600	3639	3707	3823	3988	4214
Guizhou	478	500	527	568	625	710	768	827	889	921	964	1011	1065	1124
Yunnan	1194	1213	1228	1233	1264	1300	1341	1391	1464	1522	1614	1817	2065	2300
Tibet	24	24	25	27	35	49	56	62	69	77	87	102	122	150
Shaanxi	775	824	872	911	1010	1184	1432	1625	1823	1970	2091	2213	2360	2550
Gansu	660	652	649	658	689	755	837	925	995	1065	1130	1197	1249	1281
Qinghai	168	172	187	194	201	220	235	256	278	290	310	327	345	367
Ningxia	294	293	301	314	332	371	413	455	483	498	515	540	564	594
Xinjiang	412	454	516	588	671	766	857	930	1029	1120	1253	1419	1643	1896

1994	1995	1996	1997	1998	1999	2000	2001	2002	2003	2004	2005	2006	2007	2008	2009
3867	4613	5246	5889	6690	7433	8237	9173	10388	11951	13668	15560	17604	19869	21337	23394
2245	2497	2794	3129	3530	3888	4283	4765	5340	6109	6987	8070	9403	11095	13351	16681
4346	5026	5916	6967	8122	9349	10498	11652	12829	14337	16303	18977	22164	25921	30689	36152
2468	2613	2770	3000	3362	3741	4131	4567	5106	5822	6787	8033	9572	11408	13311	16171
1591	1783	1942	2139	2351	2577	2829	3133	3647	4635	6075	8177	10638	13680	17194	22046
6138	6592	7021	7448	7942	8440	9076	9812	10669	11910	13876	16250	19348	23154	29522	34590
2147	2397	2664	2855	3087	3379	3744	4145	4632	5247	6061	7281	9326	12201	15953	19969
3402	3739	4131	4545	5069	5539	6019	6588	7217	7872	8660	9617	10880	12533	14473	17486
4830	5961	7235	8469	9626	10706	11828	13012	14369	15832	17577	19633	22017	24770	27239	30570
8441	9707	11137	12718	14585	16563	18762	21089	23659	27432	31834	37518	43819	50534	58025	67471
1110	1282	1488	1756	2126	2620	3273	4037	4955	5844	6794	12411	18181	24182	29956	36212
2964	3412	3898	4405	4914	5401	5926	6503	7150	7940	9100	10451	12063	14020	16325	19030
2868	3351	3899	4498	5221	5927	6632	7332	8089	9054	10300	11966	14063	16735	20025	23845
1838	2061	2323	2647	2977	3320	3680	4118	4779	5691	6787	8057	9495	11055	12661	14983
9184	10101	11221	12534	14093	15881	18038	20349	23161	26660	31194	36998	43671	50605	58147	68612
4540	5177	5949	6812	7795	8784	9825	10930	12220	13762	15703	18702	22817	28196	34502	42833
3278	3903	4672	5476	6352	7243	8171	9186	10206	11258	12554	14153	16287	18848	21671	25365
3627	3967	4419	4871	5398	5997	6658	7410	8250	9221	10385	11958	13841	16129	19155	22897
7008	8540	9977	11338	12997	14986	16962	19131	21740	25044	28693	33346	38551	44471	50386	58499
2081	2354	2650	2919	3256	3648	4039	4448	4931	5508	6294	7389	8854	10740	12897	16721
1017	1133	1204	1262	1326	1408	1488	1571	1671	1794	1937	2114	2331	2570	2933	3407
				2388	2803	3239	3766	4421	5309	6350	7603	8975	10526	12572	14507
4529	5008	5557	6185	6970	7723	8602	9567	10691	12045	13576	15453	17847	20779	24212	28152
1208	1309	1440	1610	1830	2097	2393	2790	3242	3749	4272	4859	5532	6303	7181	8279
2547	2836	3169	3546	4044	4530	4938	5353	5833	6480	7273	8330	9649	11190	12413	14357
187	254	307	363	428	495	544	553	599	684	788	909	1054	1219	1397	1623
2757	2992	3232	3492	3849	4250	4759	5297	5907	6790	7664	8841	10361	12321	15047	18144
1318	1375	1468	1594	1741	1931	2164	2466	2815	3211	3675	4197	4781	5464	6458	7382
391	422	472	539	619	707	813	958	1128	1315	1511	1730	1966	2227	2508	2933
626	657	688	727	784	862	959	1083	1232	1455	1703	1984	2302	2676	3173	3907
2235	2512	2792	3074	3415	3740	4120	4526	5043	5718	6444	7310	8291	9375	10352	11486

Table A.12 Provincial labour forces (unit: 10 thousand workers)

Provinces	1980	1981	1982	1983	1984	1985	1986	1987	1988	1989	1990	1991	1992	1993	1994
Beijing	484	512	535	552	556	567	573	580	584	594	627	634	649	628	664
Tianjin	395	413	421	436	447	456	467	471	465	470	470	480	486	503	513
Hebei	2183	2264	2347	2489	2534	2555	2626	2726	2808	2858	2955	3040	3106	3171	3210
Shanxi	1003	1032	1062	1080	1117	1154	1190	1223	1257	1282	1304	1332	1364	1384	1404
Inner Mongolia	698	731	762	799	828	857	875	891	910	910	925	963	976	1008	1033
Liaoning	1442	1505	1572	1639	1681	1769	1799	1835	1859	1875	1897	1938	1958	2006	2009
Jilin	715	755	850	848	867	930	988	1033	1106	1142	1169	1195	1235	1238	1250
Heilongjiang	1081	1118	1133	1196	1248	1290	1324	1333	1359	1395	1436	1482	1483	1500	1515
Shanghai	731	750	764	769	770	776	783	788	792	785	788	798	807	787	786
Jiangsu	2821	2911	2993	3057	3159	3263	3350	3430	3503	3520	4225	4273	4315	4340	4363
Zhejiang	1856	1955	2022	2141	2249	2319	2386	2445	2503	2523	2554	2579	2600	2616	2641
Anhui	2002	2078	2160	2234	2311	2421	2496	2563	2666	2724	2808	2877	2986	3157	3120
Fujian	964	1002	1028	1057	1102	1152	1189	1238	1281	1302	1348	1437	1490	1531	1554
Jiangxi	1356	1410	1434	1498	1537	1585	1623	1668	1723	1760	1817	1875	1870	1904	2008
Shandong	3118	3192	3270	3795	3564	3561	3651	3766	3887	3940	4043	4219	4303	4379	4382
Henan	2929	3039	3146	3289	3346	3520	3598	3782	3916	3943	4086	4216	4332	4400	4448
Hubei	1987	2045	2108	2145	2203	2238	2298	2349	2407	2433	3040	3083	3119	3158	3197
Hunan	2400	2449	2541	2594	2673	2729	2809	2904	2999	3091	3158	3222	3279	3346	3400
Guangdong	2368	2424	2521	2570	2637	2731	2812	2911	2995	3041	3118	3259	3367	3434	3493
Guangxi	1550	1605	1668	1713	1776	1831	1896	1961	2012	2046	2109	2171	2217	2275	2336
Hainan											305	318	322	333	336
Chongqing															
Sichuan	3260	3348	3468	3565	3643	3743	3886	3967	4090	4180	4265	4425	4521	4557	4588
Guizhou	1110	1153	1207	1234	1285	1335	1383	1436	1501	1571	1652	1701	1739	1779	1828
Yunnan	1404	1480	1544	1583	1620	1672	1731	1778	1827	1881	1923	1990	2033	2072	2109
Tibet	101	100	103	102	105	106	107	108	107	108	108	110	111	112	114
Shaanxi	1158	1202	1250	1285	1337	1375	1409	1449	1494	1529	1576	1640	1672	1708	1720
Gansu	796	842	870	994	1047	1081	1099	1140	1179	1214	1292	1302	1306	1418	1439
Qinghai	158	166	171	174	177	183	189	194	198	201	241	245	249	253	257
Ningxia	147	150	156	163	170	177	183	191	197	204	211	219	226	229	233
Xinjiang	506	524	535	546	559	566	575	585	594	600	618	638	647	656	658

1995	1996	1997	1998	1999	2000	2001	2002	2003	2004	2005	2006	2007	2008	2009
665	660	656	622	619	619	629	679	703	854	878	920	943	1174	1255
515	512	513	508	508	487	488	493	511	528	543	563	614	503	507
3252	3300	3324	3367	3322	3386	3409	3435	3470	3517	3569	3610	3665	3652	3900
1425	1441	1439	1398	1402	1392	1400	1403	1470	1475	1500	1561	1596	1583	1600
1029	1039	1050	1050	1057	1062	1067	1086	1005	1026	1041	1051	1082	1103	1142
2028	2032	1967	1959	1994	2052	2069	2025	2019	2097	2120	2128	2181	2098	2190
1271	1257	1238	1131	1120	1164	1167	1187	1203	1222	1239	1251	1266	1144	1185
1543	1558	1648	1700	1654	1601	1593	1603	1614	1681	1749	1784	1828	1670	1687
794	851	847	836	812	828	792	830	855	978	969	1005	1024	896	929
4385	4387	4389	4390	4391	4418	4434	4458	4469	4483	4510	4565	4618	4384	4536
2621	2625	2620	2613	2625	2726	2797	2859	2919	2992	3101	3172	3405	3692	3825
3207	3258	3322	3379	3399	3451	3463	3501	3545	3605	3670	3741	3818	3595	3690
1567	1594	1613	1622	1631	1660	1678	1711	1757	1814	1868	1950	2015	2080	2169
2101	2107	2121	2094	2089	2061	2055	2131	2168	2214	2277	2321	2370	2223	2244
5207	5228	5256	5288	5315	5442	5475	5527	5621	5728	5841	5960	6081	5352	5450
4509	4638	4820	5000	5205	5572	5517	5522	5536	5587	5662	5719	5773	5835	5949
3233	3276	3311	3328	3358	3385	3415	3443	3476	3507	3537	3564	3584	2876	3024
3467	3514	3560	3603	3601	3578	3608	3645	3695	3747	3801	3842	3883	3811	3908
3551	3641	3702	3784	3796	3989	4059	4134	4396	4682	5023	5250	5403	5478	5643
2383	2417	2454	2499	2515	2566	2578	2589	2601	2632	2703	2760	2769	2807	2863
334	333	342	327	327	335	338	350	360	371	379	382	400	412	431
			1711	1699	1690	1680	1655	1635	1624	1612	1605	1621	1837	1878
4619	4627	4641	4651	4654	4658	4665	4668	4684	4691	4702	4715	4731	4874	4945
1812	1783	1797	1844	1833	1866	2068	2106	2145	2186	2220	2235	2280	2302	2341
2149	2186	2224	2241	2244	2295	2323	2341	2353	2401	2461	2518	2574	2679	2730
115	118	120	120	124	124	126	130	133	137	144	148	158	160	169
1748	1776	1792	1788	1808	1813	1785	1874	1912	1941	1976	2011	2041	1947	1919
1483	1521	1530	1540	1489	1476	1489	1501	1511	1520	1391	1401	1415	1389	1407
262	266	270	275	279	284	287	291	295	297	298	304	312	277	286
241	245	257	255	272	276	279	282	291	298	300	308	310	304	329
676	684	715	681	694	673	685	701	721	744	792	812	830	814	829

Table A.13 Provincial energy consumption (unit: 10 thousand tce)

Provinces	1980	1981	1982	1983	1984	1985	1986	1987	1988	1989	1990	1991	1992	1993	1994
Beijing	1908	1903	1920	1985	2144	2211	2400	2476	2613	2653	2710	2872	2987	3265	3386
Tianjin	1550	1600	1650	1700	1750	1800	1850	1900	1950	2000	2038	1990	2141	2293	2345
Hebei	3121	3628	3929	4186	4475	4549	5080	5517	5962	6169	6124	6472	6866	7862	8169
Shanxi	3394	3269	3487	3478	3628	4000	4201	4394	4799	4719	4710	4802	5034	5473	5200
Inner Mongolia	1650	1700	1750	1800	1850	1871	1857	1967	2036	2250	2424	2505	2555	2676	2812
Liaoning	5272	5033	4111	5348	5454	6325	6360	6476	6825	7000	7171	7218	7192	8696	9205
Jilin	1930	1820	2117	2313	2496	2659	2772	3081	3284	3393	3523	3573	3615	3794	3856
Heilongjiang	3716	3796	3994	4231	4557	4581	4688	4942	5140	5371	5540	5656	5531	5078	5745
Shanghai	2300	2350	2400	2450	2500	2546	2854	2988	3049	3129	3265	3543	3736	3987	4170
Jiangsu	3900	3950	4000	4050	4100	4123	4382	4922	5508	5586	5509	5781	6297	6626	7358
Zhejiang	1700	1800	1900	2000	2100	2200	2300	2400	2500	2600	2733	3123	3484	4044	4497
Anhui	1800	1900	2000	2100	2200	2300	2400	2500	2600	2700	2769	2911	3138	3320	3672
Fujian	710	729	780	861	930	1043	1114	1215	1363	1404	1458	1531	1624	1848	1954
Jiangxi	1100	1150	1200	1250	1300	1344	1494	1615	1756	1769	1732	1793	1871	1946	2072
Shandong	3800	4100	4400	4700	5000	5300	5600	5900	6200	6500	6830	7200	7500	8000	8400
Henan	3389	3612	3560	4035	4474	4618	4709	5006	5292	5112	5206	5363	5583	5862	6225
Hubei	2011	2192	2389	2563	2756	3094	3291	3590	3870	4040	4002	4163	4472	4779	5239
Hunan	3300	3350	3400	3450	3500	3550	3600	3650	3700	3750	3821	3845	3870	3895	3920
Guangdong	1813	1920	2050	2200	2350	2495	2704	3028	3529	3944	3936	4520	5019	5590	6480
Guangxi	730	717	769	810	854	1008	1023	1136	1160	1200	1308	1387	1549	1809	2048
Hainan												159	179	215	242	279
Chongqing																
Sichuan	2755	2807	2936	3103	3244	3470	3553	3901	4230	4376	4239	4357	4475	4598	4743
Guizhou	786	740	860	942	1068	1248	1366	1649	1820	2063	2129	2313	2519	2546	2820
Yunnan	946	948	1021	1095	1226	1298	1399	1533	1623	1707	1954	1962	2017	2090	2283
Tibet	73	76	76	84	88	101	103	104	118	126	137	163	173	187	177
Shaanxi	1221	1246	1333	1508	1634	1776	1847	1989	2076	2161	2239	2359	2441	2476	2599
Gansu	1500	1575	1650	1700	1750	1790	1857	1938	2022	2128	2175	2311	2349	2508	2684
Qinghai	269	276	300	317	333	349	380	410	440	470	504	474	499	560	625
Ningxia	811	793	777	810	872	973	960	998	1032	1051	707	694	705	716	741
Xinjiang	1026	1060	1070	1174	1228	1411	1446	1461	1649	1761	1924	2071	2261	2497	2606

1995	1996	1997	1998	1999	2000	2001	2002	2003	2004	2005	2006	2007	2008	2009
3533	3735	3719	3808	3907	4144	4229	4436	4648	5140	5522	5904	6285	6344	6402
2569	2500	2452	2502	2553	2794	2918	3022	3215	3697	4115	4525	4944	5364	5783
8892	8938	9033	9151	9379	11196	12114	13405	15298	17348	19745	21690	23490	24226	24962
6574	6804	6695	6614	5573	5788	6967	8098	9038	9551	10117	11196	12135	12472	12809
3268	3144	3709	3440	3635	3938	4453	5190	6613	8602	10765	12806	14649	16268	17887
9382	9418	9192	8874	8870	9877	10357	10334	11431	12454	13592	15058	16593	17768	18943
3954	4033	4177	3627	3693	3528	3713	4209	4768	5207	5958	6622	7350	8095	8839
6461	6271	6636	6695	6390	5663	5831	6204	6310	7515	7620	7657	7958	8348	8738
4478	4611	4747	4855	5162	5500	5895	6249	6796	7406	8312	8967	9768	10314	10861
8047	8111	7991	8118	8164	8612	8881	9609	11061	13652	16895	18742	20604	21776	22947
4851	5165	5447	5657	5960	6560	7253	8280	9523	10825	12032	13223	14533	15117	15700
4194	4516	4405	4575	4683	4879	5118	5316	5457	6017	6518	7096	7752	8342	8931
2280	2452	2499	2579	2772	2943	3163	3615	4063	4528	6157	6812	7574	8238	8903
2392	2155	2132	2028	2123	2505	2628	2933	3426	3814	4286	4660	5054	5376	5698
8780	8900	9000	9012	9035	12513	13779	16150	18196	21398	25105	28250	30596	32225	33854
6473	6654	6711	7244	7380	7919	8367	9005	10595	13074	14624	16235	17841	18784	19728
5655	5731	5960	5917	5988	6269	6352	6713	7645	9120	9851	10797	11861	12603	13345
3945	3970	3995	4020	4045	4071	4622	5045	5562	7134	9110	9879	10797	11355	11914
7062	7456	7670	8083	8425	9080	9775	10862	12414	14488	17272	19059	21143	22288	23433
2257	2301	2328	2418	2473	2669	2899	2982	3421	4308	4981	5515	6137	6648	7159
309	352	390	409	431	480	520	602	684	742	819	911	1016	1089	1161
			2119	2278	2331	2464	2563	2738	3168	3882	4235	4782	5092	5401
4967	5230	5675	5925	6369	6517	6809	7510	9203	10699	11300	12538	13685	14558	15431
3079	3443	3739	4088	3854	4279	4438	4470	5534	6021	6429	7045	7696	7964	8231
2641	2819	3429	3364	3288	3468	3741	4131	4450	5210	6024	6641	7173	7578	7982
171	179	181	187	180	181	186	186	198	216	222	221	239	236	244
2869	3001	3069	2925	2584	2617	3034	3448	3919	4693	5424	6069	6639	7219	7799
2738	2803	2581	2687	2917	3012	3068	3174	3525	3908	4368	4743	5109	5373	5637
688	698	707	739	939	897	939	1019	1123	1364	1670	1903	2095	2257	2418
775	809	814	817	823	875	916	1041	1530	2283	2480	2775	3029	3189	3348
2733	3045	3208	3280	3215	3316	3496	3622	4064	4785	5506	6047	6576	7069	7563

Notes

1 Introduction

1 All raw data to construct the sub-industrial panel data, if not mentioned in the body of the text, comes from the China Statistical Yearbook (1983–2011), the China Industry Economy Statistical Yearbook (1988–2011), the China Energy Statistical Yearbook (1991–2011) and the China Compendium of Statistics 1949–2009 (2010), officially provided by the National Bureau of Statistics of China.

2 It's easier to see the preliminary heterogeneity of industry if we divide all the sectors into light and heavy industry. But the definition of light and heavy industry is not consistent, and it is difficult to differentiate the light and heavy industry at the two-digit level because there is an overlap between light and heavy sub-sectors within some two-digit sectors. Thus, in the literature, the light and heavy industry is usually divided according to their characteristics. For example, heavy industry is normally characterized by high investment, high energy consumption and heavy environmental emissions, and so on, while light industry is just the opposite. Besides the criterion of energy consumption, others used to divide light and heavy industry also include environmental emission, energy intensity, emission intensity, the ratio of capital to labor, labour and capital productivity and so on.

3 All raw data to construct the provincial panel data also comes from the China Statistical Yearbook (1983–2010), the China Energy Statistical Yearbook (1991–2010) and the China Compendium of Statistics 1949–2009 (2010), among others. The database includes all the 31 provinces, the rest of Hong Kong, Macao and Taiwan. The sample period of Hainan and Chongqing is from 1990 and 1998, respectively.

4 To build the regional classifying index, we chose five sub-indices (such as energy intensity, electricity intensity, CO_2 intensity, SO_2 intensity and COD intensity) in 2000 and 2008, respectively. To remove the influence of different units, all the sub-indices are divided by their mean to obtain their normalized value, the mean of which is now 1. The classifying index in 2000 and 20008, respectively, is obtained by simple averaging five normalized sub-indices in each year, the weighted average of which is just the final classifying index we need, the weights being 04 and 0.6 for the years of 2000 and 2008, respectively.

5 Energy intensity is the reciprocal of energy productivity.

6 The values are calculated by the author drawing on the raw data from the Statistical Review of World Energy in 2007 (in Chinese).

2 Industrial and regional composition of energy-induced CO_2 emissions

1 End-use energy consumption is only used in Chapter 2 for China as a whole. In other chapters of this book, we use the total energy consumption for both industrial and

provincial analysis. Total energy consumption includes three parts: end-use energy consumption, losses during the process of energy conversion and energy losses. But the end-use consumption accounts for the majority of total energy consumption. For example, in 2007, the proportion of end-use energy consumption to total one reaches 95.6 per cent.

2   Here, the agricultural industry includes agriculture, forestry, animal husbandry and fishery and water conservancy. The second industry consists of industry and construction. The transport industry is in fact transport, storage and post. The commerce industry includes wholesale and retail trades, hotels and catering services.

3   Similar to the classification in Chen and Zhang (2009), all the province in China are divided into eastern, middle and western regions. Specifically, the eastern region includes the 11 provinces of Beijing, Tianjin, Shanghai, Zhejiang, Jiangsu, Fujian, Guangdong, Liaoning, Shandong, Hebei and Hainan; the middle region consists of the 8 provinces of Shanxi, Jilin, Heilongjiang, Anhui, Jiangxi, Henan, Hubei and Hunan; the 11 provinces in the western region are Inner Mongolia, Guangxi, Sichuan, Chongqing, Guizhou, Yunnan, Shaanxi, Gansu, Qinghai, Ningxia and Xinjiang. Because of missing data, Tibet is not included in this study. Thus, the analytical target in this study totals 30 provinces, as described above.

4   Following Ang and Liu (2001), this chapter chooses the multiplicative logarithmic mean Divisia index (LMDI) decomposition method to analyse the influential factors that drive the change of CO_2 emission in China looking at three dimensions of 6 industries, 30 provinces and 3 energy types. The detailed method is not addressed here.

5   The six pillar industries established by the Shanghai government during the period of the tenth Five-Year Plan are petroleum and fine chemicals, iron and steel, bio-pharmaceuticals, electronic information,and the complete equipment and car industry.

3   How to reduce industrial CO_2 emission intensity

1   The industrial CO_2 intensity in Figure 2.3 and the industrial energy intensity in Figure 2.7 are the weighted mean of respective intensity values of 38 industrial sectors, the weight being the share of industrial gross output value of each sector

2   The whole period is divided into four sub-periods in terms of the different patterns of industrial carbon intensity characterized in Figure 2.3

3   According to the environmental protection programme of the tenth and eleventh national Five-Year Plans and the report *Environmental Protection in China (1996–2005)* released by the State Council on 5 June 2006, for the first time the Chinese government closed down about 84,000 small energy- and emission-intensive enterprises during the ninth Five-Year Plan (1996–2000). In the period of the tenth Five-Year Plan (2001–2005), the government shut down 33,000 small enterprises that caused heavy pollution, but the environmental policy was not implemented as strictly as in the ninth Five-Year Plan, leading to the reappearance of heavy industrialization in this phase.

4   Measurement of CO_2 shadow price

1   Murty and Kumar (2002) estimated the stochastic frontier output distance function to calculate the shadow price.

2   The methodology to calculate the carbon dioxide shadow price in this chapter can also be easily extended to calculate the shadow price of other environmental pollutions such as SO_2, NO_2, COD, waste gas, waste water and waste solids, and so on.

3   The results under the assumption of constant return to scale are also calculated here and are similar to those under NIRS.

4 Except the four traditional productivity indices described above, there is another type of Malmquist productivity index that is based on the data envelopment analysis (DEA) and does not need the price information of inputs and outputs. It will be computed later in this book.

5 Energy and environmental policies and factors driven industrial growth

1 In Chapter 7, we will discuss the difference of TFP estimates when the carbon emission variable is treated in a different way.

6 Structural change, factors reallocation and industrial growth

1 The weight is the output share of each sector.
2 If the structural effect is taken into account, the efficiency change can be ignored and it is not necessary to use the stochastic frontiers. Even if the efficiency change needs to be considered, a nonparametric method such as the DEA and Malmquist indices are better than the parametric stochastic frontier method with its priori assumption of an efficiency varying pattern. The extension in this chapter is to remove the efficiency term from the parametric frontier function.
3 This is different from another studies by decomposing the structural effect from the energy (or emission) or its intensity; see Zhang (2003), Fisher-Vanden *et al.* (2004, 2006), Wang *et al.* (2005) and Zhang (2009), among others.
4 In this study, the actual costs of labour, capital and energy are measured by the total wages, the depreciation of fixed assets and energy input cost for different sectors over time. Because the unit of energy consumption is coal equivalent in this study, we first obtain the market price of raw coal and derive the price of the coal equivalent according to the equality of 1 ton raw coal = 0.7143 ton coal equivalent, and then calculate the energy input cost. All value-type variables in this chapter are depreciated at the 1990 price level.
5 The specification of the functional form at least surpasses the priori assumption of constant return to scale and technical neutrality.
6 Zheng *et al.* (2009) define extensive growth as the higher growth rate of input factors than productivity.
7 To save space, we do not report the sectoral growth accounting here.
8 In this figure, the TFP, TC, SE, and SCE for aggregated industry, low and high emission group are all the weighted mean of 38 sectoral values. The weights are gross output share.
9 To save space, we do not report the optimal marginal output share and actual share of each factor here. The finding is consistent with many studies. Qin and Song (2009) find that the tendency of over-investment typical of centrally planned economies, the so-called investment hunger, remains in China today. By showing decelerating growth in total factor productivity and diminishing investment returns during the 1990s, Zhang (2003) suggests that China's overall fixed-asset investment has gone too far, especially with regard to its labour resources.
10 Since the late 1990s, industry in China has undergone a massive labour force reduction due to the policy of furlough (xiagang) and 'grasp the large and let go of the small' (zhuadafangxiao). Official employment data show that the number of workers employed by state industrial enterprises fell from 44.0 million in 1995 to 15.5 million in late 2002, a 65 per cent decline. Urban industrial collectives see an equally severe decline in employment, from 14.9 million in 1995 to 3.8 million in 2002. Moreover, the

13.8 million workers added to the payrolls of private and foreign-funded industrial firms do not compensate for the 39.6 million jobs lost in industrial firms of the state and collective enterprises (Frazier, 2006).
11 Dessus *et al.* (1995) and Akkemik (2005) demonstrate that the labour reallocation effect is substantially higher than the capital reallocation effect because labour is a scarce and very important resource for Taiwan and Singapore, respectively. For China, the reverse holds.

7 Undesirable output, environmental TFP and industrial economic transformation

1 They argue that if the contribution of TFP growth on output growth exceeds that of the quantitative inputs of all factors, it can be said that the development model has already been transformed into an intensive and sustainable one; otherwise, it's still extensive and unsustainable in the sense that the output growth is driven mainly by the quantitative expansion of input factors. Thus, TFP is the core variable used to evaluate the transforming extent of the economic development model.
2 For instance, the Chinese elasticity coefficient of energy consumption has exceeded 1 for three consecutive years from 2003. The IEA has also said that China, with Hong Kong included, overtook the US as the world's biggest energy consumer in 2009, consuming 2.2 billion tons of oil equivalent.
3 Though CO_2 is used as the example here, the analytical framework introduced below could be easily extended to the analysis of the other pollutions.
4 Here, the entire reform period is classified into three sub-periods according to respective reform characteristics: the trial phase (1978–1992), the decisive reform phase (1992–2001) and the assessment and adjustment period (2001–2008), as described in Chapter 6.
5 Due to the application of the nonparametric method, it's not convenient to calculate the contribution share of each input factor. Following Wu (2008), in Table 7.3, we just calculate the contribution ratio of TFP growth on output growth, 100 minus the contribution of all the input factors; the comparison between them makes the growth accounting analysis possible.
6 Tu and Xiao (2005) find that TFP growth becomes the main driver of LME growth at the turn of the century, and they vividly refer to the significant driving force of TFP on industrial growth as the industrial productivity revolution in China. Liu and Zhang (2008) concluded that the economic development model in China has become more and more sustainable since 1998. Chen *et al.* (2011) have also found that the Chinese industrial development model has transformed from being extensive to intensive after 1992.
7 Similar to the DDF used in this chapter, Watanabe and Tanaka (2007) also argue that the productivity that allows for desirable and undesirable output simultaneously is the preferable indicator to evaluate the development sustainability.
8 The classification criterion of light and heavy industry has been introduced at the first paragraph of Section 7.5 of this chapter. All the series are the weighted average, in which the weight is the respective share of GIOV for each sub-industry.
9 The signs, *, **, *** represent the significance level at the 10 per cent, 5 per cent and 1 per cent levels, respectively.
10 The finding by Zheng *et al.* (2003) is the rare case similar to ours. That is, in the 1980s, the productivity growth of light industry is higher than that of heavy industry; from the beginning of the 1990s, however, the productivity growth of heavy industry begins to exceed that of light industry.
11 Jeon and Sickles (2004) reach a similar conclusion to ours by cross countries analysis. That is, the carbon abating policies implemented in Finland, France and Sweden caused

improvement in productivity; while such countries as Canada, Japan, Ireland, Italy and Spain, with more rapid growth of carbon emission, have lower productivity growth, showing the opposite change between TFP and carbon emission.

8   Evaluation of regional low-carbon economic transformation: multiple emissions

1   In the standard DDF-AAM approach, the assumption that desirable output rises and undesirable output falls at the same proportion is called the radial assumption. The DDF values should be computed from the angle of either output or input. Unlike this, the SBM-DDF-AAM approach used in this chapter has the non-radial, non-angle assumption, modelling the reality better.
2   As for the evaluation indicator constructed in this chapter, the value of 0.5 is the threshold that is not influenced by any normalization and reflects the absolute meaning of the turning point of low-carbon transformation. The normalization will not change the relative ranking of one DMU among all DMUs at one time point and the same DMU at different time points. Due to this relative and absolute nature, the evaluation indicators could be compared either horizontally or vertically.

9   Energy-saving and emission-abating regulations and win-win development simulations

1   The denominator of energy intensity, according to its definition, is GDP or value-added. In this chapter, the gross industrial output value is chosen as the denominator to calculate energy intensity; at all events, the calculated growth of energy consumption based on the decreasing rate of energy intensity, 3 per cent, and the growth of output, either value-added or gross industrial output value, is the same.
2   In another example, Lin (2004) asserts that China is very likely to maintain an approximate 8 per cent GDP growth rate, like the middle level of our specification, for another twenty or thirty years, by adapting technological know-how from advanced countries at a lower cost. The growth of output and energy consumption is necessary for the economic development but their combination, leading to an approximate 3–4 per cent declining rate of energy intensity annually, is the most important for the sustainable development in China, as shown in the study.
3   In this case, the value of β is greater than zero, which tells us the sizes of inefficiencies for the unit.
4   For example, a varying emission abating rate can be included in our design, as opposed to most studies in which only several fixed abating rates are set for scenario simulations.
5   For the convenience of the report, Table 9.1 does not report the result of the group with 12 per cent of GIOV growth, in which the same trend as the rest of Table 9.1 is seen. Particularly, the potential net values of loss in this group are greater than the former four GIOV groups; thus, there exists no optimal energy-saving and emission-abating path in this group.
6   Another reason to choose the path with the lowest net loss in the group of 6 per cent GIOV growth, instead of four paths with lower net loss in the group of 4 per cent GIOV growth, is that if the sub-optimal path could lead to win-win opportunities, the optimal path will be more likely to cause win-win development; this has already been confirmed in this study but not reported to save space.
7   For convenience, only 35 sub-industries are included in this figure, excluding the production and supply of water (heavy industry), the manufacture of chemical fibres (light industry), and other industries. The win-win forecasting prospect of the three sub-industries is also in accordance with the conclusion of this chapter.

10 Double dividend forecasting and environmental taxation reform: carbon tax case

1 Schwartz and Repetto (2000) challenge this, arguing that the health benefits from reduced pollution will interact with pre-existing taxes and may cause the optimal environmental tax to exceed marginal damages. Williams (2003) reinforces this general notion, in the presence of pre-existing distortionary taxes, that tax-interactions tend to raise the costs of an environmental tax, and thus that the optimal environmental tax is less than the marginal environmental damages. However, West and Williams (2007) estimate the parameters necessary to calculate the optimal second-best gasoline tax and again find that the optimal gasoline tax is significantly higher than marginal damages – the opposite of the result suggested by the bulk of the earlier literature. In this chapter, we still regard the marginal abatement cost, or marginal environmental damage, as the pricing basis of an environmental tax rate.

2 Gao *et al.* (2004) use the quadratic polynomial model to fit the relationships between MAC and the abating rate.

3 The classification of light and heavy industry in Figure 10.4 is same as Figure 10.3.

4 To save space, the results are not illustrated.

5 If calculated based on carbon rather CO_2, the carbon tax rate can be obtained by multiplying the CO_2 tax rate measured in this chapter by 3.67.

References

Aghion, P. and P. Howitt (1998) *Endogenous Growth Theory*. Cambridge, MA: MIT Press.

Aigner, D. and S.F. Chu (1968) On Estimating the Industry Production Function. *American Economic Review*, 58 (2): 226–239.

Aiken, Deborah Vaughn and Carl A. Pasurka Jr (2003) Adjusting the Measurement of US Manufacturing Productivity for Air Pollution Emissions Control. *Resource and Energy Economics*, 25 (4): 329–351.

Akkemik, K.A. (2005) Labor Productivity and Inter-Sectoral Reallocation of Labor in Singapore (1965–2002). *Forum of International Development Studies*, 30 (9) see http://www.gsid.nagoya-u.ac.jp/bpub/research/public/forum/index-en.html.

Albrecht, J., D. Francois and K. Schoors (2002) A Shapley Decomposition of Carbon Emissions without Residuals. *Energy Policy*, 30 (9): 727–736.

Ambec, S. and P. Barla (2002) A Theoretical Foundation of the Porter Hypothesis. *Economics Letters* 75 (3): 355–360.

Anderson, Jonathan (2007) Solving China's Rebalancing Puzzle. *Finance and Development* 44 (3): 32–35.

Andrews, D.W.K. (1991) Heteroskedasticity and Autocorrelation Consistent Covariance Matrix Estimation. *Econometrica*, 59 (3): 817–858.

Andrews-Speed, P. (2009) China's Ongoing Energy Efficiency Drive: Origins, Progress and Prospects. *Energy Policy*, 37(4): 1331–1344.

Ang, B.W. (1993) Sector Disaggregation, Structural Change and Industrial Energy Consumption: An Approach to Analyse the Interrelationships. *Energy*, 18 (10): 1033–1044.

Ang, B.W. (2004) Decomposition Analysis for Policymaking in Energy: Which is the Preferred Method? *Energy Policy*, 32 (9): 1131–1139.

Ang, B.W. and K.H. Choi (1997) Decomposition of Aggregate Energy and Gas Emission Intensities for Industry: A Refined Divisia Index Method. *Energy*, 18 (3): 59–73.

Ang, B.W. and F.Q. Zhang (2000) A Survey of Index Decomposition Analysis in Energy and Environmental Studies. *Energy*, 25 (12): 1149–1176.

Ang, B.W. and F.L. Liu (2001) A New *Energy* Decomposition Method: Perfect in Decomposition and Consistent in Aggregation. *Energy*, 26 (6): 537–548.

Ang, B.W., F.Q. Zhang and K.H. Choi (1998) Factorizing Changes in Energy and Environmental Indicators through Decomposition. *Energy*, 23 (6): 489–495.

Ang, B.W., F.L. Liu and E.P. Chew (2003) Perfect Decomposition Techniques in Energy and Environmental Analysis. *Energy Policy*, 31 (14): 1561–1566.

Au, C.C. and J.V. Henderson (2006) How Migration Restrictions Limit Agglomeration and Productivity in China. *Journal of Development Economics*, 80 (2): 350–388.

Ayres, R.U., L.W. Ayres and B. Warr (2003) Energy, Power and Work in the US Economy, 1900–1998. *Energy,* 28 (3): 219–73.

Bai, Chong-En, Jiangyong Lu and Zhigang Tao (2009) How does Privatization Work in China? *Journal of Comparative Economics*, 37 (3): 453–470.

Bailey, I. (2002) European Environmental Taxes and Charges: Economic Theory and Policy Practice. *Applied Geography*, 22 (3): 235–251.

Baranzini, A., J. Goldemberg and S. Speck (2000) A Future for Carbon Taxes. *Ecological Economics*, 32 (3): 395–412.

Barker, T. (1997) Taxing Pollution Instead of Jobs: Towards More Employment Without More Inflation Through Fiscal Reform in the UK. In T. O'Riordan (ed.), *Ecotaxation*. London: Earthscan.

Barrett, S. (1994) Strategic Environmental Policy and International Trade. *Journal of Public Economics*, 54 (3): 325–338.

Battese, G.E. and D.S.P. Rao (2002) Technology Gap, Efficiency and a Stochastic Meta-frontier Function. *International Journal of Business and Economics*, 1 (2): 87–93.

Beaumont, N.J. and R. Tinch (2004) Abatement Cost Curves: a Viable Management Tool for Enabling the Achievement of Win–win Waste Reduction Strategies? *Journal of Environmental Management*, 71 (3): 207–215.

Bellenger, Moriah J. and Alan T. Herlihy (2009) An Economic Approach to Environmental Indices. *Ecological Economics*, 68 (8–9): 2216–2223.

Berthelemy, Jean-Claude (2001) The Role of Capital Accumulation, Adjustment and Structural Change for Economic Take-Off Empirical Evidence from African Growth Episodes. *World Development*, 29 (2): 323–343.

Bhaumik, Sumon Kumar and Saul Estrin (2007) How Transition Paths Differ: Enterprise Performance in Russia and China. *Journal of Development Economics*, 82 (2): 374–392.

Borensztein, E. and J.D. Ostry (1996) Accounting for China's Growth Performance. *American Economic Review*, 86 (1): 224–228.

Bosquet, B. (2000) Environmental Tax Reform: Does it Work? A Survey of the Empirical Evidence. *Ecological Economics*, 34 (1), 19–32.

Bossier, F. and T. Bréchet (1995) A Fiscal Reform for Increasing Employment and Mitigating CO_2 Emissions in Europe. *Energy Policy*, 23 (9): 789–798.

Bosworth, Barry and Susan M. Collins (2008) Accounting for Growth: Comparing China and India. *Journal of Economic Perspectives*, 22 (1): 45–66.

Bovenberg, A.L. and R.A. De Mooij (1994) Environmental Levies and Distortionary Taxation. *American Economic Review*, 84 (4), 1085–1089.

Bovenberg, A.Lans and Sjak Smulders (1995) Environmental Quality and Pollution-augmenting Technological Change in a Two-sector Endogenous Growth Model. *Journal of Public Economics*, 57 (3): 369–391.

Bovenberg, A.L. and L.H. Goulder (1996) Optimal Environmental Taxation in the Presence of Other Taxes: General Equilibrium Analysis. *American Economic Review*. 86 (4): 985–1000.

Bovenberg, A.L. and R.A. De Mooij (1997) Environmental Tax Reform and Endogenous Growth. *Journal of Public Economics*, 63 (2): 207–237.

Bovenberg, A.L. and L.H. Goulder (2002) Environmental Taxation and Regulation. In A.J. Auerbach and M. Feldstein (eds.), *Handbook of Public Economics, Vol. 3*. Amsterdam: North-Holland, pp. 1471–1545.

Boyd, G.A. and J.D. McClelland (1999) The Impact of Environmental Constraints on Productivity Improvement in Integrated Paper Plants. *Journal of Environmental Economics and Management,* 38 (2): 121–142.

Boyd, G.A., J.F. McDonald, M. Ross and D.A. Hanson (1987) Separating the Changing Composition of US Manufacturing Production from Energy Efficiency Improvements. *Energy*, 8 (2): 77–96.

Boyd, G., J.C. Molburg and R. Prince (1996) Alternative Methods of Marginal Abatement Cost Estimation: Nonparametric Distance Functions. *Proceedings of the USAEE/IAEE Seventeenth Conference*, pp. 86–95.

Boyd, Gale A., George Tolley and Joseph Pang (2002) Plant Level Productivity, Efficiency, and Environmental Performance of the Container Glass Industry. *Environmental and Resource Economics*, 23 (1): 29–43.

Brännlund, R. and J. Nordström (2004) Carbon Tax Simulations Using a Household Demand Model. *European Economic Review*, 48 (1): 211–233.

Brock, W. and M.S. Taylor (2005) Economic Growth and the Environment: a Review of Theory and Empirics, In P. Aghion and S. Durlauf (eds.), *Handbook of Economic Growth II*. North Holland Elsevier. pp. 1749–1821.

Brummer, B., T. Glauben and W. Lu (2006) Policy Reform and Productivity Change in Chinese Agriculture: A Distance Function Approach. *Journal of Development Economics*, 81 (1): 61–79.

Bruvoll, A. and B.M. Larsen (2004) Greenhouse Gas Emissions in Norway: do Carbon Taxes Work? *Energy Policy*, 32 (4): 493–505.

Cai, F., Y. Du and M. Wang (2008) The Political Economy of Emission in China: Will a Low Carbon Growth Be Incentive Compatible in Next Decade and Beyond? *Economic Research Journal*, 6: 4–11.

Callan, T., S. Lyons, S. Scott, R.S.J. Tol and S. Verde (2009) The Distributional Implications of a Carbon Tax in Ireland. *Energy Policy*, 37 (2): 407–412.

Carraro, C., M. Galeotti and M. Gallo (1996) Environmental Taxation and Unemployment: some Evidence on the Double Dividend Hypothesis in Europe. *Journal of Public Economics*, 62 (1–2): 141–181.

Carter, Anne P. (1974) Energy, Environment, Economic Growth. *The Bell Journal of Economics and Management Science*, 5 (2): 578–592.

Caselli, F. (2005) Accounting for Cross-country Income Differences. In Phillipe Aghion and Steven N. Durlauf (eds.), *Handbook of Economic Growth*. New York: North-Holland, 641–672.

Caves, Douglas W., R. Christensen Laurits, and W. Erwin Diewert (1982) The Economic Theory of Index Numbers and the Measurement of Input, Output and Productivity. *Econometrica*, 50 (6): 1393–1414.

CEACER (2009) *2050 China Energy and CO_2 Emission Report*. Beijing: Science Press.

Cerin, P. (2006) Bringing Economic Opportunity into Line with Environmental Influence: A Discussion on the Coase Theorem and the Porter and van der Linde Hypothesis. *Ecological Economics*, 56 (2): 209–225.

Chambers, R., R. Färe and S. Grosskopf (1996a) Productivity Growth in APEC Countries. *Pacific Economic Review*, 1 (3): 181–190.

Chambers, R., Y.H. Chung and R. Färe (1996b) Benefit and Distance Function. *Journal of Economic Theory*, 70 (2): 407–419.

Chang, Y.F., C. Lewis and S.J. Lin (2008) Comprehensive Evaluation of Industrial CO_2 Emission (1989–2004) in Taiwan by Input–Output Structural Decomposition. *Energy Policy*, 36 (7): 2471–2480.

Chen, E.K.Y. (1997) The Total Factor Productivity Debate: Determinants of Economic Growth in East Asia. *Asian-Pacific Economic Literature*, 11 (1): 18–38.

Chen, Jiagui and Qunhui Huang (2003) Industrial Modernization: Indicators and Evaluation. *Social Science in China*, 3, 18–28.

Chen, K., H.C. Wang, Y.X. Zheng, G.H. Jefferson and T.G. Rawski (1988) Productivity Change in Chinese Industry: 1953–1985. *Journal of Comparative Economics*, 12 (4): 570–591.

Chen, K.H., Y.J. Huang and C.H. Yang (2009) Analysis of Regional Productivity Growth in China: A Generalized Metafrontier MPI Approach. *China Economic Review*, 20 (4): 777–792.

Chen, P.C., M.M. Yu, C.C. Chang and S.H. Hsu (2008) Total Factor Productivity Growth in China's Agricultural Sector. *China Economic Review*, 19 (4): 580–593.

Chen, S. (2011) Reconstruction of Sub-industrial Statistical Data in China (1980–2008). *China Economic Quarterly*, 10 (3): 735–776.

Chen, Shiyi and Jun Zhang (2009) Empirical Research on Fiscal Expenditure Efficiency of Local Governments in China. *Social Sciences in China*, 30 (2): 21–34.

Chen, S.Y., W.K. Härdle and K. Jeong (2010) Forecasting Volatility with Support Vector Machines Based GARCH Model. *Journal of Forecasting*, 29 (4): 406–433.

Chen, S., G.H. Jefferson and J. Zhang (2011) Structural Change, Productivity Growth and Industrial Transformation in China. *China Economic Review*, 22 (1): 133–150.

Chen, W., P. Gao and J. He (2004) Impacts of Future Carbon Reductions on the Chinese GDP Growth. *Journal of Tsinghua University (Science and Technology)*, 44 (6): 744–747.

Chenery, H.B., S. Robinson and M. Syrquin (1986) *Industrialization and Growth: A Comparative Study*. New York: Oxford University Press.

Chow, G.C. (1993) Capital Formation and Economic Growth in China. *Quarterly Journal of Economics*, 108 (3): 809–842.

Chow, G.C. (2008) Another Look at the Rate of Increase in TFP in China. *Journal of Chinese Economics and Business Studies*, 6 (2): 219–224.

Chow, G.C. and A. Lin (2002) Accounting for Economic Growth in Taiwan and Mainland China: A Comparative Analysis. *Journal of Comparative Economics*, 30 (3): 507–530.

Chung, H.S. and H.C. Rhee (2001) A Residual-Free Decomposition of the Sources of Carbon Dioxide Emissions: A Case of the Korean Industries. *Energy*, 26 (1): 15–30.

Chung, Y. (1996) *Directional Distance Functions and Undesirable Outputs*. PhD dissertation Southern Illinois University, Carbondale.

Chung, Y.H., R. Färe and S. Grosskopf (1997) Productivity and Undesirable Outputs: A Directional Distance Function Approach. *Journal of Environmental Management*, 51 (3): 229–240.

Coggins, Jay S. and John R. Swinton (1996) The Price of Pollution: A Dual Approach to Valuing SO_2 Allowances. *Journal of Environmental Economics and Management*, 30 (1), 58–72.

Collins, S. M. and B.P. Bosworth (1996) Economic Growth in East Asia: Accumulation versus Assimilation. *Brookings Papers on Economic Activity*, 2: 135–203.

Cole, Matthew A., J.R. Elliott Robert and Kenichi Shimamoto (2005) Industrial Characteristics, Environmental Regulations and Air Pollution: An Analysis of the UK Manufacturing Sector. *Journal of Environmental Economics and Management*, 50 (1): 121–143.

Considine, T.J. and D.F. Larson (2006) The Environment as a Factor of Production. *Journal of Environmental Economics and Management*, 52 (3): 645–662.

Cremer, H., F. Gahvari and N. Ladoux. (2003) Environmental Taxes with Heterogeneous Consumers: An Application to Energy Consumption in France. *Journal of Public Economics*, 87 (12): 2791–2815.

Cuesta, Rafael A., C.A. Knox Lovell and José L. Zofío (2009) Environmental Efficiency Measurement with Translog Distance Functions: A Parametric Approach. *Ecological Economics*, 68 (8–9): 2232–2242.

Dean, Judith M. (1999) Testing the Impact of Trade Liberalization on the Environment. In Per G. Fredriksson (ed.), *Trade, Global Policy, and the Environment*, World Bank Discussion Paper, No. 402.

Dessus, S., J-D. Shea and M-S. Shi (1995) *Chinese Taipei: The Origins of the Economic Miracle*, Development Center, Long-Term Growth Series. Paris: OECD.

Diebold, F.X. and R.S. Mariano (1995) Comparing Predictive Accuracy. *Journal of Business and Economic Statistics*, 13 (3): 253–265.

Doblin, C.P. (1988) Declining Energy Intensity in the US Manufacturing Sector. *Energy*, 9 (2): 109–135.

Dorfman, J.H. and G. Koop (2005) Current Developments in Productivity and Efficiency Measurement. *Journal of Econometrics*, 126 (2): 233–240.

Dowrick, S. and N. Gemmel (1991) Industrialization, Catching-Up and Economic Growth: A Comparative Study Across the World's Capitalist Economies. *The Economic Journal*, 101 (405): 263–275.

Du, J., L. Liang and J. Zhu (2010) A Slacks-based Measure of Super-efficiency in Data Envelopment Analysis: A Comment. *European Journal of Operational Research*, 204 (3): 694–697.

Ekins, P. and S. Speck (2000) Proposals of Environmental Fiscal Reforms and the Obstacles to their Implementation. *Journal of Environmental Policy and Planning*, 2 (2): 93–114.

Fabricant, S. (1942) Employment in Manufacturing 1899–1939. NBER Working Paper.

Fagerberg, Jan (2000) Technological Progress, Structural Change and Productivity Growth: A Comparative Study. *Structural Change and Economic Dynamics*, 11 (4): 393–411.

Fan, G. (2002) Progress in Ownership Changes and Hidden Risks in China's Transition. *Transition Newsletter*, 13 (3): 1–5.

Fan Gang, Xiaolu Wang, Liwen Zhang and Hengpeng Zhu (2003) Marketization Index for China's Provinces. *Economic Research Journal*, 3: 9–18.

Fan, Shenggen, Xiaobo Zhang and Sherman Robinson (2003) Structural Change and Economic Growth in China. *Review of Development Economics*, 7 (3): 360–377.

Fan, Ying, Hua Liao and Yi-Ming Wei (2007a) Can Market Oriented Economic Reforms Contribute to Energy Efficiency Improvement? Evidence from China. *Energy Policy*, 35 (4): 2287–2295.

Fan, Y., L. Liu, G. Wu, H. Tsai and Y. Wei (2007b) Changes in Carbon Intensity in China: Empirical Findings from 1980–2003. *Ecological Economics*, 62 (3–4): 683–691.

Färe, R. and S.C. Grosskopf (1998) Shadow Pricing of Good and Bad Commodities. *American Journal of Agricultural Economics*, 80 (3): 584–590.

Färe, R. and S. Grosskopf (2000) Theory and Application of Directional Distance Functions. *Journal of Productivity Analysis*, 13(2): 93–103.

Färe, R. and S. Grosskopf (2010) Directional Distance Functions and Slacks-based Measures of Efficiency. *European Journal of Operational Research*, 200 (1): 320–322.

Färe, R., S. Grosskopf, K. Lovell and C. Pasurka (1989) Multilateral Productivity Comparisons When Some Outputs are Undesirable: A Nonparametric Approach. *Review of Economics and Statistics*, 71 (1): 90–98.

Färe, Rolf, Shawna Grosskopf, C. A. Knox Lovell and Suthathip Yaisawarng (1993) Derivation of Shadow Prices for Undesirable Outputs: A Distance Function Approach. *Review of Economics and Statistics*, 75 (2): 374–380.

Färe, R., S. Grosskopf, M. Norris and Z. Zhang (1994) Productivity Growth, Technical Progress and Efficiency Change in Industrialized Countries. *American Economic Review*, 84 (1): 66–83.

Färe, R., S. Grosskopf and C.A. Pasurka Jr (2001) Accounting for Air Pollution Emissions in Measures of State Manufacturing Productivity Growth. *Journal of Regional Science*, 41 (3): 381–409.

Färe, Rolf, Shawna Grosskopf, Dong-Woon Noh and William Weber (2005) Characteristics of a Polluting Technology: Theory and Practice. *Journal of Econometrics*, 126 (2): 469–492.

Faucheux, S and I. Nicolaï (1998) Environmental Technological Change and Governance in Sustainable Development Policy. *Ecological Economics*, 27 (3): 243–256.

Feichtinger, G., R.F. Hartl, P.M. Kort and V.M. Veliov (2005) Environmental Policy, the Porter Hypothesis and the Composition of Capital. *Journal of Environmental Economics and Management*, 50 (2): 434–446.

Fisher-Vanden, K. and M.S. Ho. (2007) How do Market Reforms affect China's Responsiveness to Environmental Policy? *Journal of Development Economics*, 82 (1): 200–233.

Fisher-Vanden, K. and G.H. Jefferson (2008) Technology Diversity and Development: Evidence from China's Industrial Enterprises. *Journal of Comparative Economics*, 36 (4): 658–672.

Fisher-Vanden, Karen, Gary H. Jefferson, Hongmei Liu and Quan Tao (2004) What is Driving China's Decline in Energy Intensity? *Resource and Energy Economics*, 26 (1): 77–97.

Fisher-Vanden, Karen, Gary H. Jefferson, Jingkui Ma and Jianyi Xu (2006) Technology Development and Energy Productivity in China. *Energy Economics*, 28 (5–6): 690–705.

Fleisher, B.M. and D.T. Yang (2003) Labor Laws and Regulations in China. *China Economic Review*, 14 (4): 426–433.

Floros, N. and A. Vlachou (2005) Energy Demand and Energy-related CO_2 Emissions in Greek Manufacturing: Assessing the Impact of a Carbon Tax. *Energy Economics*, 27 (3): 387–413.

Frazier, M.W. (2006) State-Sector Shrinkage and Workforce Reduction in China. *European Journal of Political Economy*, 22 (2): 435–451.

Fukuyama, H. and W.L. Weber (2009) A Directional Slacks-based Measure of Technical Inefficiency. *Socio-Economic Planning Sciences*, 43 (4): 274–287.

Fung, H.G., D. Kummer and J. Shen (2006) China's Privatization Reforms. *Chinese Economy*, 39 (2): 5–25.

Gao, P., W. Chen and J. He (2004) Marginal Carbon Abatement Cost in China. *Journal of Tsinghua University (Science and Technology)*, 44: 1192–1195.

Garbaccio, R.D., M.S. Ho and D.W. Jorgenson (1999a) Controlling Carbon Emissions in China. *Environment and Development Economics*, 4 (4): 493–518.

Garbaccio, R.D., M.S. Ho and D.W. Jorgenson (1999b) Why Has the Energy-Ouptut Ratio Fallen in China? *Energy*, 20 (3): 63–91.

Gassebner, Martin, Noel Gaston and Michael J. Lamla (2008) Relief for the Environment? The Importance of an Increasingly Unimportant Industrial Sector. *Economic Inquiry*, 46 (2): 160–178.

Gollop, F.M. and M.J. Roberts (1985) Cost-minimizing Regulation of Sulphur Emissions: Regional Gains in Electric Power. *Review of Economics and Statistics*, 67 (1): 81–90.

Gong, G. and J.Y.F. Lin (2008) Deflationary Expansion: An Overshooting Perspective to the Recent Business Cycle in China. *China Economic Review*, 19 (1), 1–17.

Greaker, M. (2006) Spillovers in the Development of new Pollution Abatement Technology: A New Look at the Porter-hypothesis. *Journal of Environmental Economics and Management*, 52 (1): 411–420.

Greening, L.A., W.B. Davis and L. Schipper (1998) Decomposition of Aggregate Carbon Intensity for the Manufacturing Sector: Comparison of Declining Trends from 10 OECD Countries for the Period 1971–1991. *Energy Economics*, 20 (1): 43–65.

Groom, B., P. Grosjean, A. Kontoleon, T. Swanson and S. Zhang (2010) Relaxing Rural Constraints: a 'Win-win' Policy for Poverty and Environment in China? *Oxford Economic Papers*, 62 (1): 132–156.

Guan, D.B., K. Hubacek, C.L. Weber, G.P. Peters and D.M. Reiner (2008) The Drivers of Chinese CO_2 Emission from 1980 to 2030. *Global Environmental Change*, 18 (4): 626–634.

Hailu, A. and T.S. Veeman (2000) Environmentally Sensitive Productivity Analysis of the Canadian Pulp and Paper Industry, 1959–1994: An Input Distance Function Approach. *Journal of Environmental Economics and Management*, 40 (3): 251–274.

Harberger, A.C. (1998) A Vision of the Growth Process. *American Economic Review*, 88 (1): 1–32.

He, J., K. Shen and S. Xu (2002) Carbon Tax and CGE Model of CO_2 Reduction. *Quantitative and Technological Economic Study*, 10: 39–47.

Heckman, J. (2005) China's Human Capital Investment. *China Economic Review*, 16 (1): 50–70.

Hoel, M. (1996) Should a Carbon Tax be Differentiated Across Sectors? *Journal of Public Economics*, 59 (1): 17–32.

Hoffmann, W.G. (1958) *The Growth of Industrial Economies*, translated from German by W.H. Henderson and W.H. Chaloner. Manchester: Manchester University Press.

Holz, C.A. (2006) Measuring Chinese Productivity Growth, 1952–2005. Available at SSRN: http://ssrn.com/abstract=928568 or http://dx.doi.org/10.2139/ssrn.928568

Hsieh, Chang-Tai (2002) What Explains the Industrial Revolution in East Asia? Evidence from the Factor Markets, *American Economic Review*, 92 (3): 502–526.

Hu, A., J. Zheng, Y. Gao and N. Zhang (2008) Provincial Technology Efficiency Ranking with Environment Factors (1999–2005). *China Economic Quarterly*, 7 (3): 933–960.

Hueting, R. (1991) Correcting National Income for Environmental Losses: A Practical Solution for a Theoretical Dilemma. In Robert Costanza (ed.), *Ecological Economics: The Science and Management of Sustainability*. New York: Columbia University Press. pp. 194–213.

Irmen, Andreas (2005) Extensive and Intensive Growth in a Neoclassical Framework. *Journal of Economic Dynamics and Control*, 29 (8): 1427–1448.

Islama Sardar, M.N., Mohan Munasingheb and Matthew Clarke (2003) Making Long-term Economic Growth More Sustainable: Evaluating the Costs and Benefits. *Ecological Economics*, 47 (2–3): 149–166.

Jaffe, A., S. Peterson, P. Portney and R. Stavins (1995) Environmental Regulation and the Competitiveness of US Manufacturing: What Does the Evidence Tell Us? *Journal of Economic Literature*, 33 (1): 132–163.

Jansen, H. and G. Klaassen (2000) Economic Impacts of the 1997 EU Energy Tax: Simulations with Three EU-wide Models. *Environmental and Resource Economics*, 15 (2): 179–197.

Jefferson, G.H. and J. Su (2006) Privatization and Restructuring in China: Evidence from Shareholding Ownership, 1995–2001. *Journal of Comparative Economics*, 34 (1): 146–166.

Jefferson, Gary and Ping Zhang (1998) Structure, Authority, and Incentives in Chinese Industry. In Gary Jefferson and Inderjit Singh (eds.), *Enterprise Reform in China: Ownership, Transition, and Performance*. New York: Oxford University Press.

Jefferson, G.H., T.G. Rawski, L. Wang and Y. Zheng (2000) Ownership, Productivity Change, and Financial Performance in Chinese Industry. *Journal of Comparative Economics*, 28 (4): 786–813.

Jefferson, G.H., T.G. Rawski and Y. Zhang (2008) Productivity Growth and Convergence across China's Industrial Economy. *Journal of Chinese Economic and Business Studies*, 6 (2): 121–140.

Jeon, B.M. and R.C. Sickles (2004) The Role of Environmental Factors in Growth Accounting. *Journal of Applied Econometrics*, 19 (5): 567–591.

Jorgenson, D.W. and P.J. Wilcoxen (1993) Reducing US Carbon Emissions: An Econometric General Equilibrium Assessment. *Resource and Energy Economics,* 30 (3): 433–439.

Jorgenson, Dale W. and Kevin J. Stiroh (1999) Information Technology and Growth. *American Economic Review (Papers and Proceedings)*, 89 (2): 109–1567.

Jorgenson, Dale W., Frank Gollop and Barbara Fraumeni (1987) *Productivity and US Economic Growth*. Cambridge, MA: Harvard University Press.

Jorgenson, Dale W. and Kevin J. Stiroh (2000) US Economic Growth at the Industry Level. *American Economic Review (Papers and Proceedings)*, 90 (2): 161–167.

Jung, Tae Yong, Emilio Lebre La Rovere, Henryk Gaj, P.R. Shukla and Dadi Zhou (2000) Structural Changes in Developing Countries and their Implication for Energy-Related CO_2 Emissions. *Technological Forecasting and Social Change*, 63 (2–3): 111–136.

Kahn, J.R. and D. Franceschi (2006) Beyond Kyoto: A Tax-based System for the Global Reduction of Greenhouse Gas Emissions. *Ecological Economics*, 58 (4): 778–787.

Kahrl, F. and D. Roland-Holst (2009) Growth and Structural Change in China's Energy Economy. *Energy*, 34 (7): 894–903.

Kander, Astrid and Lennart Schön (2007) The Energy-capital Relation – Sweden 1870–2000. *Structural Change and Economic Dynamics*, 18 (3): 291–305.

Kaneko, Shinji, Hidemichi Fujii, Naoya Sawazu and Ryo Fujikura (2010) Financial Allocation Strategy for the Regional Pollution Abatement Cost of Reducing Sulphur Dioxide Emissions in the Thermal Power Sector in China. *Energy Policy*, 38 (5): 2131–2141.

Karvonen, M. (2001) Natural versus Manufactured Capital: Win-lose or Win-win? A Case Study of the Finnish Pulp and Paper Industry. *Ecological Economics*, 37 (1): 71–85.

Kasahara, Hiroyuki and Joel Rodrigue (2008) Does the Use of Imported Intermediates Increase Productivity? Plant-level Evidence. *Journal of Development Economics*, DEVEC-01353: 13.

Kerkhof, A.C., H.C. Moll, E. Drissen and H.C. Wilting (2008) Taxation of Multiple Greenhouse Gases and the Effects on Income Distribution: A Case Study of the Netherlands. *Ecological Economics*, 67 (2): 318–326.

Kim, J. and L. Lau (1994) The Sources of Economic Growth of the East Asian Newly Industrialized Countries. *Journal of Japanese and International Economies*, 8 (3): 235–271.

Krugman, Paul (1994) The Myth of Asia's Miracle. *Foreign Affairs*, 73 (6): 62–78.

Kumar, Surender (2006) Environmentally Sensitive Productivity Growth: A Global Analysis Using Malmquist-Luenberger Index. *Ecological Economics*, 56 (2): 280–293.

Kumar, S. and R. Russell (2002) Technological Change, Technological Catch up, and Capital Deepening: Relative Contributions to Growth and Convergence. *American Economic Review*, 92 (3): 527–548.

Kumbhakar, S.C. (1990) Production Frontiers, Panel Data, and Time-Varying Technical Inefficiency. *Journal of Econometrics*, 46 (1–2): 201–211.

Kummel, Reiner, Julian Henn and Dietmar Lindenberger (2002) Capital, Labor, Energy and Creativity: Modeling Innovation Diffusion. *Structural Change and Economic Dynamics*, 13 (4): 415–433.

Kuosmanen, Timo and Mika Kortelainen (2007) Valuing Environmental Factors in Cost-benefit Analysis using Data Envelopment Analysis. *Ecological Economics*, 62 (1): 56–65.

Kuosmanen, T., N. Bijsterbosch and R. Dellink (2009) Environmental Cost-benefit Analysis of Alternative Timing Strategies in Greenhouse Gas Abatement. *Ecological Economics*, 68 (6): 1633–1642.

Kuznets, S. (1979) Growth and Structural Shifts. In W. Galenson (ed.), *Economic Growth and Structural Change in Taiwan. The Postwar Experience of the Republic of China*. London: Cornell University Press. pp. 15–131.

Lansink, Alfons Oude and Elvira Silva (2003) CO_2 and Energy Efficiency of Different Heating Technologies in the Dutch Glasshouse Industry. *Environmental and Resource Economics*, 24 (4), 395–407.

Lee, Chien-Chiang and Chun-Ping Chang (2007) The Impact of Energy Consumption on Economic Growth: Evidence from Linear and Nonlinear Models in Taiwan. *Energy*, 32 (12): 2282–2294.

Lee, Chien-Chiang and Chun-Ping Chang (2008) Energy Consumption and Economic Growth in Asian Economies: A more Comprehensive Analysis using Panel Data. *Resource and Energy Economics*, 30 (1), 50–65.

Lee, Chien-Chiang, Chun-Ping Chang and Pei-Fen Chen (2008) Energy-income Causality in OECD Countries Revisited: The Key Role of Capital Stock. *Energy Economics*, 30 (5): 2359–2373.

Lee, C.F., S.J. Lin, C. Lewis and Y.F. Chang (2007) Effects of Carbon Taxes on Different Industries by Fuzzy Goal Programming: A Case Study of the Petrochemical-related Industries, Taiwan. *Energy Policy*, 35 (8): 4051–4058.

Lee, D.R. and W.S. Misiolek (1986) Substituting Pollution for General Taxation: Some Implications for Efficiency in Pollution Taxation. *Journal of Environmental Economics and Management*, 13 (4): 339–347.

Lee, Jeong-Dong, Jong-Bok Park and Tai-Yoo Kim (2002) Estimation of the Shadow Prices of Pollutants with Production/environment Inefficiency taken into Account: A Nonparametric Directional Distance Function Approach. *Journal of Environmental Management*, 64 (4): 365–375.

Lee, Myunghun (2005) The Shadow Price of Substitutable Sulphur in the US Electric Power Plant: A Distance Function Approach. *Journal of Environmental Management*, 77 (2): 104–110.

Lewis, W.A. (1954) Economic Development with Unlimited Supplies of Labour. *Manchester School of Economics and Social Studies*, 22: 139–191.

Li, H. and S. Rozelle (2000) Savings or Stripping Rural Industry: An Analysis of Privatization and Efficiency in China. *Agricultural Economics*, 23 (3): 241–252.

Li, K.W. (2009) China's Total Factor Productivity Estimates by Region, Investment Sources and Ownership. *Economic Systems*, 33 (3): 213–230.

Li, K.W., T. Liu and L. Yun (2005) Decomposition of Economic and Productivity Growth in Post-Reform China (working paper), Department of Economics, Ball State University.

Li, Shengwen and Dasheng Li (2008) Fluctuation of Total Factor Productivity in Chinese Industry. *Quantitative Economy and Technique Economy Research*, 5: 43–54.

Li, S. and J. Xia (2008) The Roles and Performance of State Firms and Non-State Firms in China's Economic Transition. *World Development*, 36 (1): 39–54.

Li, W. (1997) The Impact of Economic Reform on the Performance of Chinese State Enterprises, 1980–1989. *Journal of Political Economy*, 105 (5): 1080–1106.

Li, X. and Z. Zhu (2005) Calculation of Total Factor Productivity in Chinese Industry – based on Sub-industrial Panel Data. *Management World*, 4: 56–674.

Liang, Q., Y. Fan and Y. Wei (2007) Carbon Taxation Policy in China: How to Protect Energy- and Trade-intensive Sectors? *Journal of Policy Modelling*, 29 (2): 311–333.

Lin, J.Y.F. (1992) Rural Reforms and Agricultural Growth in China. *American Economic Review*, 82 (1): 34–51.

Lin, Justin Yifu (2004) Is China's Growth Real and Sustainable? *Asian Perspective*, 28 (3): 5–29.

Liu A., S. Yao and Z. Zhang (1999) Economic Growth and Structural Changes in Employment and Investment in China, 1985–1994. *Economics of Planning*, 32 (3): 171–190.

Liu, C.C. (2006) A Study on Decomposition of Industry Energy Consumption. *International Research Journal of Finance and Economics*, 6: 73–77.

Liu, L., Y. Fan, G. Wu and Y. Wei (2007) Using LMDI Method to Analyse the Change of China's Industrial CO_2 Emissions from Final Fuel Use: An Empirical Analysis. *Energy Policy*, 35 (11): 5892–5900.

Liu, W. and H. Zhang (2008) Structural Change and Technical Advance in China's Economic Growth. *Economic Research Journal*, 11: 4–15.

Liu, X. and Y. Wu (2009) Enterprises Productivity and Its Sources: Innovation or Demand Push? *Economic Research Journal*, 7: 45–54.

Liu, X.Q., B.W. Ang and H.L. Ong (1992) The Application of the Divisia Index to the Decomposition of Changes in Industrial Energy Consumption. *Energy*, 13 (4): 161–177.

Lu, Xuedu, Jiahua Pan and Ying Chen (2006) Sustaining Economic Growth in China under Energy and Climate Security Constraints. *China and World Economy*, 14 (6): 85–97.

Lucas, R.E. (1993) Making A Miracle. *Econometrica*, 61 (2): 251–272.

Maddison, A. (1987) Growth and Slowdown in Advanced Capitalist Economies: Techniques of Quantitative Assesment. *Journal of Economic Literature*, 25 (2): 649–698.

Managi, S. (2006) Are There Increasing Returns to Pollution Abatement? Empirical Analytics of the Environmental Kuznets Curve in Pesticides. *Ecological Economics*, 58 (3): 617–636.

Massell, B.F. (1961) A Disaggregated View of Technical Change. *Journal of Political Economics*, 69 (6), 547–557.

Mohr, R.D. (2002) Technical Change, External Economies, and the Porter Hypothesis. *Journal of Environmental Economics and Management*, 43 (1): 158–168 .

Mohtadi, H. (1996) Environment, Growth and Optimal Policy Design. *Journal of Public Economics*, 63 (1): 119–140.

Morris, G.E., T. Révész, E. Zalai and J. Fucskó. (1999) Integrating Environmental Taxes on Local Air Pollutants with Fiscal Reform in Hungary: Simulations with a Computable General Equilibrium Model. *Environment and Development Economics*, 4 (4): 537–564.

Movshuk, O. (2004) Restructuring, Productivity and Technical Efficiency in China's Iron and Steel Industry, 1988–2000. *Journal of Asian Economics*, 15 (1): 135–151.

Mukherjee, Anit and Xiaobo Zhang (2007) Rural Industrialization in China and India: Role of Policies and Institutions. *World Development*, 35 (10): 1621–1634.

Murillo-Zamorano, L.R. (2005) The Role of Energy in Productivity Growth: A Controversial Issue? *The Energy Journal*, 26 (2): 69–88.

Murty, M.N. and Kumar Surender (2002) Measuring the Cost of Environmentally Sustainable Industrial Development in India: A Distance Function Approach. *Environment and Development Economics*, 7 (3): 467–486.

Murty, M.N. and Kumar Surender (2003) Win-win Opportunities and Environmental Regulation: Testing of Porter Hypothesis for Indian Manufacturing Industries. *Journal of Environmental Management*, 67 (2): 139–144.

Murty, M.N., Kumar Surender and K. Dhavala Kishore (2007) Measuring Environmental Efficiency of Industry: A Case Study of Thermal Power Generation in India. *Environmental Resource Economics*, 38 (1): 31–50.

Nakabayashi, M. (2010) Optimal Tax Rules and Public Sector Efficiency with an Externality in an Overlapping Generations Model. *Journal of Public Economics*, 94 (11–12): 1028–1040.

Nakata, T. and A. Lamont (2001) Analysis of the Impacts of Carbon Taxes on Energy Systems in Japan. *Energy Policy*, 29 (2): 159–166.

Nanere, M., I. Fraser, A. Quazi and C. D'Souza (2007) Environmentally Adjusted Productivity Measurement: An Australian Case Study. *Journal of Environmental Management*, 85 (2): 350–362.

Nelson, R.R. and H. Pack (1999) The Asian Miracle and Modern Growth Theory. *The Economic Journal*, 109 (457): 416–436.

Newey, W.K. and K.D. West (1987) A Simple Positive, Semi-Definite, Heteroskedasticity and Autocorrelation Consistent Covariance Matrix. *Econometrica*, 55 (3): 703–708.

Nielsen, Sqren Bo, Lars Haagen Pedersen and Peter Birch Sqrensen (1995) Environmental Policy, Pollution, Unemployment, and Endogenous Growth. *International Tax and Public Finance*, 2 (2), 185–205.

Ngai, L.R. and C.A. Pissarides (2007) Structural Change in a Multisector Model of Growth. *American Economic Review*, 97 (1): 429–443.

Nugent, J.B. and C.V.S.K. Sarma (2002) The Three E's – Efficiency, Equity, and Environmental Protection in Search of 'Win–win–win' Policies: A CGE Analysis of India. *Journal of Policy Modelling*, 24 (1): 19–50.

Oda, Junichiro, Keigo Akimoto, Fuminori Sano and Toshimasa Tomoda (2007) Diffusion of Energy Efficient Technologies and CO_2 Emission Reductions in Iron and Steel Sector. *Energy Economics*, 29 (4), 868–888.

Ofer, Gur (1987) Soviet Economic Growth: 1928–1985. *Journal of Economic Literature*, 25 (4): 1767–1833.

Palmer, Karen, Wallace E. Oates and Paul R. Portney (1995) Tightening Environmental Standards: The Benefit-Cost or the No-Cost Paradigm? *The Journal of Economic Perspectives*, 9 (4): 119–132.

Park, S.H. (1992) Decomposition of Industrial Energy Consumption – An Alternative Method. *Energy Economics*, 14 (4): 265–270.

Patuelli, R., P. Nijkamp and E. Pels (2005) Environmental Tax Reform and the Double Dividend: A Meta-analytical Performance Assessment. *Ecological Economics*, 55 (4): 564–583.

Pearce, D.W. (1991) The Role of Carbon Taxes in Adjusting to Global Warming. *Economic Journal*, 101 (407), 938–948.

Peneder, Michael (2003) Industrial Structure and Aggregate Growth. *Structural Change and Economic Dynamics*, 14 (4): 427–448.

Peretto, P.F. (2009) Energy Taxes and Endogenous Technological Change. *Journal of Environmental Economics and Management*, 57 (3): 269–283.

Perkins, D.H. (1988) Reforming China's Economic System. *Journal of Economic Literature*, 26 (2): 601–645.

Perkins, D.H. and T.G. Rawski (2008) Forecasting China's Economic Growth over the next two Decades. In Loren Brandt and Thomas G. Rawski (eds.), *China's*

Great Economic Transformation. Cambridge and New York: Cambridge University Press.

Picazo-Tadeo, Andrés J., Ernest Reig-Martínez and Francesc Hernández-Sanchoa (2005) Directional Distance Functions and Environmental Regulation. *Resource and Energy Economics*, 27 (2): 131–142.

Pigou, Arthur C. (1920) *The Economics of Welfare*. London: Macmillan and Co.

Pittman, R.W. (1981) Issues in Pollution Control: Interplant Cost Differences and Economies of Scale. *Land Economics*, 57 (1): 1–17.

Pittman, R.W. (1983) Multilateral Productivity Comparisons with Undesirable Outputs. *Economic Journal*, 93 (372): 883–891.

Pollak, R.A., R.C Sickles and T.J. Wales (1984) The CES-Translog: Specification and Estimation of a new Cost Function. *Review of Economics and Statistics*, 66 (4): 602–607.

Pokrovski, V.N. (2003) Energy in the Theory of Production. *Energy*, 28(8): 769–88.

Porter, M.E. (1991) America's Green Strategy. *Scientific American*, 264 (4): 168.

Porter, M.E. and C. van der Linde (1995) Toward a New Conception of the Environment – Competitiveness Relationship. *Journal of Economic Perspectives*, 9 (4): 97–118.

Qin, Duo and Haiyan Song (2009) Sources of Investment Inefficiency: The Case of Fixed-asset Investment in China. *Journal of Development Economics*, 90 (1): 94–105.

Ramanathan, Ramakrishnan (2005) An Analysis of Energy Consumption and Carbon Dioxide Emissions in Countries of the Middle East and North Africa. *Energy*, 30 (15): 2831–2842.

Rauscher, M. (1994) On Ecological Dumping. *Oxford Economic Papers*, 46 (Special Issue on Environmental Economics): 822–840.

Reddy, B.S. and G.B. Assenza (2009) The Great Climate Debate. *Energy Policy*, 37 (8): 2997–3008.

Rezek, Jon P. and Randall C. Campbell (2007) Cost Estimates for Multiple Pollutants: A Maximum Entropy Approach. *Energy Economics*, 29 (3): 503–519.

Robins, N., R. Clover and C. Singh (2009) A Climate for Recovery: The Colour of Stimulus goes Green. *HSBC Global Research*, 25 February 2009 issue: 1–45.

Rodrik, Dani (2006) What's so Special about China's Exports? NBER working paper, No. 11947.

Rose, A. and S. Casler (1996) Input–Output Structural Decomposition Analysis: A Critical Appraisal. *Economic Systems Research*, 8 (1): 33–62.

Roughgarden, T. and S.H. Schneider (1999) Climate Change Policy: Quantifying Uncertainties for Damages and Optimal Carbon Taxes. *Energy Policy*, 27 (7): 415–429.

Schäfer, Andreas (2005) Structural change in energy use. *Energy Policy*, 33 (4): 429–437.

Schaltegger, S. and T. Synnestvedt (2002) The Link between Green and Economic Success: Environmental Management as the Crucial Trigger between Environmental and Economic Performance. *Journal of Environmental Management*, 65 (4): 339–346.

Schmidt, P. (1976) On the Statistical Estimation of Parametric Frontier Production Functions. *Review of Economics and Statistics*, 58 (2): 238–239.

Schwartz, J. and R. Repetto (2000) Nonseparable Utility and the Double Dividend Debate: Reconsidering the Tax-interaction Effect. *Environmental and Resource Economics*, 15 (2): 149–157.

Scrimgeour, F., L. Oxley and K. Fatai (2005) Reducing Carbon Emissions? The Relative Effectiveness of Different Types of Environmental Tax. *Environmental Modelling & Software*, 20 (11): 1439–1448.

Sebastián, Lozano and Ester Gutiérrez (2007) Non-parametric Frontier Approach to Modelling the Relationships among Population, GDP, Energy Consumption and CO_2 Emissions. *Ecological Economics* ECOLEC-02992:13.

Shephard, R.W. (1970) *Theory of cost and production functions*. Princeton: Princeton University Press.

Sickles, R.C. and M.L. Streitwieser (1998) An Analysis of Technology, Productivity, and Regulatory Distortion in the Interstate Natural Gas Transmission Industry: 1977–1985. *Journal of Applied Econometrics*, 13 (4): 377–395.

Solow, R.M. (1957) Technical Change and the Aggregate Production Function. *Review of Economics and Statistics*, 39 (3), 312–320.

Solow, R.M. (1993) An almost Practical Step toward Sustainability. *Resources Policy*, 19 (3): 162–172.

Steenhof, Paul A. (2006) Decomposition of Electricity Demand in China's Industrial Sector. *Energy Economics*, 28 (3): 370–384.

Sun, J.W. (1998) Changes in Energy Consumption and Energy Intensity: A Complete Cecomposition Model. *Energy Economics*, 20 (1): 85–100.

Sun, Q. and W. Tong. (2003) China Share Issue Privatization: the Extent of its Success. *Journal of Financial Economics*, 70 (2): 183–222.

Syrquin, Moshe (1984) Resource Allocation and Productivity Growth. In M. Syrquin, L. Taylor and L.E. Westphal (eds.), *Economic Structure Performance – Essays in Honor of Hollis B. Chenery*. Orlando, FL: Academic Press. pp. 75–101.

Syrquin, Moshe (1995) Patterns of Structural Change. In H. Chenery and T. N. Srinivasan, (eds.), *Handbook of Development Economics*, Vol. 1. North Holland Elsevier. pp. 203–273.

Tahvonen, O. and J. Kuuluvainen (1993) Economic Growth, Pollution and Renewable Resources. *Journal of Environmental Economics and Management*, 24 (2): 101–118.

Terkla, D. (1984) The Efficiency Value of Effluent Tax Revenues. *Journal of Environmental Economics and Management*, 11 (2): 107–123.

Thompson, Henry (2006) The Applied Theory of Energy Substitution in Production. *Energy Economics*, 28 (4): 410–425.

Timmer, Marcel P. and A. Szirmai (2000) Productivity Growth in Asian Manufacturing: The Structural Bonus Hypothesis Examined. *Structural Change and Economic Dynamics*, 11 (4): 371–392.

Tone, K. (2001) A Slacks-based Measure of Efficiency in Data Envelopment Analysis. *European Journal of Operational Research*, 130 (3): 498–509.

Tone, K. (2002) A Slacks-based Measure of Super-efficiency in Data Envelopment Analysis. *European Journal of Operational Research*, 143 (1): 32–41.

Tone, K. (2010) Variations on the Theme of Slacks-based Measure of Efficiency in DEA. *European Journal of Operational Research*, 200 (3): 901–907.

Tsutomu, Miyagawa, Yukiko Ito and Nobuyuki Harada (2004) The IT Revolution and Productivity Growth in Japan. *Journal of the Japanese and International Economies*, 18 (3): 362–389.

Tu, Z. (2009) The Shadow Price of Industrial SO_2 Emission: A New Analytic Framework. *China Economic Quarterly*, 9: 259–282.

Tu, Z. and G. Xiao (2005) China's Industrial Productivity Revolution: A Stochastic Frontier Production Function Analysis of TFP Growth in China's Large and Medium Industrial Enterprises. *Economic Research Journal*, 3: 4–15.

Tuan, C., L.F.Y. Ng and B. Zhao (2009) China's Post-Economic Reform Growth: The Role of FDI and Productivity Progress. *Journal of Asian Economics*, 20 (3): 280–293.

Tullock, G. (1967) Excess Benefit. *Water Resources*, 3 (2): 643–644.

Turner, J. (1995) Measuring the Cost of Pollution Abatement in the US Electric Utility Industry: A Production Frontier Approach. Doctoral dissertation, University of North Carolina, Chapel Hill.

Tzouvelekas, E., D. Vouvaki and A. Xepapadeas (2006) *Total Factor Productivity and the Environment: A Case for Green Growth Accounting.* Mimeo: University of Crete.

United Nations (1993) Agenda 21. United Nations, New York.

Van Heerden, J., R. Gerlagh, J. Blignaut, M. Horridge, S. Hess, R. Mabugu and M. Mabugu (2006) Searching for Triple Dividends in South Africa: Fighting CO_2 Pollution and Poverty while Promoting Growth. *The Energy Journal*, 27 (2): 113–141.

Vardanyan, Michael and Dong-Woon Noh (2006) Approximating Pollution Abatement Costs via Alternative Specifications of a Multi-output Production Technology: A case of the US Electric Utility Industry. *Journal of Environmental Management*, 80 (2): 177–190.

Verspagen, B. (1993) *Uneven Growth between Interdependent Economies: an Evolutionary View on Technology Gaps, Trade and Growth.* Aldershot, England: Avebury.

Voorspools, K.R. and W.D. D'haeseleer (2005) Modelling of Electricity Generation of Large Interconnected Power Systems: How Can a CO_2 Tax Influence the European Generation Mix? *Energy Conversion and Management*, 47 (11–12): 1338–1358.

Wan, J. and A. Yuce (2007) Listing Regulations in China and Their Effect on the Performance of IPOs and SOEs. *Research in International Business and Finance*, 21 (3): 366–378.

Wang, B., Y. Wu and P. Yan (2010) Environmental Efficiency and Environmental Total Factor Productivity Growth in China's Regional Economies. *Economic Research Journal*, 5: 95–109.

Wang, C. (2007) Decomposing Energy Productivity Change: A Distance Function Approach. *Energy*, 32 (8): 1326–1333.

Wang, Can, Jining Chen and Ji Zou (2005a) Decomposition of Energy-related CO_2 Emission in China: 1957–2000. *Energy*, 30 (1): 73–83.

Wang, Can, Jining Chen and Ji Zou (2005b) Impact Assessment of CO_2 Mitigation on Chinese Economy based on a CGE Model, *Journal of Tsinghua University (Science and Technology)*, 45: 1621–1624.

Wang, Chunhua (2007) Decomposing Energy Productivity Change: A Distance Function Approach. *Energy*, 32 (8): 1326–1333.

Wang, J., G. Yan and K. Jiang (2009) The Study on China's Carbon Tax Policy to Mitigate Climate Change. *China Environmental Science*, 29: 101–105.

Wang, J., C. Ge, S. Gao, G. Yan and Z. Dong (2009b) Framework Design of the Separated Environmental Tax for China. *China Population, Resources and Environment*, 2: 69–72.

Wang, Xiaolu, Gang Fan and Peng Liu (2009a) Transformation of Growth Model and Growth Sustainability in China. *Economic Research Journal*, 1.

Watanabe, M. and K. Tanaka (2007) Efficiency Analysis of Chinese Industry: A Directional Distance Function Approach. *Energy Policy*, 35 (12): 6323–6331.

Weber, C.L. (2009) Measuring Structural Change and Energy Use: Decomposition of the US Economy from 1997 to 2002. *Energy Policy*, 37 (4): 1561–1570.

Wei, T. and S. Glomsrod (2002) The Influence of Carbon Tax on China's Economy and Greenhouse Gas Emissions. *World Economy and Politics*, 8: 47–49.

Wei, Yi-Ming, Hua Liao and Ying Fan (2007) An Empirical Analysis of Energy Efficiency in China's Iron and Steel Sector. *Energy*, 32 (12): 2262–2270.

West, S.E. and R.C. Williams III (2007) Optimal Taxation and Cross-price Effects on Labor Supply: Estimates of the Optimal Gas Tax. *Journal of Public Economics*, 91 (3–4): 593–617.

Wier, M., K. Birr-Pedersen, H.K. Jacobsen and J. Klok (2005) Are CO_2 Taxes Regressive? Evidence from the Danish Experience. *Ecological Economics*, 52 (2): 239–251.

Williams III, R.C. (2003) Health Effects and Optimal Environmental Taxes. *Journal of Public Economics*, 87 (2): 323–335.

Wissema, W. and R. Dellink (2007) AGE Analysis of the Impact of a Carbon Energy Tax on the Irish Economy. *Ecological Economics*, 61 (4): 671–683.

Woo, W.T. (1998) Chinese TFP: The Role from Labor Reallocation in Agricultural Sector. *Economic Research Journal*, 3: 31–39.

Wood, R. (2009) Structural Decomposition Analysis of Australia's Greenhouse Gas Emissions. *Energy Policy*, 37 (11): 4943–4948.

World Commission on Environment and Development (WCED) (1987) *Our Common Future*. Oxford: Oxford University Press.

World Bank (2001) *World Development Report 2000/2001: Attacking Poverty*. New York: Oxford University Press.

Wu, Qiaosheng and Jinhua Cheng (2006) Change of Energy Intensity and its Factor Decomposition in China: 1980–2004. *Economic Theory and Management*, 10: 34–40.

Wu, Y. (1995) Productivity Growth, Technological Progress, and Technical Efficiency Change in China: A Three-Sector Analysis. *Journal of Comparative Economics*, 21 (2): 207–229.

Wu, Y. (2008) The Role of Productivity in China's Growth: New Estimates. *China Economic Quarterly*, 7 (3): 827–842.

Wu, Z. and S. Yao (2003) Intermigration and Intramigration in China: A Theoretical and Empirical Analysis. *China Economic Review*, 14 (4): 371–385.

Xepapadeas, A, and A. De Zeeuw (1999) Environmental Policy and Competitiveness: The Porter Hypothesis and the Composition of Capital. *Journal of Environmental Economics and Management*, 37 (2): 165–182.

Xepapadeas, Anastasios, E. Tzouvelekas and D. Vouvaki (2007) Total Factor Productivity Growth and the Environment: A Case for Green Growth Accounting. Working Paper No.38, Fondazione Eni Enrico Mattei.

Young, A. (1994) Lessons from the East Asian NICs: A Contrarian View. *European Economic Review*, 38 (3–4): 964–973.

Young, A. (1995) The Tyranny of Numbers: Confronting the Statistical Realities of the East Asian Growth Experience. *Quarterly Journal of Economics*, 110 (3): 641–680.

Young, A. (2003) Gold into Base Metals: Productivity Growth in the People's Republic of China during the Reform Period. *Journal of Political Economy*, 111 (1): 1220–1261.

Yusuf, S., K. Nabeshima and D.H. Perkins (2005) *Under New Ownership: Privatizing China's State-Owned Enterprises*. Stanford University Press and The World Bank.

Zhang, J. (2002) Capital Formation, Industrialization and Economic Growth: Understanding China's Economic Reform. *Economic Research Journal*, 6: 3–13.

Zhang, J. (2003) Investment, Investment Efficiency, and Economic Growth in China. *Journal of Asian Economics*, 14 (5): 713–734.

Zhang Jun, Shaohua Shi and Shiyi Chen (2003) The Industrial Reform and Efficiency Change in China: Methodology, Data, Literatures and Conclusions. *China Economic Quarterly*, October, 3 (1): 1–38.

Zhang, J., S. Chen and G.H. Jefferson (2009) Structural Reform and Industrial Growth in China.. *Economic Research Journal*, 7.

Zhang, K., J. Pan and D. Cui (2008) *Introduction to Low Carbon Economy*. Beijing: China Environmental Science Press.

Zhang, Youguo (2009) Structural Decomposition Analysis of Sources of Decarbonizing Economic Development in China; 1992–2006. *Ecological Economics*, 68 (8–9): 2399–2405.

Zhang, Youguo (2010) Supply-Side Structural Efect on Carbon Emissions in China. *Energy Economics*, 32 (1): 186–193.

Zhang, Z.X. (2003) Why did the Energy Intensity fall in China's Industrial Sector in the 1990s? The Relative Importance of Structural Change and Intensity Change. *Energy Economics*, 25 (6): 625–638.

Zhang, Z.X. (2004) Meeting the Kyoto Targets: the Importance of Developing Country Participation. *Journal of Policy Modelling*, 26 (1): 3–19.

Zhang, Z.X. and A. Baranzini (2004) What do we know about Carbon Taxes? An Inquiry into their Impacts on Competitiveness and Distribution of Income. *Energy Policy*, 32 (4): 507–518.

Zheng, J., X. Liu and A. Bigsten (2003) Efficiency, Technical Progress, and Best Practice in Chinese State Enterprises (1980–1994). *Journal of Comparative Economics*, 31 (1): 134–152.

Zhou, P., B.W. Ang and K.L. Poh (2006) Slacks-based Efficiency Measures for Modeling Environmental Performance. *Ecological Economics*, 60 (1); 111–118.

Zhou, P., B.W. Ang and K.L. Poh (2008) A Survey of Data Envelopment Analysis in Energy and Environmental Studies. *European Journal of Operational Research*, 189 (1): 1–18.

Index

For Product Safety Concerns and Information please contact our EU
representative GPSR@taylorandfrancis.com
Taylor & Francis Verlag GmbH, Kaufingerstraße 24, 80331 München, Germany

www.ingramcontent.com/pod-product-compliance
Lightning Source LLC
Chambersburg PA
CBHW050412280326
41932CB00013BA/1825